Scale Invariance

Annick Lesne • Michel Laguës

Scale Invariance

From Phase Transitions to Turbulence

 Springer

Dr. Annick Lesne
Université Paris VI
Labo. Physique Théorique de la
Matière Condensée (LPTMC)
place Jussieu 4
75252 Paris Cedex 05
France
lesne@lptmc.jussieu.fr

Michel Laguës
ESPCI
rue Vauquelin 10
75231 Paris
France
michel1lagues@gmail.com

Translated by Rhoda J. Hawkins, PhD.
The original, French edition was published under the title 'Invariances d'échelle:
Des changements d'états à la turbulence'
© 2003, 2008 Editions Belin, France

ISBN 978-3-642-44896-6   ISBN 978-3-642-15123-1 (eBook)
DOI 10.1007/978-3-642-15123-1
Springer Heidelberg Dordrecht London New York

Springer is part of Springer Science+Business Media (www.springer.com)

Foreword

The opalescence of a fluid near its critical point has been a matter of curiosity for more than 100 years. However, the organisation of a fluid as it wavers between a liquid and a vapour was only understood 40 years ago – thanks to a profound insight by Leo Kadanoff in what was a major cultural achievement.

We also know of other "self-similar" systems: fractals (structures resulting from quite simple geometrical constructions) and turbulent flows (which we still struggle to control). But the case of critical points has become the key example: difficult but accessible to reasoning and all embracing enough to manifest widely varying families of behaviours.

The techniques of calculation (the "renormalisation group") are given in many works and brought together in the fine book by Toulouse and Pfeuty. But a book giving the *panorama* was needed: this is it. It starts with liquids and gases but it also shows the multiple scales that we meet in Brownian motion, flexible polymers or percolation clusters. Furthermore, the book is bold enough to address questions that are still open: cuprate superconductors, turbulence, etc. The most controversial question, which appears in the final chapter, is that of "self-organised criticality". For some this is a whole new world, and for others a world of mere words. But in any case, this book gives an accessible picture of example cases in which uncertainties still remain. And more generally it invites the reader to think. A good plan for young people curious about science.

Paris   *Pierre-Gilles de Gennes*
August 2003

Preface

The physical description of our environment consists of relating various essential quantities by laws of nature, quantities identified by observation, experiment and a posteriori verification. One common difficulty is to relate the properties of a system observed at a particular scale to those at a larger or smaller scale. In general, we separate the scales by averaging the values at the smaller scales and treating as constant the values varying on larger scales. Such descriptions are not always valid, for example in situations where multiple physical scales play a role. Here we describe new approaches specially adapted to such situations, taking changes of states of matter as the first problems we tackle. Today we know of a wide variety of critical systems in nature. The aim of this book is to unearth the common characteristics of these phenomena. The central concept of coherence length (or time) enables us to characterise critical phenomena by the size (or duration) of fluctuations. The first examples treated are spatial critical phenomena, with their associated divergence of the coherence length, second-order phase transitions, and percolation transition. After that we describe temporal critical phenomena, or even spatiotemporal, with their associated appearance of chaos.

The new challenge is that the description of these phenomena without characteristic length/time scales should be global, affecting all scales simultaneously, and should strive to understand their mutual organisation. More precisely the way in which different length/time scales are coupled to each other becomes determinate. The key concepts of *scale invariance* and *universality* are introduced. The emergence of these ideas was a turning point in modern physics and has revolutionised the study of certain physical phenomena. Analyses of characteristics at one particular scale has given way to investigations of *mechanisms of coupling between different scales*. One essential tool, renormalisation, enables this analysis; it establishes the scaling laws, demonstrates the universal physical behaviour within a Universality class and the insensitivity to microscopic details. Universality emerges as a crucial property in which it legitimises the use of basic models. A model reduced to just the essential elements is enough to describe the common characteristics of all the systems of a universality class. Examples considered in this book include the famous Ising model, percolation, logistic maps, etc.

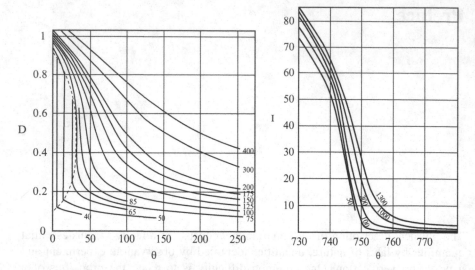

Fig. 1 Analogy, proposed by Pierre Curie as a conclusion of his PhD thesis (1895), between the density of a fluid (*left*) and the magnetization of a material (*right*), both as a function of temperature, and respectively as a function of pressure and magnetic field (for further details, see text)

To conclude this introduction, it is fitting to give tribute to Pierre Curie, who had this remarkable intuition during his work for his Ph.D. thesis in 1895, writing: "There are analogies between the function $f(I, H, T) = 0$ related to a magnetic object (Fig. 1 *right*) and the function $f(D, p, T) = 0$ related to a fluid (Fig. 1 *left*). The intensity of magnetization I corresponds to the density D, the intensity of the magnetic field H corresponds to the pressure p and the absolute temperature T plays the same role in both cases. [...]. The way in which the magnetization varies as a function of temperature close to the transition temperature, the magnetic field remaining constant, is reminiscent of the way in which the density of a fluid varies as a function of temperature near the critical temperature (while the pressure remains constant). The analogy extends to the curves $I = \varphi(T)$ that we have obtained and the curves $D = \varphi(T)$ corresponding to critical pressures. The left hand figure below, established with the experimental data from M. Amagat on carbonic acid and the right hand figure below, established with my experiments on iron, shows the idea of this analogy."

Paris   *Annick Lesne*
August 2011   *Michel Laguës*

Contents

Chapter 1
Changes of States of Matter

1.1 Introduction

It is a fact of life in our daily experience and was a real enigma for a century – pure matter dramatically changes at extremely precise temperatures. In the nineteenth century, pioneering scientists, from Gay-Lussac to Van der Waals, carried out meticulous measurements of the fluid state, paving the way for microscopic descriptions which underlie our natural sciences today. The study of properties of gases at low density enabled the introduction of absolute temperature as a measure of the kinetic energy of molecules. The striking generality of thermal behaviour and mechanical properties of gases with vastly varying chemical properties was thereby elucidated. Thermodynamics and its microscopic interpretations was born on the wave of this success. However the pioneers of fluid observations also tackled liquid–vapour transformations and discovered another elegant generality which was far from evident a priori: In a dilute gas, molecules are almost isolated and therefore one thinks their chemical properties are unimportant, but what about in a liquid where the molecules are in constant interaction?

At the dawn of the twentieth century, the study of magnetic transformations expanded the field of changes of states studied. There are also other situations where we observe a sharp change in the microscopic structure of a material; metal alloys, binary mixtures of fluids, superfluidity in helium and many others. The experimental observations beg questions such as: why do these transformations occur at such a precise temperature? And what is this origin of the surprising similarity of such changes, seemingly independent of the nature of the physical properties that transform? It was necessary to wait until the 1970s for a satisfactory answer to these questions to be proposed. But above all, why do these changes occur?

A naive reading of the second law of thermodynamics might suggest that all physical systems must evolve towards maximum disorder if the observer is patient enough. However, this argument is only valid for an isolated system. In the more

A. Lesne and M. Laguës, *Scale Invariance*, DOI 10.1007/978-3-642-15123-1_1,
© Springer-Verlag Berlin Heidelberg 2012

usual case in which the system exchanges energy with its environment, the second law of thermodynamics can be framed by the following two simple rules:

- At high temperature, the system evolves towards a highly disordered state of equilibrium, as if the system was isolated.
- At low temperature, on the other hand, the system tends to self organise to reduce its internal energy.

The challenge of studying changes of state, discussed in Chaps. 1 and 3, is to determine accurately what temperature draws the line between the two regimes and under what conditions the transition occurs.

Entropy, internal energy and transitions
The increase in entropy predicted by the second law of thermodynamics provides a criterion for the evolution of isolated systems towards thermodynamic equilibrium: the system evolves to a state of higher microscopic disorder. Valid only for isolated systems, this criterion of evolution should be modified if the system considered can exchange energy with its environment (the usual situation). At constant temperature for example, a closed system evolves to a state with free energy $F = U - TS$, which is minimal at equilibrium.

At "high" temperature, the entropic term $-TS$ dominates and the system evolves towards a state with higher entropy. On the other hand at "low" temperature, the internal energy U dominates, implying an evolution of the system towards an equilibrium state at which the internal energy is a minimum. At the transition temperature the entropic and energetic terms are of the same order of magnitude.

Molecules or electrons can decrease their energy by organising themselves in regular configurations. These configurations are established in regions of temperature sufficiently low that thermal agitation no longer dominates. In this way, in magnets (or ferromagnetic materials), a macroscopic magnetisation spontaneously appears below a temperature called the Curie temperature, which is the effect of an identical orientation of the magnetic spin at each atomic site. In the same way, when one cools a gas it liquefies or solidifies. Even though disorder reigns in the liquid state, there appears a *density order* in the sense that the molecules spontaneously confine themselves to denser regions. It is worth mentioning other physical situations where matter changes state as the temperature is lowered, in ferromagnets, liquid crystals, superconductors, superfluids, etc. As for the detailed organisation of the ordered state at equilibrium, for an entire century physicists were unable to calculate correctly what thermodynamics predicted near a change of state. To understand the reasons for their perseverance in solving this problem, we must emphasize the elegance with which the experimental studies showed a *universality* of behaviours called *critical* in the immediate vicinity of a change of state. To give an idea of the extreme precision of these measurements consider the result of William Thompson (Lord Kelvin) who, in the nineteenth century, established the variation of the melting point of ice as a function of applied pressure as $-0.00812°C$ for one atmosphere!

The study of critical phenomena initiated by Cagnard de Latour in 1822 experienced a big boost with the work of Andrews from 1867 onwards. In 1869 he observed a spectacular opalescence near the critical point of carbon dioxide.

Critical opalescence is one of the only situations where the microscopic disorder bursts into our field of view: when heating a closed tube containing a fluid at the critical density, the meniscus separating the gas and the liquid phases thickens, becomes cloudy and diffuse until it disappears. On cooling the meniscus reappears in an even more spectacular way in the midst of an opalescent cloud.[1]

Andrews correctly interpreted this opalescence as an effect of giant fluctuations in the density of the fluid, a sign of the wavering of the material between the liquid and gas states. These giant fluctuations are observed in all transitions that are called second order. The theories proposed at the beginning of the twentieth century by Einstein, Ornstein and Zernike and then Landau quantified Andrew's intuition. The predictions of these theories apply very well to certain physical situations, for example ferroelectric transitions or superconductor–insulator transitions; however, there are significant differences from that observed near the transition for most other changes of state. This is particularly so for the case of liquid–vapour transitions and magnetic transitions. The most surprising, and most annoying, is the universality shown by the obvious similarities between critical behaviours of these considerably different physical systems, which entirely escaped theoretical descriptions for a hundred years.

The first two types of changes of state to be studied in detail, corresponding a priori to very different physical situations, which will form the bulk of this introductory chapter are:

• The ferromagnetism–paramagnetism transition in a crystalline solid
• The liquid–vapour transition in a disordered fluid

The reversible disappearance of the magnetisation of iron above $770\,°C$ has been known since the Renaissance, but Pierre Curie was the first to study the variations of magnetism with temperature, during his PhD thesis work in 1895. His name is also associated with the critical temperature of the ferromagnetic–paramagnetic transition, as well as the law of variation of paramagnetism with temperature. Several physicists are well known for proposing descriptions of this transition during the first half of the twentieth century.[2] Initially the question was to describe the way in which magnetisation varies as a function of temperature and the

---

[1] The interested reader should find some pictures and film clips on the internet e.g.: http://www.youtube.com/watch?v=2xyiqPgZVyw&feature=related.

[2] Léon Brillouin, Paul Ehrenfest, Ernest Ising, Lev Landau, Paul Langevin, Louis Néel, Kammerling Onnes, Lars Onsager, Pierre Weiss to cite just a few (in alphabetical order).

applied magnetic field and then these studies gave rise to the more general theory of phase transitions. An initial description of the liquid–vapour transition was proposed by Van der Waals in 1873, also during his PhD thesis work. This description, which is the subject of Sect. 1.4.4, satisfactorily describes the change of state liquid–vapour globally but does not correctly describe the critical region. This is also the case for the description proposed in 1907 by Weiss for the appearance of magnetisation in a magnetic material, which uses the same type of approximation called *mean field*. It leads to the same behaviour and the same discrepancies in the critical region.

In this *critical region* where the temperature is near to the transition temperature, the mechanisms controlling the state of the system are complex. The fact that the material is disordered at high temperature and ordered at low temperature does not imply that the change of state should happen at a precise temperature: it could just as well be spread out over a large range of temperatures as is the case for complex mixtures or very small systems. Van der Waals' and Weiss' descriptions do predict a transition at a precise temperature, but what would happen in the case of an exact solution? This question remained unanswered until 1944 when the physicist Lars Onsager solved a 2D model of the ferromagnetic–paramagnetic transition without approximations. The results show a sharp transition for a system of infinite size (thermodynamic limit). However the mystery was far from being resolved since these exact results were in disagreement with experiment and with the mean field predictions: the plot thickened. Despite countless attempts, a century elapsed after the work of Van der Waals before a theory capable of significantly improving his descriptions and finally accounting for the universality seen in all the experiments. The power of this new "scaling" then enabled applications to very diverse fields.

1.2   Symmetry-Breaking Changes of State

Before introducing Van der Waals' and Weiss' descriptions let us specify the idea of an *order* changed during a transition. The order we are interested in in phase transitions lowers the internal energy, however as we have seen for the liquid–vapour transformation, it does not necessarily lead to a regular, crystalline organisation of the material. We need to find a more general definition.

In 1937 the Russian physicist Landau proposed a concept which allows us to unify the treatments of phase transitions. For each transition we define an *order parameter*, a physical quantity which is zero at temperatures above the critical temperature T_c, and then progressively increases as we lower the temperature below T_c up to a maximum value at zero temperature. The order can be measured by a scalar – for example the density variation in a fluid – or by a vector such as magnetisation. In general there are several different forms or orientations that can be established. For our two example cases:

- There are several possible directions for the magnetisation.
- The fluid can choose the gaseous or liquid state

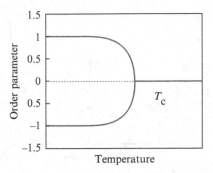

Fig. 1.1 The general characteristic of second order phase transitions is the appearance of a new type of order below a critical temperature T_c. We measure the establishment of this order by an order parameter which can in general take several different values. In the diagram above, corresponding to the example of magnetisation in a solid measured along a crystal axis, the order parameter can take two opposite values

The system orders by aligning along one of these orientations, or in one of these states. In the phase diagram (order parameter versus temperature) the transition corresponds to two paths that a system can take as the temperature is lowered below T_c (Fig. 1.1), corresponding to the respective changes of *symmetry*. Isotropic and homogeneous above the critical temperature, at low temperature in a given macroscopic region, the material locally bifurcates towards a magnetic orientation or a preferred density. This is called "bifurcation". In the absence of external excitation (in this case the applied magnetic field or gravity), the equations and boundary conditions that determine the system cannot predict which branch is chosen at a given point in space and a given moment in time. Spontaneous symmetry breaking occurs, which violates the principle established by Curie, in which the physical behaviour resulting from certain equations obeys the symmetry of these equations and the boundary conditions. According to Landau's proposal, the amount of order in a state is measured by this *deviation from the initial symmetry*.

In the case of ferromagnetic–paramagnetic and liquid–vapour transitions, this *order parameter* is respectively:

- The magnetisation (as the difference from the high temperature state of zero magnetisation).
- The difference in density between the liquid and the gas (as the deviation from the undifferentiated state of supercritical fluid).

This characterisation of a phase transition by the change in symmetry of the system under the effect of temperature is very general.

Following the terminology introduced by Ehrenfest, phase transitions can be *first* or *second order*. In a first order transition, macroscopic regions of completely different properties appear at the transition temperature, for example, ice and liquid in the case of the melting of water. From one region to another the microscopic

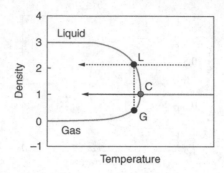

Fig. 1.2 The liquid–vapour transition is first order (*dashed arrow*) if the fluid density is other than the critical density i.e. at all points apart from C. In this case, as we lower the temperature to the coexistence curve (*point L*) gas bubbles appear of a very different density (*point G*) to that of the fluid initially. If the fluid density is at the critical density the transition is second order. At the critical point C, microscopic regions of liquid and gas form, initially at equal density and therefore they differentiate continuously when the temperature decreases

configuration changes in a discrete manner. At the macroscopic scale, this translates into a discontinuity in certain first derivatives of the thermodynamic potentials, and notably by the existence of a latent heat.

On the other hand, during second order phase transitions the first derivatives of thermodynamic potentials are continuous, whereas certain second derivatives diverge. No discontinuous change is observed at the microscopic scale, just a divergence in the size of characteristic fluctuations.

As we approach the critical temperature, extended regions of *order* multiply, interwoven within equally extended regions of *disorder*. The difference between first and second order transitions is amply illustrated by the phase diagram of a fluid (density versus temperature) (Fig. 1.2). We lower the temperature of a hypercritical fluid by keeping the volume fixed, in other words keeping the average density constant. If the fluid density is different from the critical density the liquid–vapour transition is first order (*dashed arrow*). On the other hand, if the fluid density is equal to the critical density, the transition is second order. In the following we are only interested in second order transitions.

1.3   Observations

1.3.1   *Bifurcations and Divergences at the Liquid–Vapour Critical Point*

In our environment, more often than not matter consists of mixtures and complex structures; although sometimes natural cycles drive certain substances to pure forms. Evaporation, condensation and soil filtration purify the water on which our lives depend. Water is the principal pure substance in which we see transformations in our daily lives, in particular its changes of state. In contrast to the vast majority of solids, ice contracts (by 7%) on melting. This has spectacular consequences, for example icebergs float and rocks break on freezing. Without this peculiarity, the Earth would not have had the same prospect: ice sheets would not float or play their role as effective thermal insulation, the thermal equilibrium of the planet would be profoundly different and life would not have appeared, or at least not as we know it. This anomaly of water comes from the strong electrostatic interactions between the molecules known as *hydrogen bonds*. Ordinary ice chooses a diamond-like structure that is not very dense, which makes the best use of these hydrogen bond interactions to lower its energy, that is to say increase the stability of the structure. In liquid water the hydrogen bonds also play an important role but the disorder results in a more compact structure.

Figure 1.3 represents, by isotherms, the pressure as a function of the density of water. The point C is the critical point ($p_c = 22$ bars, $\rho_c = 0.323$ kg/m^3, $T_c = 647$ K $= 374°$C). As we change the temperature of a closed vessel containing water, the liquid–vapour transformation occurs at a fixed average density, with a given latent heat. If the average density of water in the container is different from ρ_c, the latent heat is nonzero, whereas the liquid and vapour each are at the same density during the whole transformation: only the proportions of the two phases changes. The transformation we are used to at the boiling temperature $100\,°$C at atmospheric pressure is first order. In contrast however, if the average density is at the critical value the densities of the two phases are strictly equal at the critical point. As the

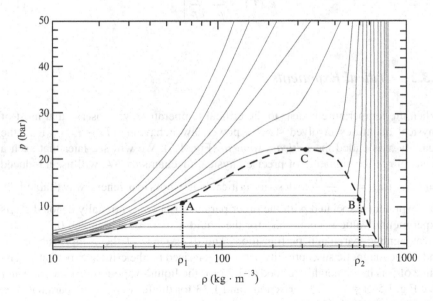

Fig. 1.3 Isotherms of water in the phase diagram pressure p as a function of density ρ (after [1])

Fig. 1.4 Latent heat of the liquid–vapour phase transition of water as a function of the transition temperature (after [1])

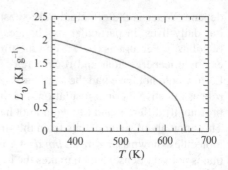

transition temperature increases, the latent heat of the liquid–vapour transformation gradually decreases, vanishing completely at T_c (Fig. 1.4).

The latent heat of water behaves qualitatively in the same way as the magnetisation of iron as a function of temperature. As the temperature is lowered, both show a similar symmetry breaking appearing at T_c. In practice, we choose the difference in density $\Delta\rho$ (Fig. 1.3) between the vapour phase (*point A*) and liquid phase (*point B*) as the order parameter of the transformation. Beneath the coexistence curve (*dashed line*), defined by these points, the corresponding homogeneous states are inaccessible for water in thermal equilibrium.

At the critical temperature the isotherm has a horizontal plateau at the critical density: the compressibility κ_c at the critical density diverges (see the example of xenon in Fig. 1.5):

$$\kappa_c(T) = \frac{1}{\rho_c} \left. \frac{\partial\rho}{\partial p} \right|_{\rho_c}. \tag{1.1}$$

1.3.2   Critical Exponents

When the temperature is near to the critical temperature, we observe that most of physical quantities involved show a power law behaviour $(T - T_c)^x$ where the quantity x is called the *critical exponent* (Fig. 1.5). We will see later that such a behaviour is the signature of precise physical mechanisms. We will use a reduced temperature $t = \dfrac{T - T_c}{T_c}$ to describe critical behaviours in a general way (Table 1.1). The exponent associated with the order parameter is conventionally denoted β. Its experimental value is about 0.32 for the liquid–vapour transition and 0.37 for the ferromagnetic–paramagnetic transition. The divergence of the compressibility κ_c and of the magnetic susceptibility χ is characterised by the critical exponent γ. The value of γ is in the neighbourhood of 1.24 for the liquid–vapour transition for water (see Fig. 1.5 or $\gamma = 1.21$ for xenon), and 1.33 for the ferromagnetic–paramagnetic transition of nickel. The exponent α conventionally characterises the divergence of specific heat, the exponent δ the variation of the order parameter as a function of

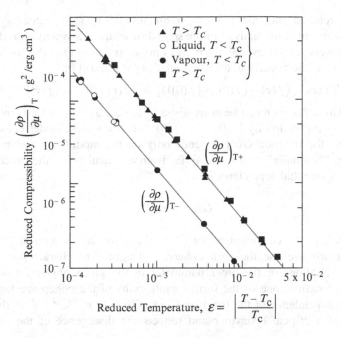

$$\text{Reduced Temperature, } \varepsilon = \left| \frac{T - T_c}{T_c} \right|$$

Fig. 1.5 Compressibility of xenon at the critical density, as a function of temperature. The compressibility diverges at the critical point. The variation obeys a power law with the exponent -1.21, with different prefactors each side of the critical point (after [9])

Table 1.1 Definition of critical exponents

Exponent	Physical property	Expression
α	Specific heat	$C \sim t^{-\alpha}$
β	Order parameter $f(T)$	$m \sim t^{\beta}$
γ	Compressibility, susceptibility, etc	$\chi \sim t^{-\gamma}$
δ	Order parameter at T_c, $f(h)$ or $f(p)$	$m(T_c, h) \sim h^{1/\delta}$
η	Correlation function	$G(T_c, r) \sim r^{-(d-2+\eta)}$
ν	Coherence length	$\xi \sim t^{-\nu}$

gravity or magnetic field h at $T = T_c$, the exponent η the spatial dependence of correlations (see later) and the exponent ν the divergence of the coherence length ξ. We introduce these last two physical quantities in Sect. 1.6, in the context of the mean field approximation.

The values of these exponents are surprisingly robust to changes in the physical system. Not only are they the same for the liquid–vapour transformation for all fluids (Fig. 1.19), but we find them in apparently very different situations. In the following paragraph we present two examples of transitions in binary fluid mixtures and metal alloys where the same values for the exponents are observed.

The correlation function $G(r)$ of a quantity $f(x)$ is a statistical measure particularly useful in the analysis of the spatial structure of a system. It is defined as the spatial average $\langle\cdots\rangle$ over all the pairs of points $(\mathbf{r}_0, \mathbf{r}_0 + \mathbf{r})$ of the product of the deviations from the average of the function $f(x)$ at $\mathbf{r}$ and at 0:

$$G(\mathbf{r}) = \langle (f(\mathbf{r}) - \langle f(\mathbf{r})\rangle)\,(f(0) - \langle f(0)\rangle)\rangle = \langle f(\mathbf{r})\,f(0)\rangle - \langle f(\mathbf{r})\rangle\,\langle f(0)\rangle \quad (1.2)$$

The correlation function can be normalised to 1, i.e., $G(\mathbf{r} = 0) = 1$, by dividing the previous expression by $\langle f(0)^2\rangle - \langle f(0)\rangle^2$. If the system under consideration is isotropic, the function $G(r)$ depends only on the modulus r and not on the direction $\mathbf{r}$. "Normally", that is to say far from a critical point, the function $G(r)$ shows an exponential dependence on r:

$$G(r) \sim e^{-r/\xi},$$

where ξ defines the characteristic length or the correlation length of the system. More generally we use the term coherence length ξ to characterise the scale of spatial variations of the order parameter in the given physical system. The correlation length is one of the formal evaluations of the coherence length. The power law dependence of the function $G(r)$, $G(r) \sim r^{-(d-2+\eta)}$, is the typical signature of a critical behaviour and reflects the divergence of the correlation length ξ.

The bifurcation that accompanies the critical point is also observed for other properties, for example dynamic properties.

Figure 1.6 illustrates the bifurcation for two dynamic properties, the coefficient of molecular self-diffusion and the relaxation time of nuclear magnetic resonance, that can be observed in ethane for which the critical point, $(T_c = 32.32°C, \rho_c = 0.207\,\text{kg/m}^3)$, is easily accessible.

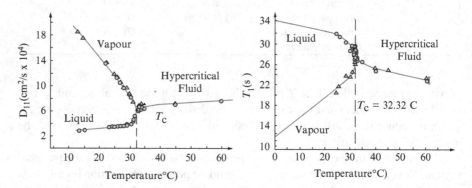

Fig. 1.6 Liquid–vapour transformation of ethane. Variation of coefficient of self-diffusion (*left*) and nuclear magnetic resonance relaxation time T_1 (*right*) near the critical temperature (after [4])

1.3.2.1 Binary Liquids and Metal Alloys

The same type of observation can be made at the critical point of mixtures of binary liquids or metal alloys: bifurcations and divergences from which we can measure the corresponding critical exponents. When their composition is equal to the critical composition, we observe a phase separation below the critical temperature.

Figure 1.7 shows the variations of the turbidity (cloudiness) of a cyclohexane-phenylamine mixture at the critical point. Some especially accurate measurements have been made on this system, allowing a detailed characterisation of its critical behaviour. In the case of binary fluids, it is also possible to observe bifurcations and divergences of transport properties (Figs. 1.8 and 1.9).

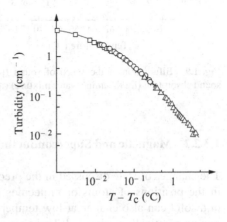

Fig. 1.7 Critical opalescence measured by the divergence of the turbidity (cloudiness) of the fluid – the inverse of the absorption depth of light – near the demixing critical point of a cyclohexane-phenylamine mixture at the critical concentration (after Calmettes et al. [2])

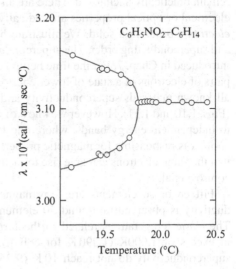

Fig. 1.8 Bifurcation of thermal conductivity of a binary mixture observed during demixing (after [4])

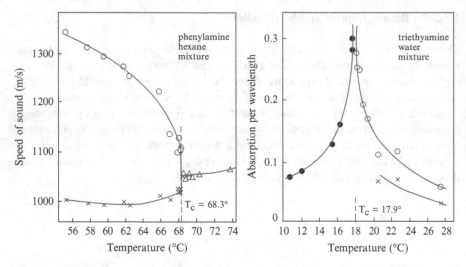

Fig. 1.9   Bifurcation of the speed of sound (phenylamine/hexane mixture) and divergence of sound absorption (triethyamine/water mixture) (after [4])

1.3.2.2   Magnetic and Superconducting Order in a Solid

The changes of state discussed in the preceding paragraphs are induced by changes in the positions of atoms or molecules in space. The configuration of electrons in a solid can also change at low temperature while the position of atomic nuclei remain practically unchanged. There are also "changes of states of matter" in which electrical or optical properties are radically changed. There exists various types of *electronic order* in a solid. We illustrate here the two main ones: magnetic order and superconducting order. The superconductor–insulator phase transition is briefly introduced in Chap. 7. For the time being keep in mind that it is the condensation of pairs of electrons in a state of lower energy than if they remained single. In almost all known materials superconductivity and magnetic order are mutually exclusive (Figs. 1.10 and 1.11). However we know of materials containing electrons belonging to independent energy bands, where one family leads to superconductivity whilst another is responsible for magnetic properties. The few cases where it is suspected that the same electrons can give rise to the two properties simultaneously are highly controversial.

Fifteen or so elements are ferromagnetic in the solid state whilst superconductivity is observed in around 50 elements, more than half the stable elements. The Curie temperatures related to the ferromagnetic–paramagnetic transition are in excess of 1,000K (1,390 K for cobalt), whereas the critical temperatures for superconductivity do not reach 10 K (9.25 K for niobium). For the elements, the characteristic energies of magnetism are therefore more than a hundred times greater

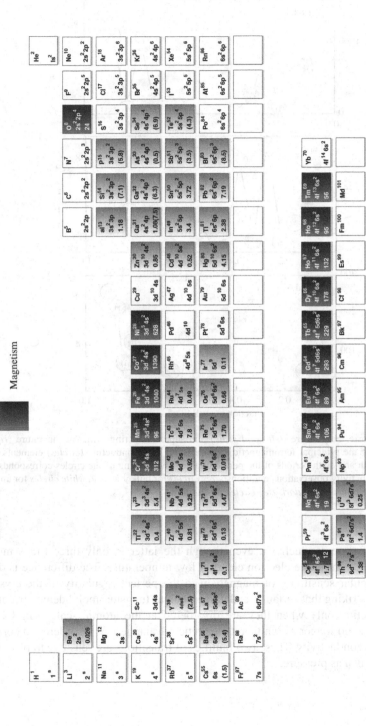

Fig. 1.10 Electronic properties of elements at low temperature. *Dark grey*: magnetic materials with the Curie temperature indicated at the bottom of the squares. *Light grey*: superconducting elements with the transition temperature indicated at the bottom of the squares. Where the temperature is in parentheses, superconductivity is only observed under pressure. Elements such as copper, silver and gold show no magnetic or superconducting properties at low temperature (after [3] *Concise encyclopedia of magnetic & superconducting materials*)

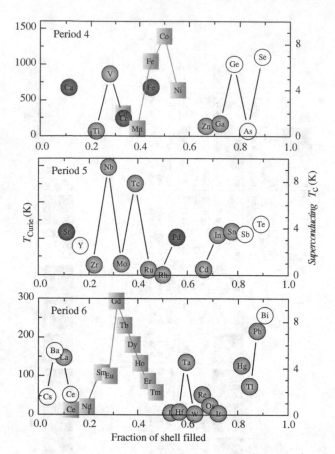

Fig. 1.11 Curie temperature (*left hand scale*) and superconducting critical temperature (*right hand scale*) of the principal ferromagnetic (*squares*) and superconducting (*circles*) elements for the fourth, fifth and sixth periods of the periodic table. The shading of the circles corresponds to the superconductivity observation conditions: *grey circles* for bulk material, *white circles* for under pressure measurements and *dark grey circles* for thin films

than that of superconductivity, even though the latter is only three times more frequently chosen by the electron cloud at low temperature. The difference is due to the particular sensitivity of magnetism to the perfect regularity of the crystal lattice. It is striking that certain metal alloys such as tungsten-molybdenum become superconducting only when they are disordered at the atomic scale (Fig. 1.12): Magnetism is no longer a significant energetic advantage in a disordered material, unlike superconductivity. These two families of transitions are sensitive to physical conditions such as pressure.

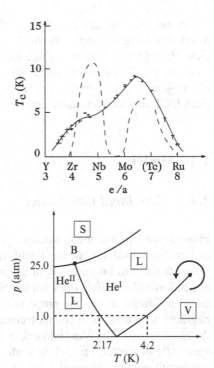

Fig. 1.12 Superconductivity of metal alloys as a function of the number of electrons in the outer shell (valence electrons). *Dashed line*: crystalline alloys; *continuous line*: disordered alloys in the form of metallic glasses/amorphous metals (after [3] *Concise encyclopedia of magnetic & superconducting materials*)

Fig. 1.13 Phase diagram of helium-4 (after [8])

1.3.3 Superfluid Helium

Of all elements, helium has the most unique properties. At atmospheric pressure, it does not solidify whatever the temperature, and at zero temperature it solidifies only above 25 atmospheres (Fig. 1.13). This peculiarity is due to the fact that helium atoms interact with each other only very weakly, like all the noble gases. Furthermore, being the lightest of the noble gases, the "zero point" atomic fluctuations predicted by quantum mechanics are sufficiently large that the solid state is unstable at absolute zero temperature.

However, the essential characteristic of the helium-4 isotope is that, below 2.17 K, it has two liquid phases, a normal liquid phase and a *superfluid* phase in which the viscosity is zero. The other stable isotope ^{3}He of helium also has a superfluid phase but at a temperature of below 2.7 milikelvins, a thousand times lower than the superfluid phase transition of ^{4}He. These two phase transitions correspond to different physical mechanisms, although they are both related to superconductivity (see Chap. 7).

Why are we so interested in physics that, while certainly rich, concerns only a single element? One major reason is that we now have a considerable body of extremely accurate experimental results on the superfluid phase transition, collected since the 1930s owing to the ability to perfectly purify the gas using powerful cryogenics techniques. Helium itself is the base of all cryostats. Another reason lies in the *quantum* nature of the superfluid phase transition: the order parameter is a complex wavefunction. Superfluidity and superconductivity are the best physical examples of the *XY model* where the order parameter is a quantity with *two real components*. Traditionally, situations where the number n of components of the order parameter is 1 or 3 are respectively called the *Ising model (n = 1)* (see Sect. 1.4.2) and the *Heisenberg model (n = 3)*.

1.4   Models

1.4.1   The Ideal Gas Model

Phase transitions are due to microscopic interactions between spins for magnetism or between molecules for the liquid–vapour transition. It is very difficult to take these interactions into account in a rigorous manner, particularly because they are not additive (their effect does not double if the size of the system is doubled). It is useful to briefly review the properties of the ideal gas model in which we totally neglect interactions between molecules. Since ancient times, gas – Geist (spirit), the name proposed by Van Helmont in the seventeenth century – has represented the ideal state of matter. From their studies of the gaseous state, the physicists of the nineteenth century deduced a simple and efficient model of an *ideal* gas. This model however supposes the molecules simultaneously possess two contradictory properties:

- The molecules interact rarely, in the sense that their interaction energy is negligible compared to their kinetic energy.
- The molecules interact often such that the ideal gas is at equilibrium at each instant in time.

By means of this conceptual leap, the ideal gas model represents the basis of all of thermodynamics. The ideal gas model equation of state is written $pv = kT$ where v is the average volume occupied by each molecule, p the pressure, T the absolute temperature and k the Boltzmann constant. Based on this empirical law, Gay-Lussac, in 1802, proposed the existence of an absolute zero of temperature, absolute zero $-273.15\,°C$ the value of which was found very accurately by Lord Kelvin half a century later. For most properties of real gases and dilute solutions this simplified description is sufficiently accurate. Let us now consider the paramagnetic properties of a magnet following Curie's law, which neglects all interactions between the spins. The average magnetisation M is related to the applied magnetic field H and the temperature T by the relation:

$$M = \frac{C}{T} H. \tag{1.3}$$

This relation can be made to resemble that of the equation of state for an ideal gas if we express the volume occupied by a molecule in the ideal gas equation of state as a function of the molecular density $n = 1/v$:

$$n = \frac{1}{kT}p. \tag{1.4}$$

In both cases the intensive variable, M and n respectively (proportional to the external applied field – the pressure p for the gas and the magnetic field H for the magnet), is *inversely proportional to the temperature*. As we show below, this proportionality is due to the fact that the models neglect *all interactions* between the gas molecules or magnetic spins. The form of these laws expresses the average thermal properties calculated for a single particle and then multiplied by the number of particles. For example, in the case of a system of N spins, $\mu_j = \pm\mu$, the average magnetisation per spin is obtained from the definition of the average of a thermodynamic variable:

$$m = \frac{<M>}{N} = \frac{1}{N}\frac{\sum\limits_{i} M_i e^{-E_i/kT}}{\sum\limits_{i} e^{-E_i/kT}}, \tag{1.5}$$

where E_i is the total energy of the set of spins $\{\mu_j\}$ in the configuration i. By showing these spins explicitly we obtain:

$$m = \frac{1}{N}\frac{\sum\limits_{\{\mu_j\}}\left[\sum\limits_{j}\mu_j\right]e^{\sum\limits_{j}\mu_j H/kT}}{\sum\limits_{\{\mu_j\}}e^{\sum\limits_{j}\mu_j H/kT}} \tag{1.6}$$

The calculation is very simple in this case since we neglect all interactions between the spins. In fact m is therefore simply the average value of the magnetisation of an isolated spin:

$$m = \frac{\mu e^{\mu H/kT} - \mu e^{-\mu H/kT}}{e^{\mu H/kT} + e^{-\mu H/kT}} = \mu\tanh(\mu H/kT) \tag{1.7}$$

If the excitation energy μH remains small compared to the thermal energy kT (the most common situation), to first order in $\mu H/kT$, the *response* M of N spins is proportional to the cause:

$$M = N\mu\frac{\mu H}{kT} \tag{1.8}$$

This response for weak excitation energies $E_{ex} = \mu H$ is shown in the following general form:

$$\text{Response} = \text{Maximum response} \cdot \frac{E_{ex}}{kT} \qquad (1.9)$$

For ideal gases, where the volume occupied by one molecule is v_m, we find the law $n = p/kT$:

$$n = \frac{1}{v_m}\frac{pv_m}{kT} \qquad (1.10)$$

Another example of an application of this law (1.9) concerns the *osmotic pressure* which causes biological cells to swell proportionally to the concentration of salt they contain. The relationship between the salt concentration and the pressure is exactly the same as that of an ideal gas as long as the concentration is not too high. Another example is that of entropic elasticity L of a long polymer chain to which we apply a traction force F. If we suppose that the N monomers (links in the chain) of length a are independent (without interactions) we can simply evaluate the relationship between L and F:

$$L = Na\frac{Fa}{kT} \qquad (1.11)$$

This leads to an elasticity $L = \frac{C}{T}F$ of entropic origin, where the stiffness is proportional to the temperature. A surprising result of this is that under a constant force a polymeric material, for example rubber, contracts on heating. This can be verified experimentally. You can do this yourself by touching your lips with a stretched rubber band and feeling it contract as your lips warm it. However, contrary to the case of gases, the above model is insufficient to quantitatively describe the elasticity of polymers for which the interactions between monomers cannot be neglected. We discuss this question of *excluded volume of polymers* later in Chap. 6.

1.4.2 Magnetism and the Ising Model

To go further, i.e. to try to take into consideration interactions between particles or spins, we need to construct and solve candidate models and discuss their relevance in representing reality. Traditionally physicists made their initial modelling attempts within the field of magnetism and most of the models used today came from magnetism.

Ferromagnetism, a magnetisation in the absence of external excitation, has many applications: significantly it is thanks to ferromagnetism that we can store vast quantities of information on our "hard disks". But the reason for the special place of magnetism in modelling changes of state is that they are simple microscopically, much simpler than fluids for example. They are simple because atomic spins are considered fixed and regularly spaced. The model of an ideal magnet that we have presented above predicts no spontaneous magnetisation: with each spin being independent of its neighbour, nothing can cause a spin to orient in one direction rather than another in the absence of an applied magnetic field. It is necessary to model interactions in order to be more realistic.

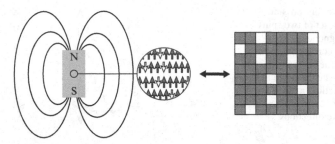

Fig. 1.14 The magnetisation of a material is the sum of the magnetisations due to the electron spins of each atom. Quantum physics states that value of spin magnetisation along a given direction can take only a few very precise values. In the simplest case, that of a spin 1/2 system, there are just two opposite values that can be observed. Ising proposed a simple model where each spin is placed on a regular lattice, represented here by a draughtboard, and can take the values +1 (*black*) or −1 (*white*). The same model can be applied to metal alloys and binary fluids

The *Ising* model, proposed by Lenz in 1920 and solved in one dimension by Ising in 1925, is the simplest model of interactions one can imagine. Despite that, it enables the rigorous description of magnetisation in solids and magnetic phase transitions in the critical region, as well as a great many changes of state in systems which have nothing to do with magnetism.

The magnet in Fig. 1.14 is at low temperature: the set of spins is globally orientated but thermal excitation orients a few spins in the opposite direction to the majority. Ising described the simplest case of a regular lattice of spins of spin 1/2, where the projection of the magnetisation of a spin in a given direction in space can take one of only two opposite values. We choose units such that the measurable values are +1 and −1. With this model we can represent, in a simple way, the state of a magnet containing a very large number of spins, for example with the aid of draughtboard in which the squares are black if the spins are in the state +1 and white if they are in the state −1 (Fig. 1.14). In this form the Ising model can be applied to the study of binary metal alloys consisting of two sorts of atoms, black or white. Named "lattice gas", this model also leads to useful conclusions when applied to the liquid–vapour phase transition: a square represents a point in space, black if it is occupied by a molecule, white if it is empty.

In ferromagnets there is an attraction between two neighbouring spins if they are of the same orientation: the Ising model assumes that their energy is therefore *lowered by the value* J (Fig. 1.15). In antiferromagnets, neighbouring spins repel each other if they have the same orientation (in the Ising model this means the value of J is negative). The quantity J can be calculated from a quantum mechanical description of the material. Iron, cobalt, nickel and many other metals (Figs. 1.10 and 1.11) are examples of ferromagnetic metals which produce a spontaneous magnetisation below their critical temperature.

Let us see how two spins behave as we change the temperature if they are isolated from the rest of the universe. At very low temperature, thermal energy does not play a role and the two spin system is in one of the lowest energy states where

Fig. 1.15 The four possible
scenarios of a set of two spins
with ferromagnetic
interaction according to the
Ising model: the spins lower
their energy when they are in
the same direction, and
increase their energy if they
are in opposite directions

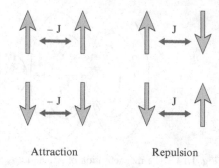

the magnetisation is maximum, that is a state in which both spins are in the same
direction. In contrast, at very high temperature the thermal energy dominates and all
states have the same occupation probability. We obtain the general expression for the
average magnetisation per spin m_F by applying (1.5). However, we are interested
in the modulus m of the magnetisation and not the direction the system chooses at
random at equilibrium; m can be evaluated by taking the average of the absolute
value of the magnetisation for each state:

$$m = \frac{1}{N} \cdot \frac{\sum\limits_{\{\mu_j\}} \left| \sum\limits_j \mu_j \right| e^{J/kT \sum\limits_{j,k} \mu_j \mu_k}}{\sum\limits_{\{\mu_j\}} e^{J/kT \sum\limits_{j,k} \mu_j \mu_k}}. \tag{1.12}$$

In the case of two spins of value $+1$ or -1, the magnetisation takes its maximum
value of 1 for the two states with both spins aligned (both up or both down) and
0 for the two states with spins in opposite directions. Summing over the four
configurations drawn in Fig. 1.15, we obtain:

$$m = \frac{1}{2} \cdot \frac{2 \cdot 2x}{2x + 2x^{-1}} = \frac{1}{1 + x^{-2}} \tag{1.13}$$

where $x = e^{J/kT}$ is the Boltzmann factor corresponding to two neighbouring spins
of the same orientation. The magnetisation m is equal to $1/2$ at high temperature
($x = 1$). This result is surprising because we expected zero spontaneous magnetisation at high temperature. It is due to the fact that we took the average of the modulus
of magnetisation over an extremely small system size: there are not enough spins for
the magnetisation to tend to zero at high temperature. To check this, we can perform
the same calculation for a slightly bigger micro-magnet – a system of 4 spins.

Figure 1.16 represents the 16 possible states for such a 4 spin micro-magnet.
At low temperature, the magnetisation is maximal and the magnet occupies one
of two states where all the spins are aligned. This example illustrates an essential

Microscopic states of a magnet containing 4 spins
(for an applied magnetic field H = 0)

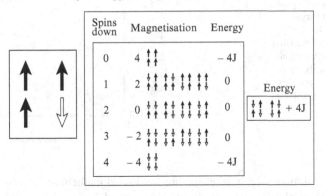

Fig. 1.16 A system of 4 spins can be found in $2^4 = 16$ distinct microscopic states if each spin can be either in state +1 (*black*) or −1 (*white*). Here these states are arranged in five families in decreasing order of magnetisation: when all the spins are in the +1 state the magnetisation takes its maximum value of 4. If 1, 2, 3 or 4 spins are in the −1 state, the magnetisation is 2, 0 −2 or −4 respectively. If we count the the interaction energies according to Fig. 1.15, we identify three families: *First*: the 2 states with all the spins aligned where the energy takes its minimum value of −4J; *middle*: 12 states where the energy is zero because there are an equal number of attractive and repulsive interactions; *right*: 2 states where the energy takes the maximum value of 4J in which the spins +1 and −1 are alternating leading to all repulsive interactions

difference taking into account the interactions makes: since the resultant energy of these interactions is not additive the usual tools of thermodynamics do not work (the energy due to interactions grows faster than the size of the system). At high temperature, thermal agitation gives an equal chance to each state. Weighting each state by the absolute value of its corresponding magnetisation, we calculate the average of the absolute value of the magnetisation m relative to the maximum magnetisation ($=4$) in the same way as for two spins. We therefore obtain:

$$m = \frac{1}{4} \cdot \frac{4 \cdot x^4 + 2 \cdot 8x^0 + 4 \cdot x^4}{2x^4 + 12x^0 + 2x^{-4}} = \frac{x^4 + 2}{x^4 + 6 + x^{-4}} \tag{1.14}$$

Figure 1.17 shows the comparison of the magnetic phase transition as a function of temperature for the micro-magnets of 2 and 4 spins. The transition occurs gradually over the range of temperature T such that kT is of the order of the interaction energy J. Here the transition is spread out over temperature from 1 to 10 in units of J/k, but the observed changes of state in real systems occur at very precise temperatures. This sharpness of the transition comes from the large number of spins involved. Before the 1970s, the only rigorous calculation was that of the Ising model on a two dimensional square lattice (by Lars Onsager in 1944). Figure 1.17 shows that in this 2D Ising model the transition occurs over a very narrow temperature window compared to the case of micro-magnets. We will return

Fig. 1.17 The magnetisation of two micro-magnets of 2 and 4 spins shows a gradual transition with varying temperature. The magnetisation of the Ising model in two dimensions (calculated exactly by Onsager) on the contrary shows a very clean sharp transition

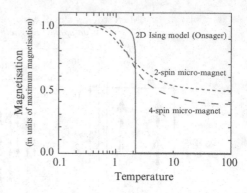

to this exact solution, which, after having fascinated mathematicians by its formal complexity, threw the physics community into turmoil. Before that, we introduce the "classical" description of phase transitions in the framework of the approximation called *mean field*.

1.4.3 A Minimal Phase Transition Model

Two ingredients are indispensable in all models of order–disorder transitions:

- The existence of interactions between the particles
- A maximum value of the order parameter

This second point is as intuitive in magnets (all the spins aligned) as in gases (all the molecules "touch each other" when the liquid state is reached). This idea was clearly expressed in 1683 by Bernoulli who was opposed to the Boyle-Mariotte law, according to which the volume of a gas can be reduced to zero by a very large pressure. Its density would therefore be infinite, remarked Bernoulli. He showed that, on the contrary, the density saturates at the value it has in the liquid state, which is practically incompressible. If we use these two properties, what is the most economical idea which predicts the appearance of a transition within a large system of particles?

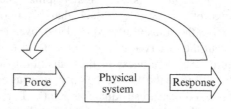

Response $= A \times [\text{Force} + a \,(\text{Response})]$

The idea of *mean field* is to improve the model of linear response in a minimal manner to take into account the interactions: we assume that part of the force applied on the system (the magnetic field or the pressure) comes from the system itself.

This feedback force, exerted on each particle by the set of all other particles, is the *mean field*.

For a gas this idea rests on the fact that the higher the fluid density the more the interactions contribute to an attractive force between the molecules: the effects of the density (response) must be added to the effects of the external pressure. In the same way for a magnet, the effects of the average magnetisation of the neighbouring spins must be added to the applied magnetic field. Rewriting the linear response in this way is in general no longer linear! A completely new property appears: *even in the absence of an external field, a self-maintained response can exist.* For this to occur, all that is needed is that the weight of the feedback a creates a mean field of sufficient intensity. Let us see the results this method gives for gases and then for magnets.

1.4.4 Van der Waals Fluid

In 1873, Johannes Van der Waals proposed the first application of this idea to gases: he started from Gay-Lussac's law and added to the pressure an internal pressure as a "mean field". He assumed that this internal pressure depends on the density:

$$n = \frac{1}{kT}\{p + a(n)\}. \tag{1.15}$$

This relation shows that the fluid density n can remain high even if the pressure is very low: *a condensed state exists at low temperature whatever the pressure.* In this case, the internal pressure replaces the external applied pressure. In practice, Van der Waals chose an internal pressure proportional to the density squared $a(n) = a \times n^2$. His reasoning for this is that the internal pressure is proportional to the number of molecules per unit volume n, multiplied by the influence of all the neighbouring molecules on each molecules. This influence being also proportional to the density n, we find Van der Waals' result. In order for the model to lead to a phase transition, it needs to be made more realistic by introducing a maximum limit to the density $1/b$. Van der Waals' equation of state is known in the equivalent form, to first order:

$$(p + a/v^2)(v - b) = kT \tag{1.16}$$

where $v = 1/n$ is the average volume occupied by one molecule. This equation of state correctly describes the liquid–vapour phase transitions, apart from close to the critical point as we will see later. The equation predicts the existence of a critical point, where the corresponding isotherm has a point of inflection of zero slope, for the following values:

Fig. 1.18 The isotherms
predicted by Van der Waals'
equation in the
neighbourhood of the critical
temperature $t = \frac{T-T_c}{T_c} = 0$

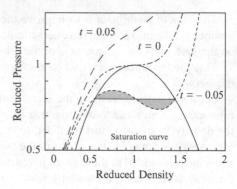

$$v_c = 3b \qquad\qquad p_c = 8a/27kb \qquad\qquad T_c = a/27b^2 \qquad (1.17)$$

Van der Waals found that by scaling the parameters by their respective critical values, curves from different substances collapse onto a general curve. If we use the reduced parameters, π, ϕ, θ, corresponding respectively to p, v, T relative to their critical values, we obtain a universal equation of state (Fig. 1.18):

$$(\pi + 3/\phi^2)(3\phi - 1) = 8\theta \qquad (1.18)$$

Most real gases obey a universal behaviour (known as the *law of corresponding states*, according to which their equation of state in reduced parameters in universal) very well. Figure 1.19 shows the coexistence curves of eight different gases, plotted in reduced coordinates $(1/\phi,\ \theta)$. These curves are remarkably superimposed. However the "universal" coexistence curve obtained from Van der Waals equation (1.18) does not fit the experimental data at all.

1.4.5 Weiss Magnet

In 1906, Pierre Weiss, after a decade of experiments on magnets, proposed modifying the Curie law in the way in which Van der Waals had modified the Gay-Lussac law:

$$M = \frac{C}{T}\{H + a(M)\} \qquad (1.19)$$

Weiss chose the "mean field" $a(M)$, also called the *molecular field*, to be simply proportional to the magnetisation M. This Weiss law describes very well the magnetic transitions, observed notably by Pierre Curie, apart from close to the critical point. Weiss' predictions surprised physicists. In the 15 years that followed, the measurements of magnetisation at low temperature by Kammerling Onnes, the amplification of crackling noise due to the reversal of magnetic domains discovered

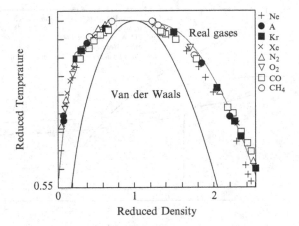

Fig. 1.19 The region of liquid–vapour coexistence is limited by the vapour saturation curve on the temperature/density graph. This is drawn here for different substances in such a way that their critical points are superimposed (after Guggenheim [5]). The *curves* lie very close to each other. Numerical study of this universal saturation curve leads to "Guggenheim's empirical law" with exponent 1/3 (*continuous line* connecting the experimental points). The saturation curve deduced from Van der Waals' law which uses the mean field approximation is presented for comparison. Its exponent of 1/2 does not correspond to experiment

by Backhausen and the direct observations of magnetic domains by Bitter showed Weiss to be absolutely right.

By using the mean field and a more precise relationship between M and H than that of Curie, proposed by Langevin, Weiss was able to calculate the spontaneous magnetisation as a function of temperature in the mean field approximation.

More quantitatively, taking (1.7) for an isolated spin, this is rewritten using Weiss' hypothesis that spins are on a lattice where each spin has q neighbours with whom it interacts with an energy J:

$$m = \tanh(qJm/kT) \qquad (1.20)$$

We introduce the *reduced coupling constant* $K = J/kT$. The above equation can be written as $m = \tanh(4Km)$ for a square lattice and leads to the reduced magnetisation shown in Fig. 1.20. To first order with respect to the temperature difference from the critical temperature $t = \frac{T-T_c}{T_c}$, the value of m is:

$$m \sim \sqrt{3} \cdot (-t)^{1/2}$$

Figure 1.20 compares the variation of this magnetisation to that calculated exactly by Onsager for a magnet in two dimensions. Despite its simplicity the mean field approximation reproduces well the essential characteristics of a change of state:

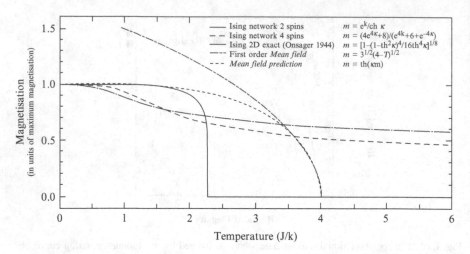

Fig. 1.20 Magnetisation calculated by Weiss using the idea of mean field (*short dashed curve* and *long-short dashed curve* to first order near the transition). The magnetisation of an Ising magnet in two dimensions calculated by Onsager in 1944 (*continuous curve*). For comparison the thermal behaviour of an Ising system of two spins (*longdash-dot line*) and four spins (*dashed line*). The parameter K is the reduced coupling constant $K = J/kT$ and q the number of first neighbours

- Maximum magnetisation at low temperature
- Zero magnetisation at high temperature
- Transition at a very precise critical temperature

Good Agreement at low Temperature

In addition, at low temperature there is excellent agreement with experiments as soon as the reduced coupling constant $K = J/kT$ departs significantly from the critical value $K_c = J/kT_c$: under these conditions the majority of spins are aligned. To reverse the direction of a spin in this perfect alignment on a square lattice, four interactions need to be changed from $-J$ to $+J$ at a total energetic cost of $8J$. To first order, the reduced magnetisation can be expressed using the Boltzmann factor as:

$$m = 1 - 2e^{-8K} \tag{1.21}$$

Figure 1.21 shows that the mean field approximation and the exact solution for the 2D Ising model coincide with that given by (1.21) as soon as K is greater than or equal to 1. However we observe that the description is not correct around the critical (Curie) temperature: neither the temperature itself nor the form of the variation are faithfully reproduced.

We have briefly explored with Johannes Van der Waals and Pierre Weiss, the idea that the effect of interactions between particles can sum up to give a "mean field", which applies in an identical way to each particle. Finally, this method gives

Fig. 1.21 Variation from the maximum magnetisation: mean field compared to the exact solution to the Ising model in two dimensions. As soon as the coupling $K = J/kT$ is of order greater than 1, the mean field approximation gives excellent agreement with the magnetisation calculated exactly for the 2D Ising model

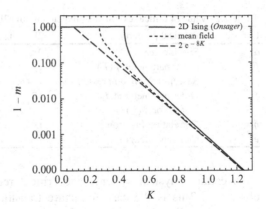

excellent results far from the critical point but poor results near the critical point. For almost a century physicists were faced with the challenge of finding a better tool.

1.4.6 Universality of the "Mean Field" Approach

No doubt Van der Waals and Weiss were not aware at this point that their approaches were so similar in principle. More importantly they did not suspect that the majority of attempts to extend their work, in fact *all* attempts until the 1970s, would lead to the *same critical behaviour* and therefore the same exponents. This is what we will later define as one and the same *universality class*. The universality of critical behaviour which follows from the mean field type approach, and its inadequacy in describing real phase transitions, comes from the following fact: the *correlations* are assumed to be *short ranged*. In Sect. 1.6 we introduce this concept of correlation which measures the way in which the values of the order parameter at two different points in the system are related. In practice, whether the range of correlations (called the *coherence length*) is taken to be zero or finite does not change the critical behaviour obtained. In fact the range of correlations diverges at the critical point and there always exists a region near T_c where the correlation range is longer than we assumed. In this region, the *critical region*, mean field approaches do not correctly describe the exponents. The physicist Ginzburg proposed a criterion to quantitatively evaluate the size of the critical region. We present this in Sect. 1.1.6.

Exponent	α	β	γ	δ	η	ν
Value predicted by mean field	0	1/2	1	3	0	1/2

Table 1.2 Value of critical exponents predicted by the 2D Ising model and mean field

Exponent	Physical property	2D Ising model exponent	Mean field exponent
α	Specific heat	0	0
β	Order parameter $f(T)$	1/8	1/2
γ	Susceptibility/Compressibility	7/4	1
δ	Order parameter at T_c $f(h)$ or $f(p)$	15	3
η	Correlation function	1/4	0
ν	Coherence length	1	1/2

In certain physical systems, the critical region is so small that we cannot observe it. This is the case for phase transitions in superconducting elements, ferromagnets, liquid crystals etc., for which the exponents predicted by the mean field approximation correspond well to the measured values.

1.4.7 2D Ising Model

In 1944 an article appeared in volume 65 of *Physical Review* (p. 117) by Lars Onsager making an essential step in the description of phase transitions: the exact solution of the Ising model in two dimensions. A skillful and complex formal piece of work granted a rigorous description of the critical behaviour of all the physical quantities for this model situation (Table 1.2).

The critical exponents calculated for the 2D Ising model are different from those predicted by the mean field approximation (Table 1.2). These results constituted a challenge for physicists. Many labs attempted the extremely precise measurements needed to distinguish between these approaches. In parallel various numerical approaches or formalisms also aimed to evaluate critical exponents. By the end of the 1960s a considerable quantity of very accurate results could be compared to the predictions, but without great success.

1.5 Universality of Critical Behaviour

Over a quarter of a century, the measurement of critical behaviours became a discipline in its own right, with its own specialists, schools and conferences. These numerous works established that the observed critical behaviours were characterised by a *universality*, that is to say possessed reproducible critical exponents, often the same ones for very different physical situations.

1.5.1 Extremely Precise Measurements

Table 1.3 collects together representative results from some of the most precise experiments, corresponding to families of diverse transitions. These three families

Table 1.3 Observed critical exponents for three different families of transitions, compared to values predicted by the 2D Ising model and the mean field model. $n(p)$ is the density difference between the liquid and gas or the density of superfluid helium as a function of the pressure

n =number of order parameter components		Models		Experiments		
		1	any	1	2	3
Exponent	Physical property	2D Ising model exponent	Mean field exponent	Liquid–vapour transition	Superfluid helium	Ferromagnetic transition (iron)
α	Specific heat	0	0	0.113 ± 0.005	-0.014 ± 0.016	-0.03 ± 0.12
β	Order parameter $m(T)$	1/8	1/2	0.322 ± 0.002	0.34 ± 0.01	0.37 ± 0.01
γ	Susceptibility/ Compressibility	7/4	1	1.239 ± 0.002	1.33 ± 0.03	1.33 ± 0.15
δ	Order parameter at T_C $m(h)$ or $n(p)$	15	3	4.85 ± 0.03	3.95 ± 0.15	4.3 ± 0.1
η	Correlation function	1/4	0	0.017 ± 0.015	0.021 ± 0.05	0.07 ± 0.04
ν	Coherence length	1	1/2	0.625 ± 0.006	0.672 ± 0.001	0.69 ± 0.02

are classified by number n of components to the order parameter, a number which, as we will see, plays an important role in the classification of families of transitions:

- $n = 1$ for the liquid–vapour phase transition. The order parameter is the density difference between the two phases and therefore a scalar. The order parameter is also a scalar for other families of transitions which have been well studied: binary fluids and metal alloys (Sect. 1.3.3). Within the precision of these experimental values the critical exponents of these three families overlap. The corresponding models are called *Ising models*.
- $n = 2$ for all "quantum" phase transitions where the order parameter is a complex wavefunction – superfluidity, superconductivity, but also for all the classical cases where there are just two degrees of freedom which is the case for example for nematic liquid crystals. The corresponding models are called *XY-models*.
- $n = 3$ for ferromagnetic–paramagnetic or ferroelectric transitions in an isotropic medium. The corresponding models are called *Heisenberg models*.

The experimental results can be classified in two categories (Fig. 1.22). In the first, we find for example superconductivity of metals, order–disorder transitions in liquid crystals and ferroelectric–paramagnetic transitions. For these transitions, the mean field type description leads to observed values for the critical exponents. In the other case, the critical exponents are not well described by either of the two models. The Ginzburg criterion, presented in Sect. 1.6, explained the difference between these two categories: for the first the "critical" region of temperature around the transition is too small to be experimentally observed (for example, less than millikelvin).

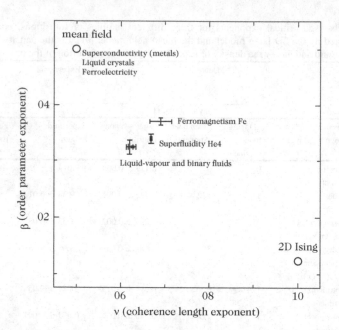

Fig. 1.22 Critical exponents β and ν: experimental results obtained for seven difference families of transitions, compared to values predicted by the mean field and 2D Ising models

When this critical region is accessible to measurement, which is the case in the second category of transitions, the critical exponents that we measure have values that we can account for today using the idea of *scale invariance* of the critical state. This idea will be the common thread in the approaches described in this book.

1.5.2   Inadequacy of Models and Universality of Exponents

We introduce and develop further the concept of scale invariance. It is useful to expand on its origin and effectiveness. The essential formal difficulty which descriptions of the critical state run into is the divergence of the characteristic length over which correlations are exerted, which we call the coherence length ξ. An initial quantitative discussion of this point is presented in Sect. 1.6. One way to interpret this divergence is to say that at the critical point:

> nothing important is modified in the physics of the critical
> state if we change the scale of observation.

For example, as we decrease the magnification of an imaginary microscope, as soon as we no longer see the microscopic details, the image of the physical system remains statistically the same. This property of scale invariance of the critical state was highlighted and used in the 1960s by Kadanoff who had

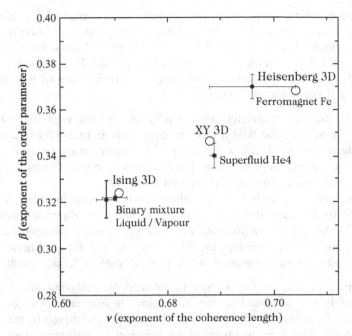

Fig. 1.23 Critical exponents measured for four families of transitions, compared to values predicted by the three corresponding models that take into account the scale invariance of the critical state

the intuition that this would be the key to an effective description of critical phenomena. In fact in 1970 several physicists, notably Wilson, proposed a series of methods called "renormalisation group" which enabled the calculation of critical behaviours drawing out the physical consequences of scale invariance. One of these consequences is that critical behaviours do not greatly depend on microscopic physical details that are "averaged out" at large scales. However they depend strongly on the geometric characteristics of the system – the spatial dimension and the number n of components of the order parameter.

Zooming in on Fig. 1.22 and adding on the predictions given by the renormalisation group for transitions in three dimensional space gives Fig. 1.23. The three models shown, Ising ($n = 1$), XY ($n = 2$) and Heisenberg ($n = 3$) predict exponents in excellent agreement with experiment.

1.6 Limits of the Mean Field Approximation

1.6.1 Landau–Ginzburg Theory

In 1937, Lev Landau proposed a general description of "mean field" type approaches [7]. Magnetic transitions are described in terms of a local free energy $f(r)$, which is expressed as a function of the order parameter (*magnetisation*), the conjugate field $h(r)$ and the temperature T. The local free energy f integrated over the whole volume gives the total free energy F, the minimum of which leads to the equilibrium values of $m(r)$ and $h(r)$ for the applied boundary conditions. This simple framework qualitatively accounts for the two observed characteristic types of transition for magnetism and many other phase transitions:

- If $T < T_c$, the order parameter spontaneously takes a finite value in the absence of an external magnetic field h. As we apply such an external excitation and gradually reverse its direction the value of the order parameter (for example the magnetisation) switches in a discontinuous manner, abruptly changing orientation. This is a "first order phase transition".
- As T increases and tends to T_c, we observe that the jump in order parameter produced by inverting the field decreases to the point where it vanishes at $T=T_c$. The behaviour as a function of h and T becomes continuous but contains singularities. In the terminology established by Paul and Tatiana Ehrenfest, this is a "second order transformation" at what we call today a "critical point".

The expression for the free energy is obtained by analysing the symmetry properties of the system around the transition point. The first terms in the expansion around $T = T_c$ are directly determined by the symmetries obeyed by the system transformations. The free energy must be invariant to transformations of the symmetry group of the system studied. We show here the simplest case of a scalar order parameter (so the sign of the order parameter below T_c can be positive or negative). In this case the function f is even with respect to m if $h = 0$:

$$f(m, h = 0, T) = f(-m, h = 0, T) \tag{1.22}$$

Considering just the two simplest terms respecting this symmetry and an "elastic" term $|\nabla m|^2$ opposing spatial variations in m, we obtain:

$$f(m, h, T) = \widetilde{a}m^2 + \frac{b}{2}m^4 + c\,|\nabla m|^2 - hm \tag{1.23}$$

where the coefficients $\widetilde{a}$, b and c can depend on temperature a priori. The first two terms of equation (1.23) were initially proposed by Landau by assuming the free energy f can be written as a Taylor series expanded around the critical point. This hypothesis does not take into account the fact that the transition point is itself a singular point in the thermodynamic potential. However, the powerfulness and generality of Landau's approach is due to the analysis of the symmetry of the physical system considered. Landau's hypothesis is that only $\widetilde{a}$ varies with T. Furthermore $\widetilde{a}$ changes sign at T_c and causes the transition by introducing a negative term in f (in the absence of a magnetic field this is the only negative term):

$$\widetilde{a} = a\,t = a\frac{T - T_c}{T_c} \tag{1.24}$$

We can calculate the local field h as a function of magnetisation m from the expression (1.23) for the free energy at equilibrium (i.e. when f is a minimum). To first order in m this is in agreement with the Curie–Weiss model (see Sect. 1.4.5):

$$h = 2\,a\,t\,m \quad \text{and} \quad \chi = \frac{\partial m}{\partial h} = \frac{1}{2a\,t} \text{ which diverges at } T_c. \qquad (1.25)$$

1.6.1.1 Homogeneous Solution of Landau's Theory in Zero External Field

When the system is homogeneous and the external field is zero, the equilibrium condition leads to a minimum free energy f_0 and magnetisation m_0:

$$f_0(m, h = 0, T) = at m_0^2 + \frac{b}{2} m_0^4 \qquad (1.26)$$

in one of the following three states:

- $m_0 = 0$ and $f_0 = 0$ for $t > 0$ $(T > T_c)$ (1.27)

- $m_0 = \pm\sqrt{\frac{-at}{b}}$ and $f_0 = -\frac{a^2 t^2}{2b}$ for $t < 0$ $(T < T_c)$

It is worth noting that for $t < 0$ the state $m = 0$ is in equilibrium but is *unstable* since $f = 0$ corresponds to a maximum (Fig. 1.24).

We can also calculate the specific heat $C = -T \frac{\partial^2 F}{\partial T^2}$ at zero field in this model:

- For $t > 0$ $(T > T_c)$ $C = 0$

                                               (1.28)

- For $t < 0$ $(T < T_c)$ $C = T \frac{a^2}{b T_c^2}$.

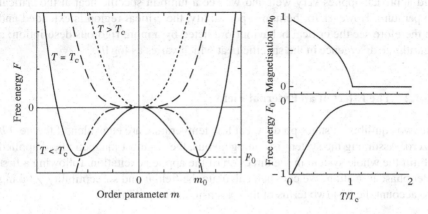

Fig. 1.24 Solutions to Landau's theory in the absence of external field for a homogeneous system

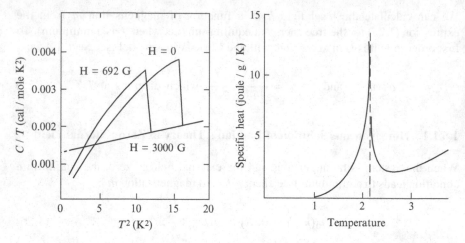

Fig. 1.25 *Left*: Jump in specific heat observed during the superconducting phase transition of tantalum for several values of magnetic field (at 3,000 Gauss superconductivity is no longer observed). *Right*: Divergence (logarithmic) of the specific heat observed during the helium superfluid phase transition (after [6])

The specific heat does not diverge – its critical exponent α is equal to zero in the mean field approximation – but it does have an abrupt jump $\Delta C = \frac{a^2}{bT_c}$ at the transition point. This jump is well observed in certain phase transitions where the mean field approximation describes the physical system well.

Figure 1.25 compares two physical situations from the same transition family (that we will define later as a *universality class*) where the order parameter is a complex wavefunction: superconductivity and superfluidity. As we will see at the end of this chapter, the difference between these two systems is in terms of the size of the *critical region*. For superconductivity of tantalum, as for all superconducting elements, the critical region is so small we cannot observe it. Therefore the mean field approach applies very well and we see a jump in specific heat at the critical temperature. However for helium superfluidity, the critical region is extended and we therefore see the critical behaviour predicted by a more rigorous description: a logarithmic divergence in the specific heat which varies as $\log |t|$.

1.6.1.2   The Effect of an External Field

The two equilibrium states predicted at low temperature are equivalent if the field h is zero. Assuming the system is homogeneous, the slightest magnetic field applied will tilt the whole system in the direction of the applied excitation, following a first order phase transition. We can then calculate the field h and susceptibility χ taking into account the first two terms of the expansion. At equilibrium:

$$h = 2atm_0 + 2bm_0^3 \quad \text{and} \quad \chi = \frac{\partial m}{\partial h} = \frac{1}{2at + 6bm_0^2}$$

- For $h = 0$ and $t > 0$ $(T > T_c)$ we obtain $\chi = \dfrac{1}{2at}$

- For $h = 0$ and $t < 0$ $(T < T_c)$ we obtain $\chi = -\dfrac{1}{4at}$. $\qquad\qquad$ (1.29)

In the mean-field approximation, the susceptibility χ varies as t^{-1}. In general the exponent describing the divergence of the susceptibility is called γ.

In practice, even at low temperature, the system is not homogeneous at equilibrium due to its finite size. In general the structure is in macroscopic regions, called domains, where the order parameter changes its magnitude or direction to minimise the surface effects. Landau's formalism is particularly powerful for studying these macroscopic spatial variations (macroscopic in the sense of at a larger scale than the range of interactions).

1.6.2   Spatial Variation of the Order Parameter

Using the general expression for the free energy (1.23), we calculate the spatial variations of h and m in the presence of a point perturbation at $r = 0$ of:

$$\delta h(r) = h_0 \delta(r) \quad \text{and} \quad m = m_0 + \delta m(r) \qquad (1.30)$$

where m_0 is the equilibrium state in absence of external magnetic excitation and δm is the small variation resulting from the perturbation δh. To calculate δm, we use the fact that $F = \int f d^3 r$ is a minimum at equilibrium:

$$\int d^3 r f(m, h, T) = \int d^3 r \left[atm^2 + \frac{b}{2}m^4 - hm \right] + \int d^3 rc|\nabla m^2| = \text{minimum}$$

$$(1.31)$$

Keeping only the terms linear in δm and using integration by parts for the second integral, we obtain the variation of F which is zero when F is minimum:

$$\int d^3 r \delta m [2atm + 2bm^3 - h - c\nabla^2 m] = 0 \qquad (1.32)$$

This calculus of *variations* leads to the minimum of F with respect to δm as for normal differential calculus. A simple solution is that f is itself minimum at each point:

$$\frac{\delta f}{\delta m} = \left[2atm + 2bm^3 - h - c\nabla^2 m \right] = 0 \qquad (1.33)$$

For $h = 0$, this equation is true for m_0, for which the gradient is zero. Therefore decomposing m gives the equation to first order in δm:

$$\nabla^2 \delta m - \frac{2}{c}(at + 3bm_0^2)\delta m = -\frac{h}{c} \tag{1.34}$$

Depending on the temperature this equation takes two related forms at equilibrium:

$t > 0 \ (T > T_c) \ m_0 = 0$	gives	$\nabla^2 \delta m - \dfrac{2at}{c}\delta m = -\dfrac{h_0}{c}\delta(r)$
$t < 0 \ (T < T_c) \ m_0 = \pm\sqrt{\dfrac{-at}{b}}$	gives	$\nabla^2 \delta m + \dfrac{4at}{c}\delta m = -\dfrac{h_0}{c}\delta(r) \quad (1.35)$

  The solution to this equation in spherical coordinates in infinite space (i.e. without boundary effects) in d dimensions leads to:

$$\delta m = \frac{h_0}{4\pi c}\frac{e^{-r/\xi}}{r^{d-2}} \tag{1.36}$$

where ξ takes the two values ξ_+ or ξ_- depending on the temperature:

- for $t > 0 \ (T > T_c) \quad \xi_+ = \sqrt{\frac{c}{2at}} = \sqrt{2}\xi_0 t^{-1/2}$

- for $t < 0 \ (T < T_c) \quad \xi_- = \sqrt{\frac{c}{-4at}} = \xi_0|t|^{-1/2}$

$$\tag{1.37}$$

where $\xi_0 = \sqrt{\frac{c}{4a}}$ is the coherence length, i.e. the range of correlations (see following paragraph), extrapolated to zero temperature.

  We note that if the initial perturbation in the field is more complex we can calculate the response of the system from (1.36) by convolution with the expression for the perturbation. We call this response to a point perturbation the *Green's function* of the problem.

Correlation Function and Coherence Length

The function $\delta m(r)$ given by (1.36) can be expressed as the product of h_0/kT and the correlation function $G(r)$ of $m(r)$ (see Sect. 3.2):

$$G(r) = \langle m(r)m(0)\rangle - \langle m(r)\rangle\langle m(0)\rangle \tag{1.38}$$

If we decompose the Hamiltonian H as $H = H_0 - \int d^d r h(r)m(r)$, the average of m is written:

$$\langle m(r) \rangle = \frac{Tr \left\{ m(r) \left[\exp \left(-H_0/kT + 1/kT \int d^d r h(r)m(r) \right) \right] \right\}}{Tr \left[\exp \left(-H_0/kT + 1/kT \int d^d r h(r)m(r) \right) \right]} \qquad (1.39)$$

Differentiating this expression with respect to h gives:

$$\delta m(r) = h_0/k_T \left[\langle m(r)m(0) \rangle - \langle m(r) \rangle \langle m(0) \rangle \right] \qquad (1.40)$$

The expression for δm (1.36) therefore leads to the correlation function:

$$G(r) = \frac{1}{4\pi c} \frac{e^{-r/\xi}}{r^{d-2}} \qquad (1.41)$$

The physical significance of the coherence length ξ is that it is the *range of correlations* of the order parameter in the system. Beyond a distance $r = \xi$ between two points the function $G(r)$ is negligible, in other words the points do not influence each other – their physical states can be considered independent of each other. Equation (1.37) show that in the mean field approximation $\xi \sim t^{-1/2}$. In general the exponent describing the divergence of the coherence length close to the critical point is called v i.e. $\xi \sim t^{-v}$.

1.6.2.1 Limits of the Mean Field Approximation and the Ginzburg Criterion

As we have seen, the mean field approach does not in general correctly describe the critical behaviour. We now know the reason for this is that it neglects the microscopic local correlations. The macroscopic term $c |\nabla m|^2$ in the free energy assumes $m(r)$ is *continuous*. This term corresponds to *perfect correlations* in a volume "small in terms of the scale of variations in m" but "large compared to the atomic scale". This corresponds well to the initial idea of the mean field calculated from *the mean m of all the spins in the system*, i.e. the idea of perfect correlation over an infinite range (see (1.19)). On the other hand, if we assume $c = 0$, we find $\xi = 0$ i.e. zero ranged correlations in (1.35). There is no middle ground in the mean field model!

In 1960, Ginzburg had the idea of quantitatively evaluating the effect of these correlations to find a criteron for the validity of mean field results:

> The mean field approach is valid if the mean amplitude of thermal fluctuations $\langle \delta m(t) \rangle$ at temperature t is less than m_0.

A Critical Dimension

Several simple arguments allow us to evaluate this limit of validity for the mean field approach. For example we can evaluate the free energy related to the order of the system in a "coherence volume" ξ^d, and if this is greater than kT at the critical point we can say the mean field model is valid:

$$|f_0 \xi^d| > kT_c \tag{1.42}$$

In other words we can neglect the effect of fluctuations inside a coherence volume within which the idea is that the order is "rigid". A coherence volume therefore corresponds to a single degree of freedom where the thermal energy is of order kT. For $t < 0$, the values of f_0, ΔC and ξ_- calculated in the previous paragraph lead to:

$$\frac{a^2 t^2}{2b} \left(\frac{c}{4a |t|} \right)^{d/2} > kT_c$$

or equivalently:

$$|t|^{2-d/2} > \frac{2k}{\xi_0^d \Delta C} \tag{1.43}$$

Since the absolute value of t is much less than 1 in the critical region, one direct consequence is that this inequality *is always valid if the exponent of t is negative.* In other words if the number of dimensions d is above a threshold known as the *critical dimension* $d_c = 4$. This conclusion, which can be justified more rigorously, is of great significance. Landau's theory, constructed without particular reference to the spatial dimension of the physical system, contains its own limits of applicability:

> The mean field description gives the correct critical behaviour for all systems of spatial dimension greater or equal to 4.

Ginzburg thereby gave the first sound argument explaining the influence of spatial dimension on phase transitions. In particular this argument could explain the conspicuous, yet hard to accept, differences between the exponents measured experimentally in 3D systems and the 2D Ising model predictions solved exactly by Onsager. Ginzburg's criterion rendered the situation in 1960 particularly frustrating – our own world of three dimensions was the only one for which there was no theoretical predictions!

A Quantitatively Predicted Critical Region

For spatial dimensions less than 4, (1.43) gives a value for the size of the critical region $|t_G| = \left| \dfrac{T_G - T_c}{T_c} \right|$:

$$|t_G| = \frac{1}{a} \left(\frac{2^{d+1} bkT_c}{c^{d/2}} \right)^{\frac{2}{4-d}} = \left(\frac{2k}{\xi_0^d \Delta C} \right)^{\frac{2}{4-d}} \tag{1.44}$$

Within the critical region $|t| < |t_G|$, the mean field description is not valid. This region can also be defined by the Ginzburg length ξ_G such that the mean field is not valid for correlation lengths $\xi > \xi_G$ because the inequality (1.40) does not hold. The Ginzburg length ξ_G is therefore given by:

$$\xi_G = \xi_0 \left(\frac{\xi_0^d \Delta C}{2k} \right)^{\frac{1}{4-d}} \tag{1.45}$$

The free energy of condensation in the ordered state in a volume ξ_G^d can be evaluated by kT_c at a distance $|t_G|$ from the critical temperature. When the correlation range ξ is less than ξ_G, the fluctuations in the order are negligible and the mean field description is valid. Since ΔC is known experimentally, we can deduce a numerical value for the size of the critical region t_G and the corresponding limit to the coherence length ξ_G for each transition (Fig. 1.26).

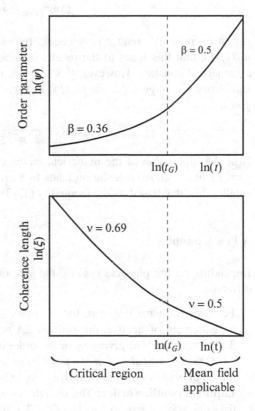

Fig. 1.26 The boundary between the critical region and the region where the mean field approximation is valid is predicted by Ginzburg's criterion. The values of critical exponents chosen here correspond to ferromagnetism

A More Rigorous Calculation Using the Amplitude of Fluctuations

Equation (1.42) is based on a crude evaluation of the effect of fluctuations. The Ginzburg criterion can be established in a more rigorous way by calculating the mean squared amplitude of fluctuations $\langle(\delta m)^2\rangle_{coh}$ over a *coherence volume* ξ^d:

$$\langle\delta m^2\rangle_{coh} = \frac{1}{(\xi^d)^2}\int_{\xi^d} d^d x d^d x' \delta m(x)\delta m(x') \tag{1.46}$$

or alternatively by taking into account the definition of the correlation function $G(r)$ and translational invariance:

$$\langle\delta m^2\rangle_{coh} = \frac{1}{\xi^d}\int_{\xi^d} d^d r G(r) \tag{1.47}$$

The Ginzburg criterion therefore is written as:

$$\langle\delta m^2\rangle_{coh} < m_0^2 \tag{1.48}$$

As an exercise the reader may check, by using the expression for $G(r)$ given in (1.41), that this leads to the results obtained in the previous paragraph, up to a numerical constant. However, this numerical coefficient can be substantial and significantly change value of (1.45). In three dimensions a rigorous calculation gives:

$$|t_G|_{3D} = \frac{1}{32\pi^2}\cdot\frac{k^2}{\xi_0^6 \Delta C^2} \tag{1.49}$$

Note the high value of the numerical factor correcting the result of the previous paragraph: the above calculation leads to a critical region about a thousand times smaller than that based on the inequality (1.40).

A Few Examples

Depending on the physical system, the size of the critical region t_G can be very different:

- **Ferromagnetism:** For iron, for example, ξ_0 extrapolated to $T = 0$ from the measurement of neutron diffusion is $2\mathring{A}$ and the jump in specific heat ΔC is 3×10^7 erg/cm^3/K, giving t_G of the order of 0.01. Given that T_c is higher than 1,000 K, *the critical region of a few tens of kelvins is easily observed over several orders of magnitude of the relative change in temperature t.*
- **Liquid crystalline order:** The *smectic A – smectic S* transition can be studied in the same way: ξ_0 extrapolated to $T = 0$ is $20\mathring{A}$ and the jump in specific heat ΔC is 10^6 erg/cm^3/K, giving t_G of the order of 10^{-5}. Since the critical temperature is of the order of 300 K, the critical region is of the order of a millikelvin. *In this*

case, the transition is well described by the mean field approximation within the normal experimental range.

- **Metal superconductors:** In the case of superconducting elements, the coherence length is long: at $T = 0$ it is of order a micron and the jump in specific heat a few $10^4 \, \mathrm{erg/cm^3/K}$. This long coherence length leads to a critical region t_G of the order of 10^{-15}, clearly impossible to observe. *In metal superconductors, the mean field approach is perfectly valid* (see Fig. 1.25). One word of caution, this is not true of high temperature superconductors (cuprates) which are two dimensional and therefore the coherence length is of order $15 \mathring{A}$ – in this case the critical region can reach tens of kelvins.

References

1. J.J. Binney, N.J. Dowrick, A.J. Fisher, M.E.J. Newman, *The theory of critical phenomena* (Oxford University Press, USA, 1992)
2. P. Calmettes, I. Laguës, C. Laj, Evidence of non zero critical exponent for a binary mixture from turbidity and scattered light-intensity measurements, Phys. Rev. Lett. **28**, 478 (1972)
3. J. Evetts, in *Concise encyclopedia of magnetic & superconducting materials,* ed. by J. Evetts (Pergamon Press, Oxford, 1992)
4. M.S. Green, J.V. Sengers, in *Critical Phenomena Proceedings of Conference NBS*, (Washington, 1965), ed. by M.S. Green, J.V. Sengers. NBS miscellaneous publications, 273, December 1966.
5. E.A. Guggenheim, The principle of corresponding states, J. Chem. Phys. **13**, 253 (1945)
6. W.H. Keesom, M.C. Désirant, The specific heats of tantalum in the normal and in the superconductive state, Physica **8**, 273 (1941)
7. L. Landau, E.M. Lifshitz, *Statistical Physics* (Pergamon Press, London, 1958)
8. G. Morandi, F. Napoli, E. Ercolesi, *Statistical Mechanics* (World Scientific, Singapore, 2001)
9. W. Smith, M. Giglio, G.B. Benedek, Correlation range and compressibility of Xenon near critical point, *Phys. Rev. Lett.* **27**, 1556 (1971)

Chapter 2
Fractal Geometry

In the previous chapter we introduced the concept of *scaling invariance* in the context of phase transitions. This invariance manifests itself in a directly observable way by the appearance of spatial structures without characteristic length scales, for example density fluctuations in a critical fluid or clusters of up-spins at $T = T_c$. Throughout this book we will meet many other scale invariant objects, which are *self-similar* in that blown-up detail is indistinguishable from the object observed as a whole. Examples include, among others, trajectories of particles performing Brownian motion, conformations of linear polymers, snowflakes, sparks, fractures, rough surfaces, percolation clusters, lungs, geographical drainage basins, blood vessels, neurons, bacterial colonies and strange attractors. We will see that the presence of self-similar structures, christened *fractals* by Mandelbrot [7, 8] is the spatial signature of critical phenomena, reflecting the divergence of the correlation length and accompanied by scaling laws for different system observables. But before we study the physical mechanisms underlying these structures, in this chapter we will look at their geometry and show that their quantitative description requires radically new concepts – that of *fractal geometry*.

2.1 Fractal Dimensions

2.1.1 Fractal Structures

Mathematicians have known of scale invariant objects (see Fig. 2.1) for a long time, but at first they considered them as oddities without any interest other than forming counter examples. Examples "invented" at the beginning of last century include continuous curves which are nowhere differentiable (Hilbert curve, Koch curve) and uncountable sets of measure-zero (Cantor set, Sierpinski gasket). At the time it was not the self-similarity of these structures that was highlighted but their mathematical consequences such as their non differentiability. It was the emergence of the concept

A. Lesne and M. Laguës, *Scale Invariance*, DOI 10.1007/978-3-642-15123-1_2,
© Springer-Verlag Berlin Heidelberg 2012

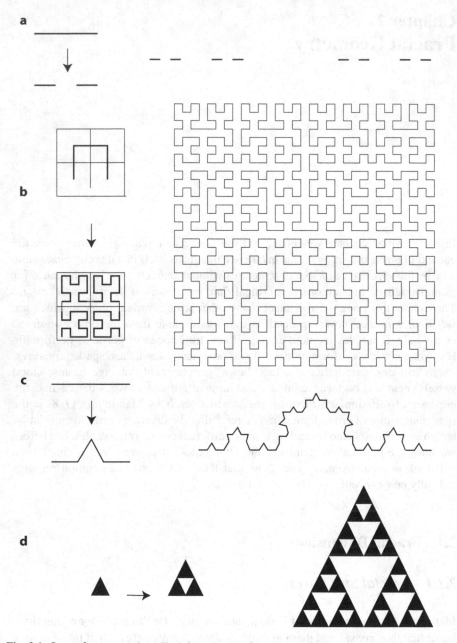

Fig. 2.1 Some famous mathematical fractals: (**a**) Cantor set [1], is an uncountable but null measure set; its fractal dimension is $d_f = \log 2/\log 3 < 1$. (**b**) Hilbert curve [5], is a space-filling curve; its fractal dimension is $d_f = 2$; (**c**) Koch curve [6], is continuous everywhere but differentiable nowhere; its fractal dimension is $d_f = \log 4/\log 3 > 1$; (**d**) Sierpinski gasket (also known as Sierpinski triangle or sieve) [10], is a set in which each point is a branch point (where the edges of two triangles meet); its fractal dimension is $d_f = \log 3/\log 2 < 2$. The figure shows the algorithm generator on the *left* and the resulting construction after three iterations on the *right*. The reader can imagine extrapolating the resulting diagram obtained ad infinitum

and the experimental realisation of scale invariance in physics that transformed this initial oversight into active interest and generated numerous works in mathematics as well as in physics. This resulted in a new geometry – fractal geometry. With the benefit of hindsight we can now see that the roots of fractal geometry are found in Perrin's work on Brownian motion (which we will go into in Chap. 4), in the work of the British biologist D'Arcy Thompson *On growth and form* [2], and in a more sporadic manner in the work of various mathematicians (Hausdorff, Hilbert, Minkowski, Bouligand). The credit goes to Mandelbrot, author of the foundational work *The fractal geometry of Nature* [8], for having shown the reality, universality and applicability of fractal geometry.

The mathematical structures shown in Fig. 2.1 are used today to define the essential concepts of fractal geometry, which we apply (in a restricted and statistical sense) to some natural structures, in the following. The construction algorithms of these mathematical structures show their self-similarity, for example, expanding the Cantor set (Fig. 2.1a) by a factor of 3 gives a structure that is exactly the same as joining two copies of the initial Cantor set. A uniform expansion of the Sierpinski gasket (Fig. 2.1d) by a factor of 2 contains 3 copies of the initial gasket. The number $n(k)$ of copies contained in the structure expanded by a factor k can be written

$$n(k) = k^{d_f} \tag{2.1}$$

which is one way of introducing the fractal dimension d_f. $d_f = d$ for a d-dimensional Euclidean structure and we find $d_f = \log 2/\log 3 < 1$ for the Cantor set and $d_f = \log 3/\log 2 < 2$ for the Sierpinski gasket. However this definition of fractal dimension is not so useful since it relies too much on the exact self-similarity character of the fractal under consideration, constructed by a deterministic algorithm iterated an infinite number of times. We will prefer a more operative approach.

2.1.2 Fractal Dimensions

Fractal structures deviate from Euclidean geometry because they are self-similar: a magnified detail is similar to the whole structure. In particular, fractal structures have details at all length scales. As a consequence, measurements we make depend on the scale at which we observe. A typical example is the coast of Britain [9], whose length varies with the scale of survey chosen, even diverging if we look at the scale of the tiniest rock crevices. The reader may check this by measuring the length of the coast from maps of different scales, from a world atlas to an Ordnance Survey map and noting that the latter gives the longest length (in km).[1] Incidentally, this

---

[1]We are, of course, talking about lengths in km, obtained by multiplying the length of the representative line on the map by the scale of the map.

property is also observed, in a more rigorous way, for the Koch curve of Fig. 2.1c. These examples show that the concept of length no longer makes sense for a fractal curve. Therefore we need to find another, more objective, indicator to quantitatively characterise fractal curves. Self-similarity enables us to define such an index; the real number d_f (in general non-integer), called the *fractal dimension*, describing the way in which the value of the measurement varies with the resolution a of the measuring apparatus: $L(a) \sim a^{1-d_f}$. It is therefore the *link* between the values $L(a)$ and $L(ka)$, obtained with different resolutions a and ka, that gives an intrinsic characteristic of the curve, the fractal dimension d_f, through the relationship:

$$L(ka) = k^{1-d_f} L(a) \qquad \text{(resolution } a). \qquad (2.2)$$

For an Euclidean ("rectifiable" in technical wording) curve we find $d_f = 1$. For $d_f > 1$, we have a convoluted curve of infinite length in the limit where $a \to 0$ (infinite resolution); the classic example being the Koch curve (Fig. 2.1c). For $d_f < 1$, we have a lacunary (from the Latin word *lacuna* meaning gap) set of zero length in the limit where $a \to 0$; the classic example being the Cantor set (Fig. 2.1a).

The length measured depends, not only on the resolution a, but also on the spatial extension l of the section of the curve considered, i.e. $L = L(a, l)$. The self-similarity of the curve can be expressed as $L(ka, kl) = kL(a, l)$, for all $k > 0$, and it follows that $L(a, l) \sim a^{1-d_f} l^{d_f}$. The length of the fractal curve is multiplied by k^{d_f} if we multiply the linear extension by k, keeping the resolution fixed:

$$L(kl) \sim k^{d_f} L(l) \qquad \text{(linear extension } l). \qquad (2.3)$$

So the fractal dimension describes the way in which our perceptions of the object at different scales of observation are linked; by considering the whole object at all scales, we can extract objective information (the value of d_f) from our subjective perceptions. We will next take the definitions and relationships above, which are restricted to fractal curves, and extend them to more general fractal objects.

2.1.3   Self-Similarity (or Scale Invariance) of Fractal Structures

Above we saw that a description of a spatial structure involves *two* subjective parameters that define the level of observation and measurement: the linear size l of the observed region (observation *field*) and the resolution a. a is actually the minimum scale, taken as a unit of measurement: if the structure is a curve, we use a *segmentation* (surveying) method in which we represent the curve by a broken line made up of segments of length a; if on the other hand the structure extends over a 2D plain or 3D space, we use a *tiling* (*box counting*) method in which we cover the surface or hypersurface with boxes (squares or cubes) of side a. If the structure is fractal all the observables describing it depend on l and a. We will use:

- The number $N(a, l)$ of "volume" elements a^d necessary to cover the structure.[2]
- The mass $M(a, l)$ given by $M(a, l) = a^d N(a, l)$.
- The density $\rho(a, l) = l^{-d} M(a, l)$ if the structure is immersed in an Euclidean space of dimension d.

For an Euclidean object of dimension d, we have

$$M(a, l) \sim l^d \qquad \text{(independent of } a\text{)} \tag{2.4}$$

$$N(a, l) \sim \left(\frac{l}{a}\right)^d \tag{2.5}$$

$$\rho(a, l) \sim \text{constant.} \tag{2.6}$$

On the other hand, for a fractal of fractal dimension d_f, we have the less trivial relationships:

$$M(a, l) \sim l^{d_f} a^{d-d_f} \tag{2.7}$$

$$N(a, l) \sim l^{d_f} a^{-d_f} \tag{2.8}$$

$$\rho(a, l) \sim \left(\frac{a}{l}\right)^{d-d_f}. \tag{2.9}$$

One critical point which should be explicitly underlined is that a fractal dimension is only well defined if the structure is *self-similar*, written (for all k, a, l) as

$$\text{self-similarity:} \qquad N(ka, kl) \sim N(a, l). \tag{2.10}$$

In other words, we cannot distinguish the structure observed in a region of size kl with resolution ka with the magnification by a factor k of the structure observed in a region of size l with a resolution a. This self-similarity, or *scale invariance*, ensures that the fractal dimension does not depend on the scale of observation. This is because the dimension d_f depends a priori on a:

$$N(a, kl) \sim k^{d_f(a)} N(a, l). \tag{2.11}$$

However, using the self-similarity property, we can also write:

$$N(a, kl) \sim N\left(\frac{a}{k}, l\right) = k^{d_f(a/k)} N\left(\frac{a}{k}, \frac{l}{k}\right) \sim k^{d_f(a/k)} N(a, l). \tag{2.12}$$

---

[2]The dimension d of elements used in the covering (rods, tiles, boxes, hypercubes, etc) can be chosen in several ways: a fractal curve could be covered with rods ($d = 1$) or tiles ($d = 2$); it can be shown that both procedures lead to the same fractal dimension.

Self-similarity therefore ensures that $d_f(a) = d_f(a/k)$. These formal manipulations are somewhat superfluous for the mathematical fractals in Fig. 2.1, however verifying self-similarity is *indispensable* in characterising the fractal nature of structures observed in the real world. We will see this in the next section which presents how the concept of fractal dimension can be used in practice.

2.2   Fractal Structures in the Natural World

2.2.1   Statistical and Limiting Properties of Fractals

The simplest method to determine the fractal dimension of an experimental structure is the box counting method that we used in the previous section to define d_f. However as soon as we leave the ideal world of mathematical fractals, self-similarity and fractal dimensions will be defined *in a statistical sense*. The useful quantity will actually be the average $\langle N(a,l) \rangle$ of the different numbers $N(a,l,O)$ obtained as we vary the origin of the box in the tiling. Using the scaling law $\langle N(a,l) \rangle \sim (l/a)^{d_f}$, rewritten as $\log\langle N \rangle \sim d_f(\log l - \log a)$, gives the fractal dimension d_f. It is not sufficient that the experimental data sufficiently accurately and reliably produces a linear section of significant length on a log-log plot[3] of N as a function of l (i.e. a linear section of the curve, as in the sequence of points, representing $\log N$ as a function of $\log l$). It is also necessary to check that the slope of this linear section, that is d_f, does not depend on the resolution a of the observation. In a similar way, if we work with the log-log plot of N as a function of a, we have to check that the slope $-d_f$ does not change if we change the observation field l. It is also necessary to observe a linear region over several decades for it to make sense to talk about a fractal dimension.

In addition, for a real fractal structure, a linear section of the log-log graph $l \mapsto N(a,l)$ with a slope independent of a (or a slope independent of l on the log-log graph of $a \mapsto N(a,l)$) is only seen in a certain range of scales for l and a. Let us consider the example of a porous rock. Although a porous rock has pores of various sizes, "at all length scales", these sizes are actually restricted between a lower bound, a_m, and an upper bound, a_M. If we observe the rock with a resolution $a < a_m$, we see the compact microscopic structure which is 3D Euclidean. If we observe the rock very roughly with a resolution $a > a_M$, we see a homogeneous rock, also three dimensional, in which the presence of pores manifests itself only in the low value of the average density. The log-log plot of $N(a)$ will therefore show two discontinuities in the slope (*crossovers* between regimes), passing from a slope of -3 for very fine resolutions $a < a_m$ to a slope of $-d_f$ (which is shallower)

---

[3]In general we use logarithms to the base 10, but the slope is independent of the choice of base for the logarithm.

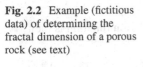
Fig. 2.2 Example (fictitious data) of determining the fractal dimension of a porous rock (see text)

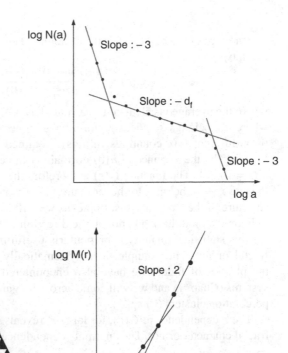

Fig. 2.3 Example of an artificial fractal of what is typically observed for a real fractal: above a certain scale the fractal becomes Euclidean again

during the region where the rock is fractal, returning to a slope of -3 for coarse resolutions $a > a_M$ (Fig. 2.2). Such crossovers are the general rule when using real experimental data (see also Fig. 2.3).

Correlation function

Another observation directly related to the fractal dimension is the spatial correlation function of the structure, which we have already introduced in Sect. 1.3.2. This point needs to be expanded on with more details in several respects: it is an effective way in practice to access the fractal dimension of the structure and the correlation functions reflect the critical character of the fractal structure.

For a tiling of d dimensional space by boxes of side a, for each site $r \in (a\mathbf{Z})^d$ we define a local observable $n(r)$ taking the value 1 or 0 depending on whether or not the box touches the structure. The correlation function is then given by[4]:

$$C(r) = \frac{\langle n(r)n(0)\rangle - \langle n(0)\rangle^2}{\langle n(0)^2\rangle - \langle n(0)\rangle^2},$$
(2.13)

where the average is calculated as a spatial average over all pairs $(r_0, r + r_0)$ where r_0 belongs to the structure.[5] We generally superimpose averages over different box countings (tilings), obtained by varying the origin O. Dividing by the variance of $n(0)$ normalises the correlation function to one: $C(r = 0) = 1$. The function $C(r)$ is therefore the conditional probability that the site $r + r_0$ belongs to the structure given that the site r_0 is found on the structure. If the structure is isotropic, as we will assume below, $C(r)$ depends only on the modulus r and not on the direction of r.

This statistical quantity is particularly useful in analysing real structures, fractal or not. For example, it is systematically calculated and studied in the physics of fluids; we have also encountered it in the analysis of spin systems (Chap. 1) and we will come across it again to quantitatively describe percolation clusters (Chap. 5).

The r dependence of $C(r)$, for large r, reveals whether the structure has a fractal character or not. The "normal" dependence of $C(r)$ is exponential:

$$C(r) \sim e^{-r/\xi},$$
(2.14)

where ξ is the characteristic length (correlation length). For a fractal structure on the other hand we have:

$$C(r) \sim \frac{1}{r^{d - d_f}}.$$
(2.15)

This replacement of an exponential dependence by a power law dependence is a typical signature of critical behaviour. It reflects the divergence of the correlation length ξ.

---

[4]There are two notations for correlation functions, C and G, with C being more common in geometric contexts (quantitative descriptions of spatial structures) and dynamic contexts (temporal correlation functions).

[5]This way of calculating the correlation implicitly assumes that the structure is statistically invariant on translation. In practice, we can vary r_0 in a sample size l, large enough to be representative but small enough to avoid meeting the edges of the structure. We can then study the l dependence of the result (finite size effects); in practice, $C_l(r) \approx C_\infty(r)$ as long as $r \ll l$.

In fact the very definition of the correlation function shows that it will depend on the scale a of tiling, which we indicate with the notation $C(a,r)$. The self-similarity of the structure is expressed by the equation $C(ka,kr) = C(a,r)$, from which we deduce the scaling law:

$$C(a,r) \sim \left(\frac{a}{r}\right)^{d-d_f}. \tag{2.16}$$

This scaling behaviour is analogous to that followed by the average density $\rho(a,r)$ of a sample of size r. Up to numerical constants,

$$\rho(a,r) \sim \frac{1}{r^3} \int_0^r C(a,r')(r')^2 dr' \tag{2.17}$$

which implies that the scaling laws describing the behaviour of $C(a,r)$ and $\rho(a,r)$ are identical.

2.2.2 The Origin of Fractal Structures

We can only understand the abundance of these scale-free structures by considering the mechanisms that led to their formation. For example, a fractal structure is often the one that best reconciles the microscopic organisation laws (molecular or cellular interactions, thermal fluctuations, diffusion etc.) and the macroscopic constraints (such as the boundary conditions, material flow etc.); fractal growth and hydrodynamic turbulent flow being the classic examples.

We can conclude that understanding the phenomena generated by scale-free structures – in practice structures with a large number of spatial or temporal scales – requires a global, dynamic and multiscale approach. We must therefore focus our studies on the relationships between different levels of organisation. More generally, what are the trade-offs between different external constraints, internal physicochemical laws and interactions between different parts of the system that need to be determined to obtain the *necessarily global* picture of the organisation and mechanism of the system?

2.3 Conclusion

We cannot measure a scale invariant object with normal Euclidean geometry. To measure such objects a new quantity had to be conceived which went beyond the subjective nature of observation at a given scale. The fractal dimension, an exponent describing how measurements made at different scales are related, fulfilled this requirement. The relationship $M(l) \sim l^{d_f}$, defining this fractal dimension d_f (where $M(l)$ is the mass of a section of the object of linear extension l), is an example of a scaling law. It shows that the only intrinsic characteristic is the exponent, which is d_f here. For this to have any sense in practice, the object must be effectively self-similar over a range of scales, in other words d_f remains unchanged if we simultaneously change the resolution of the image and the linear extension l by the same factor. It is worth noting that study of fractal geometry has two related avenues with significantly different philosophies:

• The first focuses on establishing rigorous mathematical results for properties of ideal fractals. This work led to introducing different definitions of the fractal dimension, showing relationships between them and proving links with other properties of the structures under consideration, in particular topological properties [3].

• The second, which interests us more, concerns the application of these formal concepts to structures in nature; the physical properties they manifest and information about the mechanisms which generated them that we can extract by means of these concepts [4]. The field of study therefore goes from algorithms to test self-similarity and evaluate fractal dimensions to investigating distortions that result in various physical and physicochemical phenomena when they are produced on or in fractal structures (for example liquid–vapour phase transitions, chemical reactions in porous media, diffusion in a fractal medium, exchanges across a fractal interface, etc).

Fractal structures, characterised by a scaling law defining their fractal dimension d_f, are inseparable from the phenomena that we will encounter throughout this book. These structures are the direct observable signature of scale invariance and correlation length divergence. They strongly suggest considering the system as critical and analysing it as such by abandoning a detailed description of its behaviour at a given scale and favouring a study of the link between behaviours observed at different scales. Self-similarity therefore provides tools and technical arguments replacing those applicable when there is a separation of scales (mean field, effective parameters etc).

This chapter has also prepared us to discover scale invariance more generally, far beyond the purely geometric concept associated with fractal structures. Scale invariance can be a property not only of geometric objects but also of physical processes such as dynamical behaviours, growth phenomena or various phase transitions. The fundamental idea is the same: the relevant quantities will be the exponents involved in the scaling laws expressing *quantitatively* the scale invariance of the phenomena considered.

References

1. K. Cantor, Über unendliche, lineare Punktlannigfaltigkeiten. Mathematische Annalen **21**, 545 (1883)
2. W.T. D'arcy, *On Growth and Form* (Cambridge University Press, Cambridge, 1917)
3. K. Falconer, *Fractal Geometry* (Wiley, New York, 1990)
4. J.F. Gouyet, *Physics and Fractal Structures* (Springer, Berlin, 1996).
5. D. Hilbert, Über die stetige Abbildung einer Linie auf ein Flächenstück. Mathematische Annalen **38**, 459 (1891)
6. H. von Koch, Sur une courbe continue sans tangente, obtenue par une construction géométrique élémentaire. Arkiv för Matematik **1**, 681 (1904)
7. B. Mandelbrot, *Fractals: Form, Chance and Dimension*. (Freeman, San Francisco, 1977)
8. B. Mandelbrot, *The Fractal Geometry of Nature* (Freeman, San Francisco, 1982)
9. B. Mandelbrot, How long is the coast of Britain? Statistical similarity and fractional dimension. *Science* **155**, 636 (1967)
10. W. Sierpinski, Sur une courbe cantorienne dont tout point est un point de ramification. C.R. Acad. Sci. Paris **160**, 302 (1915). Sur une courbe cantorienne qui contient une image biunivoque et continue de toute courbe donnée. C.R. Acad. Sci. Paris **162**, 629 (1916)

Chapter 3
Universality as a Consequence of Scale Invariance

In the introductory chapter (Chap. 1) we showed several examples of critical behaviours near second order phase transitions. Let us remind ourselves of the key observations.

- **When the temperature is near the critical temperature**, most physical quantities obey a power law $(T - T_c)^x$, where x is called a *critical exponent*. In this chapter we will see that this sort of behaviour is the signature of the scale invariance of the system close to the critical point.
- **Critical exponents are often not simple rational numbers**, as precise measurements have shown. For example we have seen that the exponent β associated with the order parameter is 0.32 for the liquid–vapour transition and 0.36 for the ferromagnetic–paramagnetic transition of nickel; the critical exponent γ associated with the divergence of the compressibility κ_c is 1.24 for the water liquid–vapour transition and 1.33 when it is associated with the divergence of the magnetic susceptibility χ of nickel.
- **Critical exponent values are surprisingly robust** with respect to changes of physical system. Not only are they the same for the liquid–vapour transformation of all fluids, but we find them in seemingly very different systems (demixing of binary fluids, order–disorder transitions in metal alloys, etc.)

Physicists suspected that these critical properties were linked to the *divergence* of the spatial extent of the *thermal fluctuations* (which is also the range of correlations of the order parameter, the *coherence length*), and eventually renormalisation approaches strikingly confirmed it. The difficulty was clearly identified: how can we rigorously take account of correlations at the critical temperature, due to interactions propagating from one point to another, and leading to correlations of diverging size?

The *mean field* approximation that we introduced at the end of Chap. 1 enables us to predict all the critical properties neglecting the effect of these correlations. Remember the Ginzburg criterion for the validity of the mean field description based on the powerful idea of the *critical region*:

A. Lesne and M. Laguës, *Scale Invariance*, DOI 10.1007/978-3-642-15123-1_3,
© Springer-Verlag Berlin Heidelberg 2012

> Exact calculations on a finite block are always doomed to failure in
> a small region around the critical point, the critical region, where the
> correlations have a range larger than the size of the block.

In practice there are two cases:

- For certain physical systems (metal superconductors, ferromagnets, liquid crystals, etc.), the critical region is so small that it is not observable experimentally.
  In this case, the critical behaviour obtained by the mean field approximation is
  remarkably well reproduced in experiments.
- For the numerous systems in which the critical region is experimentally observable, the critical behaviour is radically different from that predicted by the mean
  field approximation.

Physicists did not know which was the right track to tackle the problem: in the
end it turned out to be a formal calculation analysis of these unorthodox physical
mechanisms.

In Chap. 2 we introduced fractals, objects that show scale invariance, either
strictly such as the Sierpinski gasket, or statistically such as for an infinite length
random walk trajectory (see Chap. 4 on diffusion). The physicist Leo Kadanoff was
the first to suggest, in the 1960s, that the scale invariance of physical systems at the
critical point determines all the other singular properties [14]. In a few years this
idea paved the way for the pioneering methods of the *renormalisation group*.

3.1   Introduction

Let's return to the example of a ferromagnetic system at different temperatures. At
zero temperature all the spins are aligned. Figure 3.1 illustrates the situation at low
temperature (much lower than T_c): almost all the spins are aligned apart from one
or two oriented in the opposite direction grouped in little islands. The right hand
side of Fig. 3.1 represents a magnet at high temperature (much higher than T_c):
there is complete disorder, like in a system without interactions, because thermal
agitation dominates. On average there are as many spins in each orientation, isolated
or grouped in little islands. In both cases the characteristic size of the islands (that
is the fluctuations in magnetisation), measured by the coherence length ξ, is small.

Figure 3.2 shows an example of the state of spins near the critical temperature
of the ferromagnetic–paramagnetic transition (in the absence of external magnetic
excitation). For temperatures greater than or equal to the critical temperature (in
this case Curie temperature), the average magnetisation is zero. There are highly
branched islands of all sizes. While the external magnetic field is zero, there is no
preferred orientation: it is possible to change the positive and negative orientations
of the spins without changing the appearance of the system. In practice, as soon as
the temperature is detectably below the critical temperature the magnet *randomly*

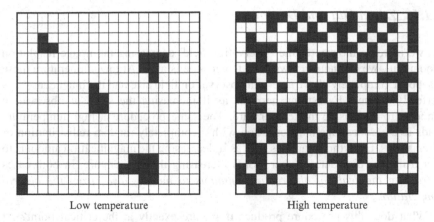

Low temperature   High temperature

Fig. 3.1 Far from the critical temperature, the arrangement of spins does not show very extended structures. At low temperature (*left*) the majority of spins are aligned in a precise direction with only a few little islands in the opposite direction. At high temperature (*right*), there is maximal disorder and the average size of islands of aligned spins is very small

Near the critical temperature

Fig. 3.2 Near the critical temperature, the spins arrange themselves on islands of varying sizes. Some islands are very big and branched

chooses an orientation (positive or negative) for the average magnetisation: the order *breaks* the symmetry imposed from outside the spin system. However the spatial structure of the fluctuations in the neighbourhood of the critical temperature is profoundly different from that which we observe at low and high temperature. Let's now look at what happens when we increase the scale of observation by *zooming out*.

3.1.1   Zoom Out and Decimation

As we increase the size of the observation field, decreasing the image resolution (zoom out), we distinguish fewer and fewer little islands. At low temperature these disappear completely above a certain observation field size and the magnet appears perfectly aligned. Everything happens as if increasing the scale of observation amounts to decreasing the temperature. The same procedure at high temperature leads to a zero average magnetisation with a completely random redistribution of orientations. Everything happens here as if decreasing the magnification amounts to increasing the disorder, in other words increasing the temperature. In both cases *increasing the scale of observation amounts to moving away from the critical temperature.*

What does this procedure produce if we are exactly at the critical point? At the critical temperature there are islands of various sizes but also very extended structures which remain whatever the increase in length scale at which we observe: we call this *scale invariance.*

Kadanoff [14] introduced a tool designed to quantify the effects of such a change in scale, a model transformation which he called decimation. He suggested that by iterating this transformation to its limit, one should be able to describe the physical properties of critical systems by revealing their scale invariance.

The etymology of the word *decimation* in Roman legions was a collective punishment consisting of grouping soldiers in packs of ten and executing one man in ten! Here in the context of renormalisation it is also used to mean grouping in packs. Figure 3.3 shows a decimation of spins in blocks of 4, which illustrates a magnification of observation scale by a factor of 2. At each block in the decimated network we place a *super-spin* in place of the original block of 4. We use a rule

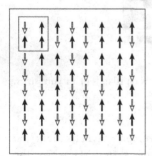 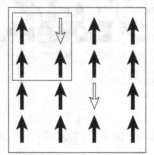 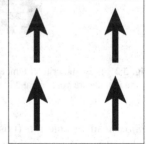

Fig. 3.3   To represent the effect of changing the scale of magnification of the microscope, Leo Kadanoff suggested grouping spins by blocks and replacing the block by a "super-spin". Then we repeat this procedure *ad infinitum*. There are different methods to attribute the orientation of the super-spin as a function of the orientations of the spins in the block (see text)

to unambiguously choose the orientation of super-spins even when the spins in the initial block are not all aligned in the same direction. When the number of spins in a block is odd, we use the simple intuitive rule of the *majority*: the super-spin is oriented in the direction of the majority of spins contained in the block. When the number of spins in a block is even, we need a rule to deal with the case when half the spins are oriented in one direction and half in the opposite direction. For these blocks we choose a super-spin orientation randomly for each configuration respecting the symmetries of the block and notably ensuring an equal number of positive and negative super-spins in total for these cases.

Figure 3.4 illustrates the effects of this transformation iterated on a system of 236,000 spins for three different initial temperatures: $1.22T_c$, T_c and $0.99T_c$. We clearly see the scale invariance at T_c, the size of clusters being limited only by the size of the system. For the other initial temperatures each iteration corresponds well to moving further away from the critical temperature.

At the critical temperature, we can, in principle, carry out an infinite number of decimations on an infinitely large system without changing the statistical properties of the system. This has an interesting consequence:

> At the critical temperature and in its immediate vicinity the microscopic details
> of the interactions between spins are not important;

a fact which connects well to the universality observed in critical behaviours. However, the method introduced by Leo Kadanoff is not sufficient to calculate the laws governing these behaviours.

It was only in 1971 that Kenneth Wilson achieved this by proposing a generalisation of this work: renormalisation methods [30, 33]. The success was considerable: these methods rapidly led to numerous results and Wilson obtained the Nobel prize in physics in 1982. The idea is to use Kadanoff's decimation not to transform the values that characterise the physical state of the model system, but to transform the form of the model itself (see the following paragraphs). At the critical point the transformation should leave the model invariant in its entirety, with all its equations and all the states accessible to the system. In formulating the conditions of this invariance, we obtain the form of model suitable for the critical system. It is also possible to classify models with regard to the physical systems they describe and then compare these predictions to numerous observations made since Van der Waals of hundreds of different systems and changes of state. The agreement between experiments and predictions obtained by renormalisation is remarkable (see for example Fig. 1.23).

The main result is that:

> the critical behaviour does not depend on the physical detais of the order
> that establishes itself at low temperature: it depends only on the number n of
> components of the order parameter and the dimension of space d.

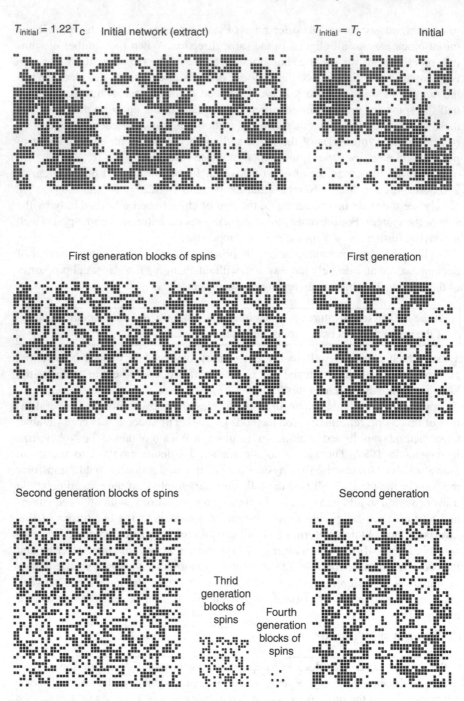

$T_{initial} = 1.22\,T_C$ Initial network (extract) $T_{initial} = T_C$ Initial

First generation blocks of spins   First generation

Second generation blocks of spins   Second generation

Thrid
generation
blocks of
spins
Fourth
generation
blocks of
spins

Fig. 3.4 (continued)

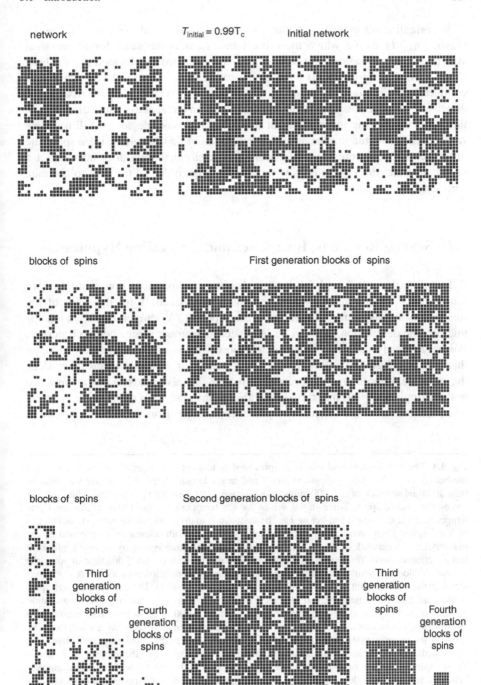

Fig. 3.4 (continued)

Renormalisation enables the classification of changes of state in *universality classes* (n, d) within which the critical behaviour is the same for all physical systems. Renormalisation rigorously confirms the conclusion obtained by Ginzburg (Sect. 1.3): the use of the mean field approximation – the results of which are completely independent of n and d – leads to the exact critical behaviour if the dimension of space d is equal to or greater than 4. In our space of three dimensions, this approximation is inaccurate, but it is even less suitable for two dimensions where thermal fluctuations have considerable weight. In one dimensional space the difference is even more blatant because the exact calculation does not predict a change of state: the system rests without long range order all the way to absolute zero temperature.

3.2   Scaling Relations, Invariance and the Scaling Hypothesis

Before describing and applying renormalisation methods to a few examples, we return to the idea of scale invariance and its consequences for the form of the free energy of a system close to the critical point. We temporarily abandon the microscopic description and will be interested in average macroscopic properties. We have already seen that experimental observations of phase transitions show that critical behaviours follow power laws such as $m \sim t^\beta$. In terms of models, the mean field approximation presented in Sect. 1.3 also leads to power laws but nothing implies a priori that this should also be the case for a rigorous description. The justification for the existence of power laws comes from a general approach in

---

Fig. 3.4 The transformation of blocks of spins applied to a network of spins several times makes the behaviour of the system appear at larger and larger length scales. A computer was used to treat an initial network of 236,000 spins; *black squares* representing up spins and white squares representing down spins. Three initial values for the temperature were taken: above the Curie temperature T_c, at T_c itself and below T_c. The first step is to divide the initial network into blocks of 3×3 spins. Then each block is replaced by a single spin with orientation determined by the majority rule. A network of blocks of spins of the first generation is thereby obtained, of which just a section is shown for reasons of readability. The procedure is restarted this time dividing the network of blocks of spins of the first generation. The second generation network thereby obtained serves as the starting point for the following transformation, and so on. There are few enough spins in the third generation network that we can draw the whole network and the fourth generation contains only 36 spins, each one representing more than 6,000 spins in the initial network. In the first step, the variations over length scales smaller than three mesh sizes were eliminated (by applying the majority rule). In the second step, variations between three and nine mesh sizes were eliminated; in the third step between nine and 27 mesh sizes, etc. When the temperature is above T_c, the spins appear more and more randomly distributed and short range variations disappear. When the temperature is less than T_c, the spins appear to have more and more a preferred uniform orientation, variations that remain being short ranged. When the initial temperature is exactly equal to T_c, large scale variations remain at every step. We say that at the Curie temperature the system is at a fixed point because each transformation of blocks of spins conserves the large scale structure of the lattice (after [33])

physics called *scaling* which justifies the establishment of general *scaling laws* for systems with many degrees of freedom.

3.2.1 A Unifying Concept

Scaling laws are a new type of statistical law on the same level as the law of large numbers and the central limit theorem. As such they go far beyond the scope of physics and can be found in other domains such as finance, traffic control and biology. They apply to global, macroscopic properties of systems containing a large number of elementary microscopic units. To obey a classical description these units:

- Must be statistically independent or weakly correlated, that is characterised by a finite coherence length ξ
- Must not fluctuate too much: the variance σ of their elementary observables must be finite so that the law of large numbers and the central limit theorem apply.

Scaling laws do appear when one of these conditions is not satisfied (infinite ξ or σ) provided that the nature of the elementary constituents and the regularity of their organisation let us predict self-similarity properties. We will illustrate this aspect during the course of this book and with a more detailed example in the next chapter (Chap. 4) on normal and abnormal diffusion.

The existence of scaling laws is very useful: it allows us, by a simple change in scale, to deduce the behaviour of a system of arbitrary size from that of a system of a given size. We can therefore describe the behaviour of a large system from that of a small system, analysed rigorously or simulated numerically. A typical example is that of polymers, which show scaling behaviours at all temperatures and not just near a critical temperature (see Sect. 6.3.4). The scale invariance of hydrodynamic equations allows us to reproduce the behaviour of a huge oil tanker by using a miniature boat a few meters long navigating in a liquid that is much more viscous than water. This enables pilots to practice navigation with the sensations they will encounter when manoeuvring a real oil tanker (see Sect. 9.5.1).

Even though such power laws had been introduced in fluid mechanics by Kolmogorov since 1941, they were not accepted as a description suited to critical behaviours until the 1960s, during which an intense activity was dedicated to their justification. Various authors established relationships between the critical exponents, inequalities from thermodynamic stability considerations and then equalities traditionally called *critical scaling relations*. Each of these relations was another argument in favour of the universality of critical behaviours. It was clear by the end of the 1960s that all the critical exponents could be deduced from two of them. In the following we will describe one of these approaches: Widom scaling.

3.2.2   Widom Scaling

Widom's basic idea [29] was to characterise the generic properties of the singular part f of the free energy[1] per unit volume, in the vicinity of the critical point. This function should depend on t, the relative distance from the critical point temperature, and on the external field h. The different physical quantities, specific heat C, order parameter m and susceptibility χ are derived from $f(t, h)$:

$$C = \left(\frac{\partial^2 f}{\partial t^2}\right)_h \qquad m = \left(\frac{\partial f}{\partial h}\right)_t \qquad \chi = \left(\frac{\partial^2 f}{\partial h^2}\right)_t. \qquad (3.1)$$

Widom's reasoning starts from the idea that if these three quantities show a critical behaviour in a power law, then f must itself have a form of power law. He supposed therefore that f has the form:

$$f(h, |t|) = |t|^{2-\alpha} g(h|t|^{-\Delta}) \qquad (3.2)$$

which applies whatever the sign of t with potentially different values of g. For zero field h, this form ensures that $C = g(0)t^{-\alpha}$ with the usual meaning of the exponent α (see Table 1.1). Similarly, the two other quantities lead to the expression for the critical exponents β and γ:

$$\beta = 2 - \alpha - \Delta \quad \text{and} \quad \gamma = -2 + \alpha + 2\Delta. \qquad (3.3)$$

Eliminating Δ, gives the Rushbrooke scaling relation, perfectly verified by measured exponents but also by the mean field exponents[2]:

$$\alpha + 2\beta + \gamma = 2. \qquad (3.4)$$

3.2.2.1   Dependence on the Field

The exponent δ, defined by $m \sim h^{1/\delta}$ at $t = 0$, was introduced by taking $h|t|^{-\Delta} = x$ in the expression (3.1) for m, using (3.2):

$$m = |t|^{2-\alpha-\Delta} g(x) = \left(\frac{h}{x}\right)^{\frac{2-\alpha-\Delta}{\Delta}} g(x). \qquad (3.5)$$

---

[1] Even if we call it f and not g, the quantity we are interested in here is strictly speaking the *Gibbs free energy*, which we could call the *free enthalpy* in so far as it depends on the intensive magnetic variable h and not on the corresponding extensive variable, the magnetisation m.

[2] For simplicity, the term *mean field exponents* refers to the exponents calculated within the framework of the mean field approximation.

By fixing for example $x = 1$, we obtain the value $\delta = \Delta/(2 - \alpha - \Delta)$. Eliminating Δ leads to a new relation, Griffith's scaling relation:

$$\alpha + \beta\delta + \beta = 2. \tag{3.6}$$

3.2.2.2 Coherence Length

One way to evaluate the *singular part* f of the free energy is to express it to order kT per degree of freedom. We can show that here one degree of freedom corresponds to each *coherence volume element* ξ^d, where d is the spatial dimension. This result can be understood by considering that *the coherence length is the smallest distance over which the order parameter could change*: everything occurs as if the space was divided up into cells of volume ξ^d within which the order parameter is constant. The value of the free energy is therefore:

$$f \sim kT\,\xi^{-d} \sim kT\,|t|^{-dv} \tag{3.7}$$

which, by comparison with the form of f given by Widom (3.2), leads to a new relation:

$$\alpha + dv = 2. \tag{3.8}$$

This scaling relation, called the Josephson scaling relation, is also called the *hyperscaling* relation because it is the only one to explicitly involve the spatial dimension d. We note that for mean field exponents $\alpha = 0$ and $v = 1/2$ this relation is only satisfied for the critical dimension $d = 4$ below which the mean field description is not valid (see the Ginzburg criterion at the end of Chap. 1). Another relation, the Fisher scaling relation, can be obtained by studying the dimensional relationships linking the spatial correlation function, characterised by the exponent η, to the susceptibility and the coherence length, characterised by the exponents γ and v:

$$\eta + \gamma/v = 2. \tag{3.9}$$

3.2.2.3 Excellent Agreement with Experiments

In order to compare their validity, we have presented the four scaling relations above in a normalised form, in which the right hand side is 2. Table 3.1 shows the experimental measurements of the exponents [2] for six different families of changes of state. It also shows the degree of validity of the above scaling relations. The table shows the value of the left hand side of the scaling relations calculated using the experimental values of the exponents. Taking into account the experimental error bars on the measurements of the exponents, the observed agreement is remarkable. The experimental confirmation of these scaling relations implies that the Widom-type formal approach is grounded. But how can we

Table 3.1 Verification of scaling relations with experimental measurements for critical exponents for a few families

	Xe	Binary mixture	Alloy	^{4}He	Fe	Ni
n number of components to the order parameter	1	1	1	2	3	3
α	0	0.113	0.05	−0.014	−0.03	0.04
β	0.35	0.322	0.305	0.34	0.37	0.358
γ	1.3	1.239	1.25	1.33	1.33	1.33
δ	4.2	4.85		3.95	4.3	4.29
η	0.1	0.017	0.08	0.021	0.07	0.041
ν	0.57	0.625	0.65	0.672	0.69	0.64
Rushbrooke $\alpha + 2\beta + \gamma = 2$	2.00	1.99	1.91	1.99	2.04	2.09
Griffith $\alpha + \beta\delta + \beta = 2$	1.82	2.00		1.67	1.93	1.93
Fisher $\eta + \gamma/\nu = 2$	2.38	2.00	2.00	2.00	1.99	2.12
Josephson $\alpha + \nu\, d = 2$	1.71	1.99	2.00	2.00	2.04	1.96

justify the physical meaning of the relation postulated by Widom? What physical mechanism could lead to such a simple form for the singular part f of the free energy at the critical point? Kadanoff proposes a simple response to this question: the divergence of the coherence length.

3.2.3 Divergence of ξ and the Scaling Hypothesis

Let us return to the simple question: why do phase transitions show critical behaviour as power laws such as $m \sim t^\beta$ when there is nothing a priori to indicate that they should? A physical argument in favour of such a type of critical behaviour follows from an essential hypothesis called the *scaling hypothesis*.[3] Among the many ways in which it has been formulated, according to Kadanoff it can be expressed in the following way:

---

[3] Scaling methods have become a standard tool for physicists today. The word *scaling* evokes a structure of nature and particular relations, but most of all a universal approach and type of analysis. It refers to an observation as much as to an action, with the idea that the this active approach molds the observation process.

The critical behaviour of a physical property $P(t)$ essentially originates from the dependence of the coherence length on temperature, which diverges at the critical temperature with a power law $\xi(t) = \xi_0 t^{-\nu}$:

$$P(t) \equiv \Pi(\xi(t)). \qquad (3.10)$$

The function Π does not explicitly depend on the temperature but only on the coherence length. In this section, we assume the power law for the divergence of the coherence length is given. We will see, in the section on *renormalisation flow*, that we can obtain its form rigorously. Kadanoff proposed this hypothesis, which we will come back to later, in the 1960s, and he was the principal craftsman of its application. If we take into account this hypothesis, transformations of the scale of a critical system naturally lead to power law behaviours. The question is the following: when we study a system after having subjected it to a change in length scale $x \to bx$, what is the temperature difference t' such that:

$$Rb\,[P(t)] = P(t') \quad \text{or alternatively} \quad \Pi(b\xi(t)) = \Pi(\xi(t')) \ ?$$

where $Rb\,[P(t)]$ is $P(t)$ transformed by the change of scale. By assuming $\xi(t) = \xi_0 t^{-\nu}$, as we observe experimentally, this approach leads to power laws of the type:

$$\Pi(\xi) \equiv \Pi_0 \xi^z \quad \text{and therefore} \quad P(t) = P_0 t^x. \qquad (3.11)$$

Unlike Widom's approach, which postulated these power laws, we now have these relations derived from a concrete physical hypothesis. The phenomenological hypothesis of scaling behaviour in the vicinity of the critical point will be proved by the renormalisation method. We now present the step proposed by Kadanoff which would lead to renormalisation when Wilson, and many others, added an additional conceptual step.

3.2.4 Scaling Relations Revisited with Decimation

Let us introduce the *decimation* operation D of an Ising network such as that drawn in Fig. 3.3: the spins in the initial network, of lattice spacing a, are grouped in blocks of size La containing $n = L^d$ spins (in Fig. 3.3 $L = 2$, $d = 2$, $n = 4$). The number of degrees of freedom is reduced by a factor n in the new network of *super-spins*. The crucial step is to reduce the 2^n states accessible to each super-spin to two states $+1$ or -1 as for each spin in the starting network. The quantities t and h are assumed to be transformed to t' and h' by the decimation. Kadanoff made the following hypothesis:

$$h' = hL^x \quad \text{and} \quad t' = tL^y. \qquad (3.12)$$

The transformation of the free energy per unit volume from f to f' during decimation can be deduced by seeing that the volume occupied by one spin in the network changes from a^d to $(La)^d$ and noting that the physical value of the free energy per spin should not change. This gives:

$$f(t,h) = L^{-d} f'(tL^y, hL^x).$$ (3.13)

As for the coherence length, it transforms as:

$$\xi(t,h) = L\xi'(tL^y, hL^x).$$ (3.14)

the thermal dependence of which we can identify with the divergence of the coherence length, $\xi(t) = \xi_0 t^{-\nu}$.

At the critical point, scale invariance imposes that this transformation corresponds to a *fixed point* where $f = f'$ and $\xi = \xi'$. In addition these relationships must apply for all L. For example, we can choose $L = |t|^{-1/y}$ in a way that the first argument is either $+1$ or -1 according to the sign of t. By putting $h = 0$ we obtain;

$$f(t,0) = |t|^{d/y} f(\pm 1, 0) \quad \text{and} \quad \xi(t,0) = |t|^{-1/y} \xi(\pm 1, 0).$$ (3.15)

By identifying the exponent d/y as $2 - \alpha$ (3.2) and $1/y$ as ν (3.7) and eliminating y, we recover the Josephson relation called *hyperscaling*:

$$d\nu = 2 - \alpha.$$ (3.16)

It is easy to also recover the Rushbrook and Griffith scaling relations by directly calculating m and χ (3.1). The derivation of the fourth relation, that of Fisher, is interesting because it directly involves the physical significance of decimation. Let us recall the definition of the exponent η linked to the dependence of the correlation function G on the field $h = 0$ given by $G(T_c, r) \sim r^{-(d-2+\eta)}$.

In general, the correlation function $G(r, t, h)$ can be expressed as:

$$G(r, t, h) = \frac{\partial^2 f}{\partial h(r) \partial h(0)}$$ (3.17)

which leads to:

$$G(r, t, h) = L^{-2d} L^{2x} G'(r/L, tL^y, hL^x).$$ (3.18)

By choosing $L = r$, $t = 0$ and $h = 0$, and by expressing x using the exponent γ characterising the susceptibility $\chi \sim \left(\frac{\partial^2 f}{\partial h^2}\right)_t$, we obtain the relation $\gamma = (2 - \eta)\nu$ which is the Fisher scaling relation.

Formally the approaches of Widom and of Kadanoff are very close but the latter is based on a physical argument, the scaling hypothesis. Here the scaling relations are the signature of the divergence of the coherence length. We have defined six exponents which are related by four scaling relations: all the critical singularities can therefore be deduced from two exponents, but how can we calculate the value of these two exponents? Let us come back to the microscopical description for this next step.

3.2.4.1   Kadanoff's Scheme

Kadanoff and other physicists used the decimation process in trying to calculate the transformed coupling constants $K' = D(K)$ in the search for a fixed point. The coupling constants are the parameters in the Hamiltonian characterising the energy of the system. We have shown an example in Chap. 1 for the Ising model. The Hamiltonian in this case is characterised by the coupling constant $K = J/kT$, where J is the interaction energy between the nearest neighbouring spins. The idea was to transform the coupling constants using the following three steps:

1. Group the spins into super-spins using one decimation step D into blocks of b^d spins, thereby obtaining a "super-lattice" of lattice spacing $D(a) = ba$.
2. Change the scale of the super-lattice by a factor $1/b$ in order to recover the lattice spacing of the initial lattice $a' = D(a)/b = a$ with a shorter coherence length $\xi' = \xi/b$.
3. Transform the coupling constants K to $K' = D(K)$.

As we iterate these three steps (together making a transformation we call R) we expect the following limits as a function of the initial temperature:

- $T > T_c$, i.e. $K < K_c$, for which we expect that the transformation entails a decrease in K, $K' < K$, and that therefore the coupling constant tends towards the limit $K_1^* = 0$, that is $T_1^* = \infty$. This "high temperature" fixed point is *stable* since the transformation drives the system to it, and corresponds to the predicted total disorder for zero coupling, that is to say non interacting spins at zero external field.
- $T < T_c$, i.e. $K > K_c$, for which we expect the transformation leads to an increase in K, $K' > K$, and that the coupling constant tends towards the limit $K_2^* = \infty$, that is $T_2^* = 0$. This "low temperature" fixed point is also *stable*. It corresponds to perfect order.
- $T = T_c$, i.e. $K = K_c$, for which, since the system is scale invariant, we expect that the transformation will not change K, $K' = K = K_c$, and that therefore the coupling constant remains equal to the critical value $K^* = K_c$, that is $T^* = T_c$. This fixed point corresponds to the critical point and is *unstable* because the smallest fluctuation in temperature leads to one of the two previous cases.

All the physics of renormalisation was already in this procedure. However for this approach to succeed in a systematic way it was missing a crucial idea: without exception, it is not just the parameters of a model that should be transformed, but the form of the model itself. Having laid down the phenomenological basis of this approach, let us return to the microscopic models which are required for its implementation.

3.3   Transitions and Model Hamiltonians

Historically microscopic models of the ferromagnetic–paramagnetic phase transition have served as the basis for most phase transition studies. Here we will also use this system to introduce renormalisation methods. The results obtained will then be applied to a wide variety of physical systems. The approach used is the usual one in statistical physics consisting of calculating the values of macroscopic physical observables (magnetisation m, susceptibility χ, etc.) at equilibrium from the Hamiltonian H which characterises the energy of the system.

The first step is to construct the partition function Z and the Gibbs free energy f:

$$Z = \sum_{\{S_i\}} e^{-H/kT} \qquad\qquad f = -kT \log(Z)$$

from which we can deduce the average values of the quantities we are interested in. In this way Hamiltonians completely characterise the physical models they correspond to. We will show the forms of the most commonly used Hamiltonians, along with the names they are known by, below.

3.3.1   Number of Components of the Order Parameter

The spins S_i are vectors of n components with constant modulus (an assumption that is often chosen but can be changed). It is necessary to distinguish between the number of components n and the dimension d of space.

- If we are talking about spins in the physical sense, in most general case, the number of components is $n = 3$. The corresponding models are characterised by Hamiltonians called *Heisenberg* Hamiltonians.
- When the number of components is $n = 2$, the current name is *XY-model*.
- The case $n = 1$ is associated with *Ising models*.

Do not forget however that we will use the ferromagnetic–paramagnetic transition as an example to describe changes of state in a general manner. Therefore n and d will be considered as parameters that a priori could take any value. We will see for example in Chap. 6 that the case $n = 0$ provides a fruitful analogy with folding polymer chains. The fictitious situation in which $n = \infty$, called the spherical model, is also useful as a reference.

3.3.2   The Interactions

The general form of the Hamiltonian that takes into account interactions between two spins S_i and S_j is written:

$$H = -\sum_{\alpha=1}^{n} \sum_{i} \sum_{j} J_{\alpha ij} S_{\alpha i} S_{\alpha j} - h \sum_{i} S_{zi}, \qquad (3.19)$$

where α is one of the n components of the spins and h is the magnetic field oriented in the z direction. The coupling constant $J_{\alpha ij}$ simplifies in the following cases:

- If the system is isotropic the coupling constant is independent of α.
- If the system is invariant under translation:

$$J_{ij} = J(r_i - r_j) = J(r_{ij}).$$

- And finally if the interactions are only present when $\mathbf{S}_i$ and $\mathbf{S}_j$ are *nearest neighbour* spins, there remains only one coupling parameter J (only one reduced coupling constant $K = J/kT$).

With these three hypotheses the Hamiltonian is written:

$$H = -J \sum_{\alpha=1}^{n} \sum_{<ij>} S_{\alpha i} S_{\alpha j} - h \sum_{i} S_{zi}, \qquad (3.20)$$

where $< ij >$ refers to the set of pairs of nearest neighbour spins $\mathbf{S}_i$ and $\mathbf{S}_j$. We will return at the end of the chapter to the relevance of this third hypothesis. In the light of the previous paragraphs, the reader will realise that interactions between second (or third) nearest neighbours are effects remaining localised at the microscopic scale which do not seriously affect the scale invariance of a system at the critical point. We will show later that renormalisation methods provide the means to discuss this point quantitatively.

3.3.3 The Ising Model

We introduced the simplest model, the Ising model, in Chap. 1. Proposed in 1920 by Lenz, the Ising model assumes, as well as the hypotheses in the previous paragraph, that spins can only be in one of two opposite orientations. We use reduced physical quantities such that the spins take the values $\sigma_i = \pm 1$. In the following we will mainly use the Ising Hamiltonian:

$$H = -J \sum_{<ij>} \sigma_i \sigma_j - h \sum_{i} \sigma_i. \qquad (3.21)$$

3.4 1D and 2D Solutions of the Ising Model

The *Ising model problem* can be posed as: can statistical physics predict an order–disorder transition in a system described by the Hamiltonian in (3.21)?

The problem was solved by Ernst Ising in 1925 for the case of a one dimensional chain of spins. We had to wait for another twenty years and an exceptional feat of mathematics by Onsager [1944] for a solution of the Ising model in two dimensions. Today we know of no exact analytical solution for dimensions 3 or above.

3.4.1 Transfer Matrices

In this paragraph we present the solution of the Ising model problem in one dimension, based on the method of *transfer matrices*. The same idea leads to a solution to the 2D problem that is simpler than Onsager's solution [26].

The Hamiltonian and the partition function for a 1D Ising chain are written:

$$H = -J \sum_i \sigma_i \sigma_{i+1} - \frac{h}{2} \sum_i (\sigma_i + \sigma_{i+1}) = \sum_i H_i,$$

$$Z = \sum_{\substack{\{\sigma_1 = \pm 1, \\ \sigma_2 = \pm 1, \ldots\}}} e^{-H/kT}. \tag{3.22}$$

where we have written the field term in a form symmetric in i and $i + 1$. We will see that it is useful to construct the matrix $\mathscr{T}$ in which the elements are the possible values of $Z_i = \exp(-H_i/kT)$ relative to the pair of spins (σ_i, σ_{i+1}):

	$\sigma_{i+1} = 1$	$\sigma_{i+1} = -1$
$\sigma_i = 1$	$\mathscr{T}_{11} = e^{(J+h)/kT}$	$\mathscr{T}_{1,-1} = e^{-J/kT}$
$\sigma_i = -1$	$\mathscr{T}_{-1,1} = e^{-J/kT}$	$\mathscr{T}_{-1,-1} = e^{(J-h)/kT}$

We now associate a vector $\sigma_i = (\sigma_i^+, \sigma_i^-)$ with the spin σ_i such that

$$\sigma_i = (1,0) \quad \text{if} \quad \sigma_i = 1 \quad \text{and} \quad \sigma_i = (0,1) \quad \text{if} \quad \sigma_i = -1.$$

The quantity $< \sigma_i| \mathscr{T} |\sigma_{i+1} >$ corresponds to the observable for Z_i :

$$< \sigma_i| \mathscr{T} |\sigma_{i+1} > = \sigma_i^+ \sigma_{i+1}^+ e^{(J+h)/kT} + \sigma_i^+ \sigma_{i+1}^- e^{-J/kT} + \sigma_i^- \sigma_{i+1}^+ e^{-J/kT}$$

$$+ \sigma_i^- \sigma_{i+1}^- e^{(J-h)/kT}. \tag{3.23}$$

in which one, and only one, of the four terms in the expression is non zero. Assuming periodic boundary conditions ($\sigma_{N+1} = \sigma_1$), (3.23) implies the expression for Z:

$$Z = \sum_{\substack{\{\sigma_1=\pm 1, \\ \sigma_2=\pm 1,\dots\}}} \prod_i Z_i$$

$$= \sum_{\substack{\{\sigma_1=\pm 1, \\ \sigma_2=\pm 1,\dots\}}} <\sigma_1|\,\mathcal{T}\,|\sigma_2> <\sigma_2|\,\mathcal{T}\,|\sigma_3> <\sigma_3|\,\mathcal{T}\,|\sigma_4> \dots <\sigma_N|\,\mathcal{T}\,|\sigma_1>.$$

$$(3.24)$$

In this way we can identify Z with the trace of $\mathcal{T}$ to the power N:

$$Z = Tr\{\mathcal{T}^N\} = \lambda_1^N + \lambda_2^N, \tag{3.25}$$

where λ_1 and λ_2 are the eigenvalues of $\mathcal{T}$, which can be calculated in the classical manner:

$$\lambda_\pm = e^{J/kT} \cosh\left(\frac{h}{kT}\right) \pm \sqrt{e^{2J/kT}\sinh^2\left(\frac{h}{kT}\right) + e^{-2J/kT}}. \tag{3.26}$$

At the thermodynamic limit $N \to \infty$, only the largest positive eigenvalue $\lambda_1 = \lambda_+$ counts, for which the modulus is the largest:

$$Z \approx \lambda_1^N. \tag{3.27}$$

Everything happens as if the system was made up of non interacting particles, for which the partition function would be $Z_1 = \lambda_1$. The transfer matrix method reduces the calculation of a chain of N interacting spins to one in terms of properties of a single fictitious particle. In the same way in two dimensions, this method reduces the calculation to a problem in one dimension.

3.4.2 Properties of the 1D Ising Model

From the expression for Z, (3.27), we directly obtain the free energy

$$G = -kT \log Z = -NkT \log \lambda_1,$$

and all the physical quantities characterising the system. The form of λ_1 (3.26), which has no singularities at finite temperature, confirms the observation of no phase transition for the 1D Ising model. The magnetisation is

$$m = -\frac{1}{N}\frac{\partial G}{\partial h} = \frac{kT}{\lambda_1}\frac{\partial \lambda_1}{\partial h} = \frac{\sinh\left(\frac{h}{kT}\right)}{\sqrt{\sinh^2\left(\frac{h}{kT}\right) + e^{-4J/kT}}} \tag{3.28}$$

and the susceptibility is

$$\chi = \frac{\partial m}{\partial h} = \frac{1}{kT} \frac{e^{-4J/kT} \cosh\left(\frac{h}{kT}\right)}{\left(\sinh^2\left(\frac{h}{kT}\right) + e^{-4J/kT}\right)^{3/2}}. \tag{3.29}$$

Therefore, in zero magnetic field $h = 0$ we obtain

$$m = 0 \quad \text{and} \quad \chi = \frac{1}{kT} e^{2J/kT}, \tag{3.30}$$

in other words the susceptibility diverges at $T = 0$, even if the spontaneous magnetisation remains strictly zero. Note that this divergence, which is often considered
as being a "zero temperature phase transition", is an exponential divergence and
not a power law. From this point of view this "transition" is very unusual: it is
not dominated by the scale invariance of the system but by the individual thermal
excitations. We can also calculate the specific heat $C = \frac{\partial^2 G}{\partial T^2}$, which shows a
maximum at a temperature slightly lower than J/k. Figure 3.5 depicts G, χ and
C as a function of reduced temperature (in units of J/k) compared to their values
in the mean field approximation.

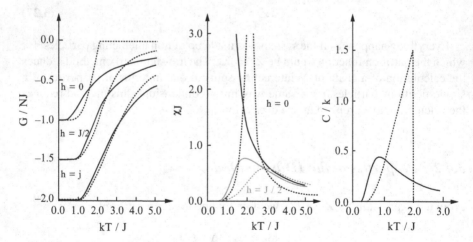

Fig. 3.5 *Left*: Reduced free energy G/NJ of the 1D Ising model as a function of reduced
temperature in units of J/k (*continuous line*) compared to the value obtained in the mean field
approximation (*dashed line*), for $h = 0$, $h = J/2$, and $h = J$. *Middle*: Reduced susceptibility χJ
of the 1D Ising model as a function of reduced temperature (*continuous line*) compared to the value
obtained in the mean field approximation (*dashed line*), for $h = 0$ and $h = J/2$. *Right*: Reduced
specific heat C/k of the 1D Ising model as a function of reduced temperature (*continuous line*)
compared to the value obtained in the mean field approximation (*dashed line*)

Remarks

- The general form of f, obtained in the mean field approximation is qualitatively good. Furthermore, it is even better if the applied field is non zero: the field h appreciably reduces the effect of thermal fluctuations.
- The reader may be surprised to see that the free energy saturates at its minimum value, NJ, for T less than about $J/3k$, while $m = 0$. This can seem contradictory because, at low temperature, we would expect $f = -mNJ$. However this result, $f = -mNJ$, is only exact for a homogeneous system: here there is no long range order until zero temperature. Switches of the type ↑↑↑↑↑↑↑↓↓↓↓↓↓↓, between regions oriented in opposite directions, are more and more rare as we reduce the temperature, but the magnetisation remains zero, oppositely oriented domains strictly compensating for each other. We will return to the discussion of this point in paragraph 3.6.3.
- On the other hand there is a considerable difference in behaviour of the susceptibility at zero magnetic field, since mean field predicts a transition at $T_c = 2J/k$. At field $h = J/2$ the qualitative agreement is much better even though the position of the maximum is clearly different from the rigorous prediction.
- The behaviour of the specific heat C is interesting: it behaves as if there was a "soft" transition around $T = J/k$. In reality, below this temperature, there exist finite but large pieces of chain that are completely ordered.

3.4.3   Properties of the 2D Ising Model

The 2D Ising model corresponds to the *only physical situation* (characterised by $n = 1$ and $d = 2$) which shows a real phase transition for which an exact solution is known. Since it was solved initially [Onsager 1944] (a remarkable mathematical achievement), the method of transfer matrices has lightened the mathematical treatment of the model. Here we present just the essential results.

- **An order–disorder transition is predicted** in which the critical temperature, for a square lattice, is $T_c \approx 2.269185 \times J/k$ (solution to the equation $\sinh(\frac{2J}{kT_c}) = 1$). This temperature is well below the $T_c = 4J/k$ predicted by the mean field approximation (see Chap. 1, Fig. 1.20).
- **Above the critical temperature T_c, the average magnetisation is zero**, while below it, it is:

$$m = \left[1 - \frac{\left(1 - \tanh^2\left(\frac{J}{kT}\right)\right)^4}{16\tanh^4\left(\frac{J}{kT}\right)} \right]^{1/8} \tag{3.31}$$

which behaves as $m \approx |t|^{1/8}$ in the vicinity of the critical point. The critical exponent of the order parameter here is $\beta = 1/8$, compared with $1/2$ predicted by the mean field approximation.

- **Similarly, the susceptibility at zero field behaves**, in the vicinity of the critical point, as:

$$\chi = |t|^{-7/4},$$

(3.32)

where the exponent $\gamma = 7/4$, compared to the value 1 predicted by mean field.

The exact solution of the 2D Ising model was a milestone. It established a fundamental point: statistical physics is capable of predicting phase transitions by rigorous calculation. However, the distance between these results and the mean field results was troubling. Numerous labs tackled extremely precise measurements to distinguish between these two approaches, but the measured exponents were most often intermediate values with error bars clearly excluding both formal predictions (see Chap. 1, Fig. 1.22). A quarter of a century later, the dilemma was finally resolved.

3.5 Renormalisation

Renormalisation methods are used in all areas of physics and are necessary as soon as one is interested in asymptotic properties of systems in which fluctuations exist at all spatial and/or temporal scales. They give access to properties that are intrinsic, universal and independent of the microscopic details of the system. Renormalisation results, based on scale invariance, are insensitive to many simplifications and shortcomings of a particular model. Renormalisation was not born in 1971 with the solution of critical behaviour [31], but has a long history. After a brief presentation of the semantics of the term "renormalisation" and the changes of meaning it has known, we will present a few principles common to all renormalisation methods, which transformed the very status of models, their relevance and use.

3.5.1 Brief History of the Concept

The concept of "renormalisation" was first used in the form of an adjective in the context of hydrodynamics. During the 19th century, the *renormalised mass* of a body moving in a fluid corresponded to its apparent mass, i.e. its inertial mass m (appearing in the expression for its kinetic energy $1/2\ mv^2$ and in its fundamental equation of motion $m\ dv/dt$ = resultant force), modified by the presence of the fluid. The movement of the mass drags fluid along with it and thereby increases its inertia. The renormalised mass m_R takes into account a contribution due to the displaced fluid:

$$1/2\ m_R\ v^2 = 1/2\ m\ v^2 + \text{ kinetic energy of displaced fluid.}$$

More generally, by extension, a *renormalised* quantity signifies the *apparent* value of this quantity obtained by adding to its intrinsic value contributions resulting from interactions, external influences or degrees of freedom not appearing explicitly in the description. For example, we distinguish the direct correlations between two particles of a fluid from renormalised correlations reflecting implicit interactions transmitted by the solvent.

3.5.1.1 The Original Renormalisation Group for Quantum Field Theories

Renormalisation in the new sense of *regularisation* appeared later in quantum field theories, as a technique to integrate over all virtual quanta with very high frequency ω, or equivalently very high energy (a situation called *UV catastrophe*, see box below) [6]. At the beginning of the 1950s, Feynman, Schwinger, Tomonaga and Dyson, overcame this difficulty using the renormalisation technique. A *renormalisable* theory is one in which the influence of phenomena can be taken into account in an implicit way by replacing the initial parameters by *effective parameters*, for modes with frequency ω lower than a cutoff value Ω. We call transformations $\mathfrak{R}_{\Omega_1, \Omega_0}$ linking initial parameters to renormalised parameters in a frequency range (Ω_0, Ω_1), *renormalisation operators*. The condition of coherence of the operation is a generalised group law. When $\mathfrak{R}_{\Omega_1, \Omega_0}$ depends only on $k = \Omega_0/\Omega_1$, we recover the usual group law $\mathfrak{R}_{k_2}\mathfrak{R}_{k_1} = \mathfrak{R}_{k_2 k_1}$. This group structure is of such great interest that we talk about the *renormalisation group*. This structure primarily enables the renormalisation to be carried out using tools from group theory (Lie algebras, representations etc.), already widely used to exploit the existence of symmetry groups. But over and above this technical argument,

> the symmetry groups of a physical system determine a large percentage of its observables: the renormalisation group appears as a group of particular symmetry and can often be used to translate the scale invariance properties of a physical system under the action of renormalisation into quantitative information.

Later, the renormalisation group was powerfully applied to statistical mechanics due to its ability to treat phenomena in which a divergence of a correlation length prevents microscopic degrees of freedom being replaced by average quantities due to a separation of length scales. The renormalisation approach determines the way fluctuations are organised at different scales and formulates scaling laws resulting from this organisation. More recently, renormalisation has been used in the context of dynamical systems to study the appearance of chaos. In this case of dynamical systems, renormalisation applies to dependencies and invariances of *temporal* scales.

Spatial Limits of a Scaling Behaviour: UV and IR Catastrophes

- **UV divergences**. We say there is an *ultraviolet catastrophe* (or divergence) in a physical
  system if there is an infinite number of possible processes at small scales (i.e. at high
  frequencies q/ω). The formal treatment of such a system requires the introduction
  of a cutoff scale a allowing the regularisation of a model M(scales $\rightarrow$ 0) by an
  effective model M(a). The expression "UV divergence" comes from quantum field
  theory in problems of divergence at the high energy limit. The same issue is met when
  constructing an ideal model of the point charge of an electron. These divergences appear
  when we extend a theory that is valid only in a finite domain, to $q = \infty$ or $\omega = \infty$. At
  the "UV" end, descriptions of critical systems are naturally limited by the atomic scale.
- **IR divergences**. *Infrared divergences* (large scale limit) appear if there are divergent
  behaviours at the macroscopic scale. This is the case for critical systems. IR divergences
  are also linked to long range correlations: effects of fluctuations add constructively
  instead of leading to an average of zero. Perturbation methods are ineffective in cases of
  IR divergences and they can only be treated with renormalisation methods.

We now introduce a few general principles common to all renormalisation
methods, justifying their shared name.

3.5.2 Renormalisation Steps

Renormalisation operates on a model M, that is on a physical representation of a
real system S. We construct a model as a function of the phenomena we want to
study: in particular we fix the *minimum scale* a that is possible to detect, related
to the distance between particles or the resolution of the measurement apparatus.
This microscopic scale determines the sub-systems of S that will be considered
as elementary constituents. The model assumes they have no internal structure.
Their physical state (position, magnetisation etc.) is described by a small number
of quantities s. A microscopic state of S is described by a *configuration* $\{s_i\}$ $\equiv$
$(s_1, \ldots , s_N)$ of N constituent elements. The ingredients of a model M are as
follows:

- The *phase space* $\mathsf{E} = \{\ \{s_i\}\ \}$ of configurations.
- The macroscopic quantities A associated with the configurations $\{s_i\} \rightarrow A(\{s_i\})$.
- A function $\phi(\{s_i\})$, which we call the *structure rule* (the backbone of the model).

At equilibrium, $\phi(\{s_i\})$ determines the statistical weight of the configuration $\{s_i\}$
within the macroscopic properties. Out of equilibrium, $\phi(\{s_i\})$ determines the
evolution of the configuration $\{s_i\}$. ϕ is for example the Hamiltonian of a system
in statistical mechanics, the law of motion for a dynamical system, or transition
probabilities for a stochastic process. The structure rule itself is written using a set
of parameters $\{K_j\}$ written in compact form as K.

The forms of $\phi(\{s_i\})$ – the Hamiltonian, in the case of phase transitions which interests us here – may be wanted in a functional ensemble $\Phi = \{\phi\}$. The elements of Φ are, for example, constructed by adding different terms, each corresponding to a parameter in the expression of the structure rule ϕ. This point is crucial: it differentiates the method proposed by Kadanoff – renormalisation within a given ensemble of parameters K – and that of the renormalisation group, which corresponds to changing the form of the original Hamiltonian, within a general functional ensemble Φ.

The novel approach brought by the renormalisation method is to iterate a *renormalisation transformation* consisting of the following steps:

1. Decimation
2. Change of scale of the system such that is superimposes on its starting structure
3. Transformation of parameters replacing them with effective parameters

and to study the action of this transformation in the *space of models*.

Let us return to the signification of these steps, which we briefly introduced in Sect. 1.1.

3.5.2.1 Decimation

The term "decimation" (coarse-graining) designates a change in resolution $a \to ba$ of the configuration $\{s_i\}$, in other words a regrouping of constituent elements in packets $\{s_i\}'$ each containing b^d original elements in a space of dimension d. A rule T_b relates the configuration $\{s_i\}'$ of these packets to the initial configuration $\{s_i\}' = T_b(\{s_i\})$. Two points of view are possible: either this operation reduces the number N of degrees of freedom by a factor b^d, or the size L of the observation field is increased keeping N constant. Decimation is accompanied by a loss of information at scales smaller than ba: the transformation T_b is chosen in a way that preserves at least the information and properties we think play a crucial role at large scales. This step is a subtle point in the renormalisation process and requires a prior understanding of the studied phenomena. If the step is performed in conjugate space (frequency and wavevector Fourier transformed space), decimation results in a change $\Omega \to \Omega/b$ in the cutoff value $\Omega = 2\pi/a$ beyond which the renormalised model no longer explicitly describes the system.

3.5.2.2 Change of Scale

The change of scale[4] restores the effective minimum scale after spatial regrouping during decimation. Distances in the decimated model are therefore reduced by the

---

[4]Note that more generally the study of changes of scale is also very useful in the case of spatio-temporal phenomena, outside the context of decimation. Take for example a random walk in discrete time (Sect. 4.3.3) obeying the law of diffusion: when we contract time by a factor b,

factor b

$$x \;\rightarrow\; x/b.$$

3.5.2.3   Effective Parameters

The combined effect of the first two steps, decimation and change of scale, must be compensated for by a transformation of the structure rule ϕ, so that we continue to describe the *same* physical system. The new rule $\phi' = R_b(\phi)$ must describe the statistics or evolution of the new configuration $T_b(\{s_i\})$. R_b, acting in the functional space Φ, is called the *renormalisation operator*. In the case of a parameterised model ϕ_K, we write $K' = R_b(K)$ paying a price for approximations, the relevance of which we will discuss later. The *renormalised* parameters K' contain, in an integrated manner, the effect of the microscopic details at scales smaller than ba.

3.5.2.4   Covariance and Invariance by Renormalisation

Coherence in these three steps is established by the fact that the transformed model $M' = R_b(M)$ must describe the *same* physical reality. This property, the *covariance*, ensures in particular that the initial model and the renormalised model conserve the physical invariants of the problem. Covariance therefore expresses the essential objective of renormalisation: to exploit the inalterability of the physical reality as we change *our* manner of observing and describing it. As we have already highlighted, a renormalisation group is nothing other than a particular symmetry group: like all other symmetry operations it leaves the system studied invariant. A stronger concept is that of *invariance by renormalisation* of a property $A = R_b(A)$. This expresses the *self-similarity* of A: with respect to this property, the system observed at the scale ba is identical to the image observed at scale a and expanded by a factor b. By construction of R_b itself, the fixed points of renormalisation are therefore associated with systems possessing an exact scale invariance for the property A.

3.5.3   Renormalisation Flow in a Space of Models

A renormalisation group can be described as a dynamical system (see Chap. 9), that is to say the evolution in time of models in a space $\mathsf{E_M}$. On iterating the action

---

space must be contracted by a factor $b^\alpha = b^{1/2}$ to obtain a statistically identical trajectory. This change of scale factor is the only one for which we obtain a non trivial limit of the diffusion coefficient D at $b \rightarrow \infty$. In the case of anomalous diffusion, with "pathological" jump statistics (Lévy flights) or diffusion on a fractal object, the required exponent α is greater than $1/2$ in the first case (Lévy flights involve superdiffusion) and less than $1/2$ in the case of a fractal space (where the dead branches lead to subdiffusion).

of the operator R_b starting from a model M_0, we obtain a trajectory corresponding to a succession of models showing the same thermodynamic properties as M_0, in particular with the *same partition function*. The set of these trajectories is called *renormalisation flow*. To the user of the model, $R_b(M_0)$ shows a correlation length $\xi' = \xi/b$ and R_b apparently reduces the critical character.

In the context of a parametrised model $M(K)$, we transfer the study of flow induced by the transformation $M \rightarrow R(M)$ to that of a flow induced in the parameter space by a transformation $K \rightarrow r(K)$ with of course $R[M(K)] = M[r(K)]$. The critical parameter values are given by the fixed point $r(K^*) = K^*$. Neglecting nonlinear terms, we show that $Dr(K^*)$ and $DR(M^*)$ have the same eigenvalues, so lead to the same critical exponents.

Space of models	$M(K) \rightarrow R(M)$		$M(K^*) = R(M^*)$
	$\updownarrow$ $\updownarrow$		$\updownarrow$ $\updownarrow$
Parameter space	$K \rightarrow r(K)$		$K^* = r(K^*)$

This is a good place to reiterate the essential step differentiating the search for a fixed point in the space of a determined set of parameters (approach proposed by Kadanoff but which does not work in general) from renormalisation, that is the search for a fixed point in the space of models.

3.5.3.1 Fixed Points

Fixed points, obtained by expressing that a property A is invariant if its value is A^* such that $A^* = R_b(A^*)$, play a crucial role in the analysis of renormalisation. By definition, all models presenting the same fixed points show the same properties on large scales. In addition:

- The characteristic scale ξ^* (typically a correlation length or time) associated with a fixed point has to obey $\xi^* = b\,\xi^*$ and is therefore either zero or infinity. If $\xi^* = 0$, the fixed point is an ideal model in which there is *no coupling* between the elementary constituents. If $\xi^* = \infty$, the fixed point corresponds to a model of *critical phenomena* because it takes into account the scale invariance and the *divergence of* ξ.
- The analysis of R_b in the region around a critical fixed point ϕ^* ($\xi^* = \infty$) determines the scaling relations describing the corresponding critical phenomena. In particular it is shown [9, 11, 17] that the *critical exponents* are simply linked to eigenvalues of the renormalisation operator linearised in ϕ^*.
- These results are *robust* and *universal* in the sense that they are identical for all models in what we call the *same universality class*. At this stage of our introduction to the renormalisation group, this property constitutes a definition of a universality class. Later we will see that each of these universality classes contains an infinite number of different models. Renormalisation results are therefore not invalidated by potential ignorance or misconstruction of the system at small scales. On the contrary, the renormalisation approach enables the

classification of the *relevant* and *irrelevant* aspects of a model, and thereby
the construction of the minimal model for each universality class. In this sense,
renormalisation methods overcome experimental limitations in certain physical
systems. In this way they mark a turning point in theoretical physics.

3.5.3.2 Models at Fixed Points

Models M^* at fixed points obey $M^* = R_b(M^*)$. They correspond to exactly scale
invariant systems, either trivial ($\xi^* = 0$), or critical ($\xi^* = \infty$). The set of models
that converge to M^* or diverge from M^*, under the action of R_b, is a hypersurface
of E_M, which we will call the universality class $C(M^*)$. The basin of attraction of a
fixed point is the sub-space of models for which the renormalisation flow *flows to* a
fixed point. The universality class $C(M^*)$ is a larger set which also contains all the
models *flowing from* the fixed point under the action of renormalisation.

Analysis of the neighbourhood of M^* predicts the (asymptotic) critical properties
of all the models of $C(M^*)$. A space tangent to $C(M^*)$ at M^* is generated by
the eigenvectors of the linearised transformation $DR_b(M^*)$ (see Sect. 9.1.1). The
eigenvalues λ with modulus $|\lambda| < 1$ correspond to *stable* directions,[5] such
that the transformation R_b results in M converging to M^*, whereas in *unstable*
directions ($|\lambda| > 1$), R_b drives M away from M^*. Situations where $|\lambda| = 1$ are
called marginal. If the fixed point shows stable and unstable directions, it is called
hyperbolic (see Fig. 3.6). These are the physically interesting fixed points.

3.5.3.3 Power Law and Calculation of Critical Exponents

We distinguish two parts within $C(M^*)$: the basin of attraction of M^* (the hyperspace V^S generated by the stable directions) and the hyperspace V^U generated by
the unstable directions. V^S corresponds to phenomena with no consequences at
macroscopic scales, i.e. to classical systems, which are in other words "non critical".
The *critical properties* we are interested in correspond to *unstable directions*,
themselves associated with relevant coupling parameters K_i. This is one possible
definition of the *relevance of a parameter*, which we will discuss below. Let us
suppose that there exists only one unstable direction, corresponding to the relevant
parameter K, along which R will perform an expansion. The essential idea is that
all the models M_K of $C(M^*)$ that project onto the space V^U in an identical manner
are equivalent (denoted $\Leftrightarrow$) because flow in the space V^S does not change their
asymptotic behaviour. So we can express the linearisation of R in the neighbourhood
of the fixed point as:

---

[5]A fixed point $M^* = (x^*, y^*, z^*, \ldots) = f(x^*, y^*, z^*, \ldots)$ is said to be *stable* in the direction
x, if f applied iteratively at point $M = (x^* + dx, y^*, z^*, \ldots)$ brings M to M^* (dx being
infinitesimal).

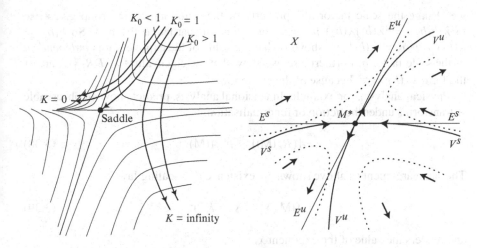

Fig. 3.6 (*Left*) Hyperbolic fixed point in the case of an essential coupling parameter K (*right*) If x^* is a hyperbolic fixed point of the transformation $x_{n+1} = f(x_n)$ in space X, written $x^* = f(x^*)$, there exists a unique pair (V^S, V^U) of stable and unstable manifolds, that are invariant under f and tangent at x^* to vector spaces E^S and E^U respectively (direct sum of X on x^*). *Dotted curves* show the appearance of discrete trajectories (after [33])

$$R(\mathsf{M}_K) \Leftrightarrow \mathsf{M}_{r(K)} \Leftrightarrow \mathsf{M}_{K^*} + \lambda_1(\mathsf{M}_K - \mathsf{M}_{K^*}), \qquad (3.33)$$

where $r(K)$ is the renormalisation transformation of parameter K and λ_1 is the positive eigenvalue associated with the unstable direction. Identifying the expansion action of R along V^U with the action of r along the direction K gives, to dominant order:

$$r(K) \approx K^* + \lambda_1(K - K^*) \quad \text{where} \quad K^* = r(K^*). \qquad (3.34)$$

If b is the change of scale factor, the behaviour of the coherence length in the vicinity of the fixed point is written:

$$\xi\left[r(K) - K^*\right] = \xi\left[\lambda_1(K - K^*)\right] = \frac{\xi(K - K^*)}{b}. \qquad (3.35)$$

This is true only if $\xi(K)$ obeys a power law:

$$\left[\lambda_1(K - K^*)\right]^{-\nu} \sim \frac{(K - K^*)^{-\nu}}{b} \qquad (3.36)$$

from which we derive the value of the exponent ν:

$$\nu = \frac{\log(b)}{\log(\lambda_1)}. \qquad (3.37)$$

The exponent ν and the other exponents are calculated from the eigenvalues of the renormalisation transformation linearised around the fixed point $DR_b(\mathsf{M}^*)$. If we change the scale factor, the property of the renormalisation group gives rise to $\lambda(b_1 b_2) = \lambda(b_1)\lambda(b_2)$ for any eigenvalue, in particular the first. So $b_1 b_2 = \lambda(b_1)^\nu \lambda(b_2)^\nu = \lambda(b_1 b_2)^\nu$, showing that the value of the exponent ν is *independent* of the scale factor b. Furthermore, we show that the eigenvalues of $DR_{b(M^*)}$ are of the form $\lambda_j(b) = b_j^\gamma$ because of the group law.

Physical analysis, for example dimensional analysis, predicts how an observable A transforms under the action of renormalisation:

$$A(R_b(M)) \sim b^\alpha A(M). \tag{3.38}$$

The same argument as above shows the existence of a scaling law

$$A(M_K) \sim (K - K^*)^x \tag{3.39}$$

and renders the value of the exponent x:

$$x = \frac{\alpha \log(b)}{\log(\lambda_1)} = \alpha \nu. \tag{3.40}$$

If the observable is local, depending on the position **r** in real space, its renormalisation transformation is written:

$$A\left(R_b(M_K), \frac{r}{b}\right) = b^\alpha A(M_K, r). \tag{3.41}$$

This is for example the case for the correlation function $G(M, r)$ whose general expression:

$$G(M, r) = \frac{e^{-r/\xi(M)}}{r^{d-2+\eta}} \tag{3.42}$$

is compatible with the scaling behaviour of ξ and A above: it obeys (3.41) when we simultaneously transform r to r/b and ξ to ξ/b.

3.5.3.4 Universality Classes

The procedure of renormalisation we have just described shows the *universality* of macroscopic properties of critical systems: all models within the same *universality class* $C(M^*)$, have the same asymptotic properties. Renormalisation enables the classification of the ingredients of each model into two categories, *irrelevant* factors, which contribute nothing to the critical behaviour, and *relevant* factors, which determine the universality class. To make this classification, it is enough to examine the *renormalisation flow*. The action of the transformation R_K, represented in the space of models by the renormalisation flow, leads to the determination of whether or not a modification δM of a model M has observable consequences on its thermodynamic behaviour (large systems limit):

- If the trajectories coming from M and M + δM approach one another under the action of R_K, the perturbation δM does not affect the macroscopic properties predicted by the model M. In this case δM is said to be *irrelevant* and it is futile to take it into account if we are interested in the critical behaviour.
- If, on the contrary, the trajectories move away from each other, the perturbation δM is *relevant*: its influence is amplified and discernible in the macroscopic characteristics. The perturbation is capable of changing the universality class of M.

This argument shows that critical phenomena are associated with hyperbolic fixed points of R_K, unstable directions being associated with relevant control parameters. Universality classes are associated with these hyperbolic fixed points, which are the ideal representatives of the models in the universality class.

3.5.4 Remarks

3.5.4.1 Role of Choice of Transformation: M* changes but Exponents do not

The form of the renormalisation flow depends directly on the definition of the renormalisation operator R. As a general rule, if we significantly change the transformation R, we expect that it will change the position of the fixed point M*. However in practice, the transformations most often used are connected by a diffeomorphism such that $R' = fRf^{-1}$. The fixed point of R' becomes $f(M^*)$ but the *critical exponents do not change*, simply because $DR(M^*)$ and $DR'[f(M^*)]$ have the same eigenvalues. This argument reinforces the universal characteristic of critical exponents.

3.5.4.2 Influence of Non Linearities in the Transformation

We could draw a parallel between the action of the transformation R_K on the family of models and its linearised action along the unstable manifold of the renormalisation flow leading to the *exact* scaling laws. This parallel is however only rigorous if we neglect the nonlinear terms in the action of the renormalisation. This means the scaling laws only give the dominant dependence with respect to K. The rigorous nonlinear analysis (very tricky) effectively shows that the nonlinear terms do not destroy the scaling law deduced from the action of the linearised renormalisation operator.[6]

---

[6]It was proved for the renormalisation describing the transition to chaos by doubling the period (see Chap. 9) [Collet and Eckmann 1980].

3.6   Phase Transitions Described by Renormalisation

3.6.1   Examples of Renormalisation

Since the 1970s many renormalisation methods have been developed, some in direct space and others in Fourier space. Here we develop a first, simple, example corresponding to the Ising model in two dimensions. We first show the calculation for a triangular lattice, then for a square lattice.

3.6.1.1   Ising Model in 2D: Triangular Lattice

Although the concept of renormalisation can seem abstract, it is possible to illustrate it with simple concrete examples. Figure 3.7 shows the first situation we will describe in detail: six spins σ_i on a triangular lattice are regrouped in two blocks of three spins. In this case, the decimation can be summarised in the following way:

- Six initial spins σ_i $\rightarrow$ 2 super-spins μ_i.
- Change in spatial scale factor $b = \sqrt{3}$.
- Majority rule:

$$\mu_i = +1 \quad \text{if} \quad \sum_{j \in Block(i)} \sigma_j > 0$$
$$\mu_i = -1 \quad \text{if} \quad \sum_{j \in Block(i)} \sigma_j < 0.$$

In the absence of an external field, the Ising Hamiltonian is written:

$$-\frac{H}{kT} = K \sum_{<ij>} \sigma_i \sigma_j, \tag{3.43}$$

where $< ij >$ means that σ_i and σ_j are nearest neighbours. The partition function is:

$$Z = \sum_{\{\sigma_i\}} e^{\displaystyle K \sum_{<ij>} \sigma_i \sigma_j}. \tag{3.44}$$

Fig. 3.7 A set of 6 spins, regrouped in 2 triangular blocks

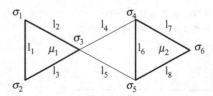

Recall that the reduced coupling constant K is linked to the coupling constant J and the temperature by $K = J/kT$. First we need to state that the decimation conserves the value of the partition function for the set of spins ($Z' = Z$). In fact, by this transformation all we are doing is arbitrarily changing the level of description of a reality which itself remains the same (covariance). The partition function Z' of the decimated system is particularly simple. It contains just two terms corresponding to the 4 states accessible to the super-spins μ_1 $\{\uparrow\uparrow, \downarrow\downarrow, \uparrow\downarrow, \downarrow\uparrow\}$. We can then express Z' as:

$$Z' = Z'_{\uparrow\uparrow} + Z'_{\downarrow\downarrow} + Z'_{\uparrow\downarrow} + Z'_{\downarrow\uparrow} = 2e^{K'} + 2e^{-K'}. \tag{3.45}$$

The calculation of Z brings into play $2^6 = 64$ states, 16 per state of the system of super-spins μ_i. We can therefore decompose Z into 4 contributions:

$$Z = Z_{\uparrow\uparrow} + Z_{\downarrow\downarrow} + Z_{\uparrow\downarrow} + Z_{\downarrow\uparrow}, \tag{3.46}$$

where the arrows in the indices refer to the state of μ_1 and μ_2. Table 3.2 shows the 16 configurations that lead to the state $\uparrow\uparrow$ for μ_1 and μ_2, i.e. situations where only one spin σ_i at most is flipped down in each block. The 16 configurations that lead to the state $\downarrow\downarrow$ for μ_1 and μ_2 are deduced directly by reversing the orientations of every spin in the table: they lead to the same values in the Hamiltonian and contribute in the same way to Z:

$$Z_{\uparrow\uparrow} = Z_{\downarrow\downarrow} = e^{8K} + 3e^{4K} + 2e^{2K} + 3e^{0K} + 6e^{-2K} + e^{-4K}. \tag{3.47}$$

Table 3.2 The 16 configurations of the system of spins that correspond to the state $\uparrow\uparrow$ for μ_1 and μ_2

	σ_1	σ_2	σ_3	σ_4	σ_5	σ_6	l_1	l_2	l_3	l_4	l_5	l_6	l_7	l_8	$-H/kT$
Order	↑	↑	↑	↑	↑	↑	K	K	K	K	K	K	K	K	$8K$
	↓	↑	↑	↑	↑	↑	$-K$	$-K$	K	K	K	K	K	K	$4K$
	↑	↓	↑	↑	↑	↑	$-K$	K	$-K$	K	K	K	K	K	$4K$
1 spin	↑	↑	↑	↑	↑	↓	K	K	K	K	K	K	$-K$	$-K$	$4K$
flip	↑	↑	↑	↓	↑	↑	K	K	K	$-K$	K	$-K$	$-K$	K	$2K$
	↑	↑	↑	↑	↓	↑	K	K	K	K	$-K$	$-K$	K	$-K$	$2K$
	↑	↑	↓	↑	↑	↑	K	$-K$	$-K$	$-K$	$-K$	K	K	K	$0K$
	↓	↑	↑	↑	↑	↓	$-K$	$-K$	K	K	K	K	$-K$	$-K$	$0K$
	↑	↓	↑	↑	↑	↓	$-K$	K	$-K$	K	K	K	$-K$	$-K$	$0K$
2 spin	↓	↑	↑	↓	↑	↑	$-K$	$-K$	K	$-K$	K	$-K$	$-K$	K	$-2K$
flips	↑	↓	↑	↓	↑	↑	$-K$	K	$-K$	$-K$	K	$-K$	$-K$	K	$-2K$
	↑	↑	↓	↓	↑	↑	K	$-K$	$-K$	K	$-K$	$-K$	$-K$	K	$-2K$
	↓	↑	↑	↑	↓	↑	$-K$	$-K$	K	K	$-K$	$-K$	K	$-K$	$-2K$
	↑	↓	↑	↑	↓	↑	$-K$	K	$-K$	$-K$	$-K$	$-K$	K	$-K$	$-2K$
	↑	↑	↓	↑	↓	↑	K	$-K$	$-K$	$-K$	K	$-K$	K	$-K$	$-2K$
	↑	↑	↓	↑	↑	↓	K	$-K$	$-K$	$-K$	$-K$	K	$-K$	$-K$	$-4K$

We leave as an exercise for the reader to show in the same way that:

$$Z_{\uparrow\downarrow} = Z_{\downarrow\uparrow} = 2e^{4K} + 2e^{3K} + 4e^{0K} + 6e^{-2K} + 2e^{-4K}. \qquad (3.48)$$

The *covariance* we would like to impose, that is the invariance of system properties under the effect of the transformation, requires $Z'_{\uparrow\uparrow}=Z_{\uparrow\uparrow}$ and $Z'_{\uparrow\downarrow}=Z_{\uparrow\downarrow}$. But one difficulty is that we have two independent relationships relating K and K'. These relationships are not identical: in Z', the product of the two terms equals 1 ($e^{2K'} \times e^{-2K'} = 1$), which is not true for $Z_{\uparrow\uparrow} \times Z_{\uparrow\downarrow}$. Here we touch on the difficulty that Kadanoff's approach ran into: the renormalisation of the coupling parameter K seems to lead to a contradiction. We have indicated the general idea enabling us to avoid such contradictions: transform the *model* and not just the value of the parameters. Here one possibility is to add a term g, per spin, to the Hamiltonian. Note that this term, equivalent to a chemical potential, has no reason to show singular behaviour at the critical point:

$$-\frac{H}{kT} = K \sum_{<ij>} \sigma_i \sigma_j + Ng. \qquad (3.49)$$

Identifying each term gives:

$$e^{K'+2g'} = \frac{Z_1}{2} = e^{6g}\left[e^{8K} + 3e^{4K} + 2e^{2K} + 3e^{0K} + 6e^{-2K} + e^{-4K}\right] \qquad (3.50)$$

and

$$e^{-K'+2g'} = \frac{Z_2}{2} = e^{6g}\left[2e^{4K} + 2e^{3K} + 4e^{0K} + 6e^{-2K} + 2e^{-4K}\right] \qquad (3.51)$$

from which we get the value of K' (and g'):

$$K' = \frac{1}{2} \log\left[\frac{e^{8K} + 3e^{4K} + 2e^{2K} + 3 + 6e^{-2K} + e^{-4K}}{2e^{4K} + 2e^{3K} + 4 + 6e^{-2K} + 2e^{-4K}}\right]. \qquad (3.52)$$

In this way we have established the renormalisation relation of K. There are three fixed points of which two are stable:

- The fixed point $K = 0$ corresponds to a coupling $J = 0$ or $T = \infty$: the spins are independent from each other and the disorder is total.
- The fixed point $K = \infty$ correspond to an *infinite* coupling or $T = 0$: the spins are aligned with perfect order.
- There exists another fixed point, which is hyperbolic and corresponds to the value $K^* = 0.45146$. It is this fixed point that corresponds the critical point.

Since we have taken into account only a finite part of the Ising network, we only expect an approximate representation of the critical behaviour. However, this should still be an improvement with respect to the mean field results. We can check this by calculating the exponent v as it is expressed in (3.37). The eigenvalue corresponding to the renormalisation transformation $r(K)$ is simply the derivative of this function at the fixed point $r'(K^*)$. The value of the exponent v is therefore:

$$v = \frac{\log(b)}{\log(\lambda_1)} = \frac{1/2 \log 3}{\log(r'(K^*))}. \tag{3.53}$$

The value obtained here is $v = 1.159$, compared with the exact value $v = 1$ calculated by Onsager and the value $v = 1/2$ predicted by mean field.

Additionally, we establish the relation $g'(K^*) = 5.60 g(K^*)$ which shows that at the fixed point the chemical potential g is $g^* = 0$. We say that this quantity is *irrelevant*, meaning that taking it into account in the Hamiltonian, facilitating the calculations, does not affect the fixed point.

3.6.1.2 Ising Model in 2D: Square Lattice

The same type of reasoning can be carried out on two blocks of a square lattice. This time the 8 spins can organise in 256 possible configurations, which makes the calculation long. Another difficulty could seem like a dilemma. Given the even number of spins in each block, we cannot use the majority rule for the six cases where the magnetisation is zero in each block: we have to randomly consider that three cases correspond to $\mu = 1$, and the other three to $\mu = -1$. Using this rule, we leave to the reader to establish in the same way as previously that the renormalisation transformation of K can be expressed as:

$$K' = \frac{1}{2} \log \left[\frac{e^{10K} + 6e^{6K} + 4e^{4K} + 14e^{2K} + 18 + 12e^{-2K} + 6e^{-4K} + 2e^{-6K} + e^{-10K}}{e^{6K} + 8e^{4K} + 10e^{2K} + 22 + 13e^{-2K} + 6e^{-4K} + 4e^{-6K}} \right]. \tag{3.54}$$

Here the non trivial fixed point $K^* = 0.507874$, leads to the value $v = 1.114$. These two calculations therefore give similar values of the exponent. Their main drawback is that they start with finite blocks that do not show the symmetry of the lattice (triangular or square). A more sophisticated calculation [20] on a finite lattice of 16 spins regrouped in blocks of 4 (Fig. 3.8) leads to exponent values extremely close to the exact values calculated by Onsager.

Table 3.3 regroups the results obtained by our renormalisation example of the Ising model. The value of the fixed point K^* of the reduced coupling constant K leads to values of the critical exponents which are universal and therefore

Fig. 3.8 Decimation into blocks of 4 spins on a Ising square lattice

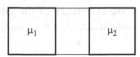

Table 3.3 Results obtained in the case of the 2D Ising model. The value of the reduced coupling constant K^* at the fixed point leads to the value of the critical temperature T_c (non universal) which is compared to the value obtained in the mean field approximation T_c (MF) (MF for mean field). K^* also leads to the value of critical exponents (universal and therefore independent of the lattice type). The example of the exponent ν characterising the coherence length is presented above

	Triangular lattice			Square lattice		
	Mean field	This calculation	Exact value	Mean field	This calculation	Exact value
K^*	1/6	0.45146	0.27465	1/4	0.507874	0.440687
kT_c/J (MF)						
(non universal)	6	2.709	3.64	4	2.0315	2.269185
ν (universal)	1/2	1.159	1	1/2	1.114	1

independent of the lattice type. As an example we have taken the exponent ν which characterises the coherence length. Renormalisation, even at the simple level we have applied it, noticeably approaches the exact value of this exponent. The value of the reduced coupling constant K^* also leads to the value of the critical temperature $T_c = J/kK^*$. Even though it is not universal (it depends in particular on the nature of the lattice) this value enables us to evaluate the effect of fluctuations if it is compared to the value obtained in the mean field approximation T_c (MF). The noticeable lowering of the critical temperature seen shows that this simple renormalisation calculation takes into account the effect of fluctuations.

3.6.2   Expansion in $\varepsilon = 4 - d$

Here we very briefly present the famous approach developed (from 1972) by Wilson and Fisher [9,10,32]. The idea is to generalise renormalisation freeing it from being saddled with specific models. Let us remind ourselves of the accomplishments of renormalisation as verified when applied to a particular model:

- **Universality.** Critical behaviour (and for example critical exponents) do not depend on the microscopic details of the model. Note that the critical temperature itself, directly determined by the value of K^*, *is not a universal quantity*: it depends on the *position of the fixed point*, which is itself sensitive to the details of the model. On the other hand, we show that the properties of the *neighbourhood of the fixed point*, characterised by the *eigenvalues*, are independent of the position of the fixed point, within the same universality class.
- **Critical exponents** are universal quantities, which principally depend only on two "relevant" geometric parameters, the dimension d of space and the number n of components of the order parameter. Examples of "irrelevant" parameters are the symmetry of the lattice and the nature of the interaction range introduced in the model (on the condition that it decays reasonably fast with distance).

- **For $d \geq 4$** the Ginzburg criterion (Sect. 3.1.3) indicates that the mean field exponents are valid. This is verified by renormalisation approaches.

Wilson and Fisher [32] use a very general model in which n and d are parameters assumed to vary continuously. They then expand the value of the exponents as a function of $\varepsilon = 4 - d$ where ε is small. For $\varepsilon = 0$, the exponents take the mean field values which we presented in Chap. 1 ($\alpha = 0, \beta = 1/2, \gamma = 1, \delta = 3, \eta = 0, \nu = 1/2$). The starting Hamiltonian chosen by Wilson is:

$$H = -\sum_{r,a} J(a)S(r)S(r+a) - h \sum_r S_\alpha(r), \qquad (3.55)$$

where the interaction energy $J(a)$ depends in general on the distance between the spins. Spins S have n components, position r is located in space of dimension d and h is the applied external magnetic field. This Hamiltonian will now undergo several transformations in order to simplify the renormalisation. The main transformation consists of switching to Fourier space allowing the sum over a to be replaced by the product $S(q) S(-q)$. Finally, the Hamiltonian can be written in the form of the following expansion:

$$H = -\frac{1}{2}\sum_q (p+q^2)S(q)S(-q) - \frac{U}{N} \sum_{q_1+q_2+q_3+q_4=0} [S(q_1)S(q_2)][S(q_3)S(q_4)]$$

$$-\frac{W}{N^2} \sum_{q_1+q_2+q_3+q_4+q_5+q_6=0} [S(q_1)S(q_2)][S(q_3)S(q_4)][S(q_5)S(q_6)]$$

$$-\cdots + hN^{1/2}S_\alpha(0) \quad (3.56)$$

where p, U, W, etc. are the new coupling constants which depend on J, a, d and n. If we take only the first terms, we obtain the model called "Gaussian" in which the critical behaviour is characterised by mean field exponents. We can write the renormalisation transformations of the coupling constants as a function of n and $\varepsilon = 4 - d$. To first order in ε it is sufficient to take into account just the first two terms: the corresponding model is usually called the S^4 model. It establishes the following expressions for the exponents to first order in ε:

$$\alpha = \varepsilon \left[1/2 - \frac{n+2}{n+8} \right] + O(\varepsilon) \qquad\qquad \beta = 1/2 - \frac{3\varepsilon}{2(n+8)} + O(\varepsilon)$$

$$\gamma = 1 + \frac{(n+2)\varepsilon}{2(n+8)} + O(\varepsilon) \qquad\qquad \delta = 3 + \varepsilon + O(\varepsilon).$$

$$(3.57)$$

The other two exponents η and ν can be derived using the scaling relations (3.8) and (3.9) (see Sect. 3.2.1). One may feel that a century of study of phase transitions finally succeeded with these explicit expressions for the critical exponents. How-

Table 3.4 Values of the exponents obtained by expansion in $\varepsilon = 4 - d$. To zero order, first order and finally the best value obtained by summing all the contributions [34]

Exponent (3D Ising)	α	β	γ	δ	η	ν
Mean field values ($\varepsilon = 0$)	0	1/2	1	3	0	1/2
Value to first order in ε	0.16666	0.3333	1.16666	4	0.0909	0.61111
Best value obtained	0.1070	0.3270	1.239	4.814	0.0375	0.6310
Experimental values (liquid $\leftrightarrow$ vapour)	0.113	0.322	1.239	4.85	0.017	0.625

ever, though the approach may be excellent, the first order results are not perfect (see Table 3.4).

An aspect even more disappointing is that the expansion in ε does not converge [34] whatever the order! However, sophisticated techniques to sum over all terms of the expansion leads to values in perfect agreement with experiments within the measurement resolution (Table 3.4). Thanks to these summation techniques, expansion in ε has become the characteristic tool of the renormalisation group. Using this tool the contours in (d, n) space for each exponent can be traced, where each point represents a universality class.

Figure 3.9 summarises the results traced in this way by Michael Fisher, for situations which do not in general have a direct physical meaning. For example n varies from -2 to infinity (spherical model)! In fact there is no particular *formal* limit in this range of values for n. In four and higher dimensions the critical behaviour is that of mean field, while in one dimension long range order is only established at strictly zero temperature in a system of infinite size. Let us look at Peierls' powerful argument on this point.

3.6.3 In a 1D Space

A simple argument due to Peierls rigorously shows that in a one dimensional system at finite temperature no long range order is allowed. Let us suppose that there exists a single defect on a 1D Ising lattice, separating up spins $\uparrow$ on the left from down spins $\downarrow$ on the right:

The free energy of such a defect, which could be located at $N - 1$ positions, has the simple value of:

$$f_{\text{defect}} = 2J - kT \log(N - 1),$$

where N is the number of spins. At the thermodynamic limit $N \to \infty$, this free energy is negative for all non zero temperatures. Therefore such defects are stable at all temperatures and they are produced spontaneously so long range order is unstable at all finite temperatures.

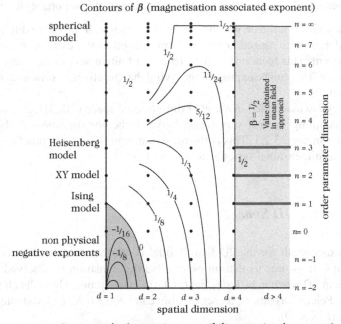

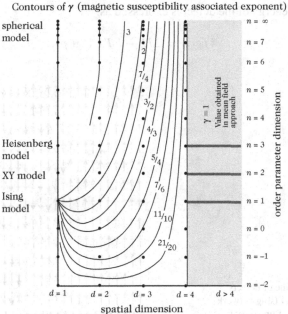

Fig. 3.9 The variation of exponents β and γ taken from the expansion in $\varepsilon = 4 - d$ (traced by Michael Fisher) as a function of the dimension d of space and number n of components of the order parameter. Each point (d, n) corresponds to a universality class. The coordinates are assumed to vary continuously but only their integer values make sense physically. The variation of critical exponents is represented as contours. At and above four dimensions, the mean field exponents are valid (after [33])

One dimensional systems are however interesting in two important situations:

- When their size is finite or, which amounts to the same, if they have dilute defects or disorder: a finite ranged order can "graft" itself to the extremities or defects of the system. This local order could lead to a finite macroscopic order at finite temperature. This can be expressed by saying that the effect of *structural disorder* is relevant.
- When they are assembled in a higher dimensional space (2D, 3D, ...) and there exists a coupling, even if very weak, between the one dimensional objects, as we discuss in Sect. 3.7.1. This is the most common situation in practice because strictly one dimensional systems do not exist.

3.6.4 And in a 2D Space?

Onsager's exact result for the 2D Ising model is entirely confirmed by renormalisation results. It assures us that an order–disorder transition is observed at finite temperature in 2D when $n = 1$, despite the large fluctuations which develop in two dimensions. Peierls' argument, presented above in Sect. 3.6.3, can be adapted to two dimensions (Fig. 3.10).

The free energy of the defect is expressed as:

$$f_{\text{defect}} = 2JL - kTL \log(p),$$

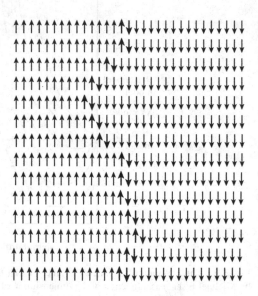

Fig. 3.10 A line of defects separating a 2D Ising network into two parts of up spins ↑, to the *left*, and down spins ↓, to the *right*

where L is the length of the line of defects measured in number of sites,[7] and p is the average number of choices of orientation of this line at each site. We do not know the exact value of p, but due to the constraint that the line does not loop back on itself, we know that the average value is between 2 and 3. We deduce that f_{defect} changes sign at temperature T_c:

$$T_c = \frac{2J}{k \log(p)}.$$

This leads to $1.82J/k < T_c < 2.88J/k$ compared to the exact (Onsager) value of $T_c = 2.2692J/k$ and that of mean field $T_c = 4J/k$. It is interesting to note how much the stability of these simple defect lines explain the reduction in the critical temperature compared to that predicted by mean field.

The same type of stability analysis of defects in two dimensional models with $n = 3$ (*Heisenberg*), $n = 4$, $n = 5$ etc. shows that the critical temperature is zero in these cases. What about the borderline case of the 2D XY model where $d = 2$ and $n = 2$? Actually a transition at finite temperature is observed. It does not have characteristics of universality and corresponds to amazing physics. We present its main characteristics in the following paragraph.

3.6.5 XY Model in 2D: The Kosterlitz–Thouless Transition

The argument of the previous paragraph is specific to the Ising model where spins can only take two opposite values. In the case of a continuous variation, the types of defects possible are far more diverse and their entropy can be high: this can profoundly change the nature of the transition. The particular case of a planar array of spins $S = (S_x, S_y)$ in two dimensions – the 2D XY model – illustrates this effect. It was the object of intense debate in the early 1970s, before Kosterlitz and Thouless proposed a famous description [15]. This transition is also often called the "KT transition".

There seemed to be a contradiction about this model. On one hand, Mermin and Wagner established that all long range order is excluded in two dimensional systems that possess a rotational symmetry, which is the case in the XY model where the spins S can take any orientation in the plane. On the other hand, numerical simulations clearly showed a very clean transition, but to a state without long range order. Here we give a quick introduction. The XY model Hamiltonian in zero field is written:

---

[7]Given that only one of these p choices progresses the line in each direction, the length L takes the value $pN^{1/2}$ where N is the total number of spins. However, since L is a factor in the expression of f_{defect}, its precise value does not play a role in the sign of f_{defect}.

$$H = -JS^2 \sum_{<ij>} \cos(\theta_i - \theta_j),$$ (3.58)

where θ_i is the orientation of spin i. This angle is a convenient order parameter. Is it legitimate to assume that it varies slowly in space on the scale of neighbouring spins? At very low temperature, certainly, and we will assume so initially. A determination by renormalisation of the KT transition is presented in Sect. 8.6.2.

3.6.5.1 Slow Spatial Variation of the Order Parameter θ

This approximation enables us to write, to first order in $\theta_i - \theta_j$:

$$H = -\frac{1}{2}qNJS^2 + \frac{1}{2}JS^2 \sum_{<ij>} (\theta_i - \theta_j)^2$$ (3.59)

or, in a continuous approximation and assuming $S = 1$:

$$H = E_0 + \frac{J}{2a^{d-2}} \int d^d r \, |\nabla \theta|^2.$$ (3.60)

This form of the Hamiltonian leads notably to excitations of the system which are called "spin waves", excitations in which the gradient of θ is constant. From this Hamiltonian all the thermodynamic properties can be calculated, notably the correlation function $G(r)$. From the general form of the correlation function we directly obtain:

$$G(r) = \langle \cos[\theta(r) - \theta(0)] \rangle,$$

where the average is performed over the set of origins $r = 0$ in the system.

$G(r)$ tends towards a constant when $r \to \infty$, if long range order exists. In particular, at zero temperature, $G(r) = 1$ for the perfectly ordered state. An evaluation of the partition function and then of $G(r)$ to first order leads to:

$$G(r) \sim \left(\frac{r}{a}\right)^{-\eta(T)}$$ (3.61)

in agreement with the definition of the exponent η in (3.42) for $d = 2$ at the critical temperature for which $\xi = \infty$. But here, the completely new result is that the exponent $\eta(T)$ depends on the temperature:

$$\eta(T) = \frac{kT}{2\pi J}.$$ (3.62)

This result is radically different from what we have observed for universal phase transitions, for which a power law $G(r) \sim (r/a)^{-(d-2+\eta)}$ is observed strictly only at the critical point (outside the critical point the dominant dependence is exponential).

> Everything happens as if the coherence length stayed infinite, as if the system
> stayed in the critical state up to zero temperature.

Do not forget that to get to this result we have assumed that we are at very low
temperature so that the order parameter varies slowly in space. At high temperature
this hypothesis is wrong since there is total disorder. How does the transition
between the two regimes take place?

3.6.5.2 Vortices and Vortex-Antivortex Pairs

When we start with a perfectly ordered system, all phase transitions are the result of
local excitations produced by thermal agitation. In the case of an Ising model, these
excitations are the flipping of isolated spins. Kosterlitz and Thouless [15] identified
the excitations responsible in the case of the XY transition in two dimensions: they
are pairs of *vortices*. The simplest vortex of an order parameter is an arrangement
of spins which is rotationally invariant around a point M (see Fig. 3.11).

If we use polar coordinates (r, φ) centred on M for the position of spins, the order
parameter θ is equal to φ up to a constant and it does not depend on r. Generally
θ in a vortex can be a positive or negative integer multiple of φ: $\theta = q\varphi + \theta_0$. We
will see below that the number q plays the role of an electric charge. From a few
atomic distances onwards, the energy of a vortex can be calculated by the continuous
approximation of (3.60). By neglecting the terms due to nearest neighbours of M
(their contribution could not be assessed by the hypothesis of continuous variation
of θ), the energy of a vortex of unit charge in a system of size L is written:

$$E_{\text{vortex}} = 2\pi J \log\left(\frac{L}{a}\right) \tag{3.63}$$

which diverges for a system of infinite size, a being, as usual, the distance between
spins. Since the entropy of a vortex is directly linked to the number of possible
positions for M (L^2/a^2 positions), the free energy is given by:

$$f_{\text{vortex}} = 2(\pi J - kT) \log\left(\frac{L}{a}\right). \tag{3.64}$$

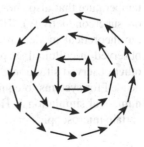

Fig. 3.11 A vortex of unit
"charge" on a 2D XY lattice

This defines a critical temperature $T_c = \pi J / k$ below which no isolated vortices in the system are stable. The energy of a dipole, a pair of vortices of charges q and $-q$ separated by r, is written:

$$E_{\text{Dip}} = 2\pi J q \cdot \log \left(\frac{r}{a}\right). \tag{3.65}$$

Since this energy is finite, while their entropy is large (varies as $\log(L/a)$), these dipoles are always present in a large system, even at low temperature. The number of vortex pairs increases with temperature. At the critical temperature, the average separation between pairs reaches the size of the system and disorder settles. Kosterlitz showed that the coherence length, all the way long infinite below T_c, diverges for $t = (T - T_c)/T_c > 0$ as:

$$\xi \sim \exp(1.5t^{-1/2}). \tag{3.66}$$

This dependence on temperature, which has nothing to do with the power laws observed for universal critical transitions, clearly illustrates the unusual nature of the KT transition. Many other properties are special to this transition, such as the thermal dependence of the exponents.

3.6.5.3   Experiments in 2D

During the 1970s, many teams attempted the challenge of experimentally observing the KT transition [16]. The systems they considered were as diverse as they were difficult to produce: liquid crystals, soap bubbles, adsorbed atomic layers, ultrathin films of superfluid helium, two dimensional superconductors, two dimensional gases of electrons, etc. Most of the attempts ran up against an apparently unsolvable problem. To find a real XY system where the order parameter really has two components, it is difficult to imagine and produce a system where the third spatial dimension, the support or environment in general, does not perturb the two dimensional nature of the system: adsorbed layers are perturbed by the substrate, 2D electron gases by the crystal lattice, etc.

However, observations of films of superfluid helium verified point by point the predictions of the KT transition [3]. Similarly for superconducting films [8] and for planar networks of Josephson junctions [12] show exponents depending on temperature that also show a discontinuity at the critical temperature (see Chap. 7 on superconductivity). The case of melting of 2D crystals is less clear despite the theoretical model proposed by Nelson and Halperin [22] KT behaviour was observed in colloidal suspensions [19] whereas melting seems to be first order in other observed situations. But effects of the support and the extremely long times needed for systems to equilibrate considerably complicate experiments (as well as numerical simulations). However it seems that this situation is not *universal* and certain microscopic details have a relevant effect on the nature of the transition.

We will see another example of the KT transition in the chapter on growth mechanisms (Chap. 8) namely the *roughening transition*, in which a solid becomes rough above a given temperature. This transition, predicted by theory, is well confirmed by numerical simulations.

Among the numerous reviews of these experiments, we recommend those published by Abraham [1], Saïto and Müller-Krumbar [28] and Nelson [21].

3.7   And in Real Situations

3.7.1   Crossover from One Universality Class to Another

Suppose we have two hyperbolic fixed points M_1^* and M_2^* in the same set of parameterised models. There exists a limit separating the zone of validity of the linear approximation around M_1^* from the corresponding zone around M_2^*. Within these zones, the trajectory of a point M under the action of renormalisation is more sensitive to the presence of the fixed point M_1^* or M_2^* respectively (see Fig. 3.12).

In practice, we could observe a *crossover* when we vary one of the parameters of the Hamiltonian: beyond a given threshold the fixed points are suddenly changed. An example of such a crossover results from the following Hamiltonian, cited by Fisher, for a system of 3 component spins with an anisotropy characterised by ε:

$$H(\sigma, \sigma') = J(\sigma_1 \sigma_1' + \sigma_2 \sigma_2' + \sigma_3 \sigma_3') + \varepsilon J(\sigma_2 \sigma_2' + \sigma_3 \sigma_3'). \qquad (3.67)$$

If $\varepsilon = 0$, the critical exponents are those associated with the fixed point M_3^* (Heisenberg spins with three components). However if $\varepsilon = -1$, the critical exponents are those associated with the fixed point M_1^* (Ising spins with one component). Finally if $\varepsilon = +\infty$, the critical exponents are those associated with the fixed point M_2^* (XY spins with two components).

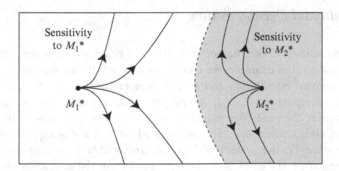

Fig. 3.12 Line separating the zones of influence of two fixed points

For which values of ε are the changes of universality classes situated? It can be shown that, for all nonzero values of the anisotropy ε we exit the universality class of 3 component Heisenberg spins:

- For $\varepsilon > 0$, the universality class is the XY model (2 component spins).
- For $\varepsilon < 0$, the universality class is the Ising model (1 component spins).

3.7.2 Establishment of Equilibrium and Dynamic Critical Exponents

Up until now, we have assumed that our systems are at equilibrium and that we have been taking an instantaneous microscopic snapshot averaged over all configurations of the system. This assumption rests on the implicit assumption of ergodicity. In practice, the system constantly evolves at the microscopic scale and its dynamic behaviour is singular at the critical point. In fact, at the critical point, fluctuations in the order parameter are of divergent spatial range, but they also evolve very slowly: we can say that their *temporal range* diverges. The fast exponential relaxation $e^{-t/\tau}$ of correlations far from the critical point, is replaced by a power law decay $t^{-\kappa}$, κ being a *dynamic critical exponent* and here t is the time. This can also be expressed by the fact that the relaxation time $\tau(T)$ diverges at T_c. We call this phenomenon, which is characteristic of critical phenomena, *critical slowing down*. Other exponents have been defined describing for example the temperature dependence of the time taken to relax to equilibrium. Critical phenomena are then characterised by a double singularity at $(q \rightarrow 0, \omega \rightarrow 0)$. Like other critical exponents, dynamic critical exponents are invariant within universality classes. They can be determined with the help of a spatiotemporal extension of the renormalisation methods used to calculate the spatial critical components [13, 18], see also Chap 8, Sect 8.6.2.

3.7.3 Spinodal Decomposition

We may also wonder about the *path* that leads to order at low temperature when we cool a system. For example, we can try to describe the evolution of a system, initially at thermal equilibrium and at high temperature $T_0 \gg T_c$ and suddenly *quenched* to a temperature $T_1 < T_c$, keeping the order parameter constant (constant magnetisation in the case of a magnetic system or constant density in the case of a fluid or alloy). Since the initial disordered state is not at equilibrium at the new temperature T_1, the system will evolve irreversibly to one of the probable configurations at T_1. If $T_1 \neq 0$, in general two phases coexist at equilibrium (liquid–vapour, positive–negative magnetisation). So we see domains, occupied by one or the other phase, growing by aggregation, fusion etc. This process of separation, which starts on a very small scale and propagates to the macroscopic scale, is called *spinodal decomposition*. Let us mention a few examples where the phenomenon has been observed experimentally or numerically:

- Growth of domains of magnetisation $\pm M_0(T)$ in ferromagnetic systems and the associated Ising model
- Spinodal decomposition in binary alloys
- Phase separation in binary fluids

Several studies have brought to light numerically and experimentally [27] that the relaxation of such a system at long times shows a remarkable scale invariance. In Chap. 2 we introduced structures that are spatially scale invariant: fractals. These structures evolve and are related to one another at successive times by changes of spatial scale. Note that the relaxation of the system acts exactly like an inverse renormalisation. Quantitatively, the morphology of a system is described by the correlation function of the order parameter m at two points:

$$G(r, t) = \langle m(r, t)m(0, t) \rangle - \langle m(r, t) \rangle \langle m(0, t) \rangle \tag{3.68}$$

(an average over the initial condition is implicit in this definition). The scale invariance is expressed by writing that the correlation function only depends on r or t via the dimensionless parameter $r/L(t)$:

$$G(r, t) = f(r/L(t)) \quad \text{when} \quad t \to \infty. \tag{3.69}$$

The scale factor $L(t)$ contains the physics of this decomposition observed at long times, $L(t)$ showing a scaling behaviour:

$$L(t) \sim t^n \quad \text{when} \quad t \to \infty. \tag{3.70}$$

We can specify the condition $t \to \infty$: the typical domain size must be larger than the correlation length for the scaling regime to be established. The scaling law is observed when $L(t) \gg \xi(T)$, where $\xi(T)$ is the correlation length at thermal equilibrium. Starting from a Langevin equation describing the dynamics (see Chaps. 4 and 8), n may be determined by a spatiotemporal renormalisation study (we describe this type of approach in the following chapter). This study leads to $T = 0$ and not to the critical fixed point, therefore the exponent n is not a dynamic critical exponent. It can be shown [4, 5] that the phenomenon shows a certain universality in the sense that n does not depend on the details of systems within a universality class. In particular, the details of the kinetics of domain growth are not involved.

Scale invariance of the spinodal structure has been well established experimentally and numerically, but it is more difficult to demonstrate theoretically, apart from in a few specific cases. Two cases need to be distinguished depending on whether the order parameter is *conserved* or not during the evolution following cooling. In the Ising model, spins may flip independently of each other in such a way that the magnetisation changes during relaxation. In contrast, in an alloy or a binary fluid, the number of molecules of each type is conserved, which is equivalent to working at constant magnetisation in a spin system.

3.7.3.1 Domain Growth with Non Conserved Order Parameter

In the first case, where the *order parameter is not fixed*, domain growth is governed by the principle of minimisation of surface energy at the boundaries between domains. A similar growth process, also governed by minimisation of surface energy, is observed in clusters of soap bubbles. The interfaces relax independently of one another and the relaxation is therefore rapid: $L(t) \sim \sqrt{t}$. Analytically, we can describe this situation with local laws. In this case, the proportion of phases is not fixed in advance and we see a competition between domains. The phase observed in the final state is the result of this competition. We stress that this result will depend on the initial condition but also on all the successive random events involved in the domain growth.

3.7.3.2 Domain Growth with Conserved Order Parameter

In the second case, deformations of different domains are all correlated due to the constraint of conservation of the order parameter. The evolution is still governed by the principle of minimisation of surface energy, but this minimisation must be global. It occurs through diffusion of the order parameter from interfaces of high curvature towards those with low curvature. The presence of correlations at all scales in the system, induced by the constraint of conservation of the order parameter and the global nature this imposes on the dynamics, will significantly slow down the evolution with respect to the first case of non fixed order parameter.

When the order parameter is conserved, we observe and calculate by renormalisation $L(t) \sim t^{1/3}$, if the order parameter is scalar, and $L(t) \sim t^{1/4}$, if the order parameter is a vector. In binary fluids the situation is more complicated because the scaling regime is truncated by the (in general dominant) influence of gravity (unless we work in microgravity or the fluids have the same density). We will then typically see three successive regimes:

$$\text{diffusive regime, in which } L(t) \sim t^{1/3}$$

$$\text{viscous regime, in which } L(t) \sim t$$

$$\text{inertial regime, in which } L(t) \sim t^{2/3}. \qquad (3.71)$$

In summary, the dynamics associated with the appearance of order by symmetry breaking observed at low temperature shows a spatiotemporal scale invariance at long times, in which the characteristic length $L(t)$ associated with the spatial scale invariance behaves like $L(t) \sim t^n$. The mechanism of domain growth is driven by minimisation of surface energy: the system evolves for as long as there are transformations lowering its interfacial energy. Relaxation occurs globally and therefore more slowly in systems where the order parameter is conserved compared to systems where it is not.

3.7.4 Transitions and Scale Invariance in a Finite Size System

3.7.4.1 Can a System of Finite Size show a Phase Transition?

We have seen that a phase transition is only well defined in the thermodynamic limit $N \to \infty$. In this sense, we can define a phase transition as a singularity of physical behaviour related to the non commutativity of the limits $T \to T_c$ and $N \to \infty$. This definition does not help us to understand what a phase transition corresponds to at the microscopic scale and hardly indicates what will be the natural extension of the notion of phase transition in a system of finite size. Let us return to the meaning of the formalism in which we define phase transitions. It it based on the identification, in the thermodynamic limit, of:

- A thermodynamic quantity A_{thermo}
- The experimentally observed quantity A_{obs}
- The statistical average $\langle A \rangle$
- The most probable value A_m

This identification is justified by the probability distribution $P_N(A)$ with relative variance proportional to $1/N$, assumed to have one single peak centred on A_m. For reasonably large N, the saddle point method (method of steepest descent) applies and ensures that $\langle A \rangle \approx A_m$. Fluctuations being negligible, we can make the identification: $\langle A \rangle \approx A_m \approx A_{obs} \approx A_{thermo}$. This approach fails in at least two cases:

- If $P_N(A)$ has several peaks.
- And close to a critical point, where giant fluctuations not decaying as $1/N$ can appear due to the divergence of the correlation length.

These two cases correspond precisely to first and second order phase transitions respectively. From a physical point of view, it seems natural to base the definition of phase transitions on the form of the probability distribution $P_N(A)$. Then the essential quantity is no longer the thermodynamic limit of the free energy but the distribution, at thermal equilibrium and finite size N, of a variable A physically characterising the phase. A will be for example the density in the case of liquid–vapour transitions or magnetisation per spin in the case of ferromagnetic–paramagnetic transitions. A first order phase transition corresponds to a bimodal distribution $P_N(A)$. The fraction of the system in each phase is given by the area of the corresponding peak, the peaks remaining well separated. The case in which two peaks merge into a single peak when we vary a control parameter X, for example the temperature, corresponds to a second order phase transition. The peak is large

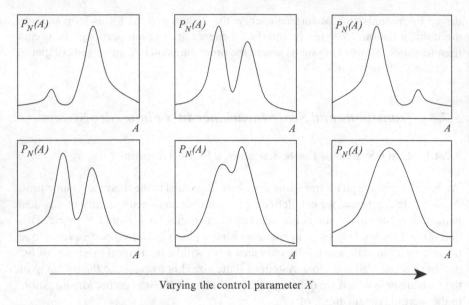

Varying the control parameter X

Fig. 3.13 The action of a control parameter X in a finite sized system, in the case of a first order transition (*top*), and a second order transition (*bottom*)

at the moment of fusion, corresponding to the presence of giant fluctuations and a divergence in the susceptibility (see Fig. 3.13).

3.7.4.2   Scale Invariance in a Finite Sized System

Here we face an apparent contradiction: how can we apply scaling approaches to a finite system when these approaches are based on scale invariance, which assumes an infinite system? The paradox is lifted when we are interested in systems that are finite but large compared with the element size. The behaviour of the system is then determined by the values of L and the coherence length ξ:

- If $L > \xi$, the system behaves as if it was infinite.
- If $L < \xi$, then the boundaries of the system "truncate" the transition.

In practice, the quantity L appears like a new "field", in the same manner as h and t, with its own critical exponents. According to the *scaling* hypothesis expressed by Kadanoff, we expect the behaviour of the system to be directly linked to the relationship between L and ξ. A given property $A(t)$ will take for example the following form $A(t, L)$, in a system of size L:

$$A(L, t) \sim t^\alpha f\left(\frac{L}{\xi(t)}\right) \sim t^\alpha f\left(t^\nu L\right), \tag{3.72}$$

where the universal function f is constant if its argument is small compared to 1. We will return to this approach of scaling in finite sized systems in the chapters treating percolation (Chap. 5), superconductivity (Chap. 7) and above all growth mechanisms (Chap. 8).

3.7.4.3 Effect of Structural Disorder

The existence of dilute static disorder – impurities, topological defects, etc. – can also profoundly change the nature of a transition. We will give just one example: that of one dimensional systems. An impurity plays the role of an extremity: the system is transformed into an assembly of finite sized systems. Impurities represent "seeds" for the order which is established on each side over a coherence length. If the distance between impurities is comparable to the coherence length, the system can become ordered on average due to the static "disorder" induced for instance by defects. In general, studies of transitions take into account the structural disorder as a "field" in the same way as temperature and magnetic field. We are often obliged to draw phase diagrams of physical systems as a function of the structural disorder intensity [24, 25].

3.8 Conclusion: The Correct Usage of a Model

It was a formidable challenge to correctly describe a physical system containing infinite relevant spatial scales. Experimental results showing the robustness of observed critical behaviours were accumulated, but no model, even very simplified ones, had been solved in three dimensions. This frustrating failure gave birth to a new type of physical approach, in which the description of properties of a particular model takes a back seat behind the comparison between the models themselves.

By striving to describe the *organisation* of phenomena rather than the phenomena themselves, renormalisation methods lead to intrinsic results, insensitive to the inevitable approximations involved in theoretical descriptions. A good picture of this approach is the description of a fractal curve: its length is a relative quantity $L(a)$ depending on the step size a with which we survey the curve. In contrast, the *connection* between the measurements $L(a)$ and $L(ba)$ obtained by choosing different minimum scales of a and ba respectively, leads to an intrinsic characteristic of the curve, the *fractal dimension* d_f (see Chap. 2). This relationship expresses the self-similarity – the scale invariance – of the fractal curve.

To construct a physical model is to choose the minimum and maximum scales, the boundaries and degrees of freedom of the system S and to simplify its interactions with the environment, etc. A model is therefore necessarily subjective and imperfect. The usual route, consisting of deriving as much information as possible about the behaviour of an ideal system described by the model, is therefore marred by uncertainties, not to mention errors. Renormalisation methods suggest a completely different approach, where the analysis moves from *phase space* to a *space of models*. The renormalisation group classifies the models into *universality classes*, bringing together each model leading to the same asymptotic properties. Note that the renormalisation group is nothing but a dynamic system in the space of models.

Therefore to correctly predict the large scale properties of a system, all we need to do is to describe the universality class the physical system belongs to, and to do that a only a basic model is necessary! The exponents appearing in the asymptotic scaling laws are the same as those of the archetypal model of the universality class. However, the crux lies in the fact that renormalisation methods are able to determine whether or not a perturbation to the model is able to change which universality class the model lies in, that is to say whether the perturbation will destroy (relevant perturbation) or not destroy (irrelevant perturbation) the macroscopic predictions obtained before perturbation. In this way we can test the *robustness* of the results with respect to modifications in the microscopic details and therefore the *validity of the model*, since many microscopic approximations (discrete–continuous, interactions restricted to nearest neighbours, type of lattice, etc.) have no consequences at the scale of observation.

References

1. F.F. Abraham, Computational statistical mechanics methodology. Applications and supercomputing. Adv. Phys. **35**, 1 (1986)
2. J.J. Binney, N.J. Dowrick, A.J. Fisher, M.E.J. Newman, *The Theory of Critical Phenomena (An Introduction to the Renormalization Group)* (Oxford University Press, 1992) p. 140
3. D.J. Bishop, J.D. Reppy, Study of the superfluid transition in 2-dimensional He-4 films. Phys. Rev. B **22**, 5171 (1980)
4. A.J. Bray, Exact renormalization-group results for domain-growth-scaling in spinodal decomposition. Phys. Rev. Lett. **62**, 2841 (1989)
5. A.J. Bray, Theory of phase ordering kinetics. Adv. Phys. **51**, 481 (2002)
6. L.M. Brown, *Renormalization. From Lorentz to Landau (and beyond)* (Springer, Berlin, 1993)
7. P. Collet, J. -P. Eckmann and O. E. Lanford, Universal properties of maps on an interval, *Com. Math. Phys.*, 76, 3, 211 (1980)
8. A.T. Fiory, A.F. Hebard, W.I. Glaberson, Superconductiong phase-transition in Indium-oxide thin-film composites. *Phys. Rev. B* **28**, 5075 (1983)
9. M.E. Fisher, The renormalization-group and the theory of critical behavior. Rev. Mod. Phys. **46**, 597 (1974)
10. M.E. Fisher, Renormalization group theory: its basis and formulation in statistical physics. Rev. Mod. Phys. **70**, 653 (1998)
11. N. Goldenfeld, in *Frontiers in Physics*. Lectures on phase transitions and the renormalization-group, vol 85 (Addison-Wesley, Reading, 1992)
12. A.F. Hebard, A.T. Fiory, Critical-exponent measurements of a two-dimensional superconductor. Phys. Rev. Lett. **50**, 1603 (1983)
13. F.C. Hohenberg, B.J. Halperin, Theory of dynamical critical phenomena. Rev. Mod. Phys. **49**, 435 (1977)
14. L.P. Kadanoff, Scaling laws for Ising models near T_c. Physics **2**, 263 (1966)

15. J.M. Kosterlitz, D.J. Thouless, Ordering, metastability and phase-transitions in 2-dimensional systems. J. Phys. C. Solid State **6**, 1181 (1973)
16. M. Laguës, Concevoir et raisonner en dimension 2. La Recherche **90**, 580 (1978)
17. A. Lesne, *Méthodes de renormalisation* (Eyrolles, Paris, 1995)
18. S.K. Ma, *Modern Theory of Critical Phenomena* (Benjamin, Reading Mass, 1976)
19. C.A. Murray, D.H. Winkle, Experimental observation of two-stage melting in a classical two-dimensional screened Coulomb system. Phys. Rev. Lett. **58** 1200 (1987)
20. M. Nauenberg, B. Nienhuis, Critical surface for square Ising spin-lattice. Phys. Rev. Lett. **33**, 944 (1974)
21. D.R. Nelson, Phase Transitions and Critical Phenomena **7** (1983).
22. D.R. Nelson, B.I. Halperin, Dislocation-mediated melting in 2 dimensions. Phys. Rev. B **19**, 2457 (1979)
23. L. Onsager Phys. Rev vol 65, p 117 (1944)
24. E. Orignac, T. Giamarchi, Effects of disorder on two strongly correlated coupled-chains. Phys. Rev. B. **56**, 7167 (1997)
25. E. Orignac, T. Giamarchi, Weakly disordered spin ladders. Phys. Rev. B **57**, 5812 (1998)
26. M. Plischke, B. Bergersen, *Equilibrium Statistical Physics* (World Scientific, Singapore, 1994)
27. T.M. Rogers, K.R. Elder, R.C. Desai, Numerical study of the late stages of spinodal decomposition. Phys. Rev. B **37**, 9638 (1988)
28. Y. Saïto, H. Müller-Krumbhaar, *Applications of the Monte-Carlo Method in Statistical Physics* (Springer, Berlin, 1984)
29. B. Widom, Equation of state in neighborhood of critical point. J. Chem. Phys. **43**, 3898 (1965)
30. K.G. Wilson, Renormalization group and critical phenomena. 1. Renormalization group and Kadanoff scaling picture. Phys. Rev. B **4**, 3174 (1971)
31. K.G. Wilson, Renormalization group and critical phenomena. 2. Phase-space cell analysis of critical behavior. Phys. Rev. B **4**, 3184 (1971)
32. K.G. Wilson, M.E. Fisher, Critical exponents in 3.99 dimensions. Phys. Rev. Lett. **28**, 240 (1972)
33. K.G. Wilson, Les systèmes physiques et les échelles de longueur. Pour la Science 16 (1979)
34. J. Zinn-Justin, *Quantum Field Theory and Critical Phenomena* (Oxford University Press, 1989)

Chapter 4
Diffusion

This chapter is devoted to diffusion phenomena, which provide many examples involving principles central to this book:

- The experimental analysis of Brownian motion by Perrin marks, amongst others, the emergence of the concept of self-similarity in physics [53].
- In particular, the trajectory of a particle performing Brownian motion and the front of an initially localised cloud of diffusing particles (diffusion front), are fractal structures, illustrating the ideas introduced in Chap. 2.
- The diffusion laws $R(t) \sim t^{\gamma/2}$, describing the temporal dependence of the root mean square displacement, $R(t)$, of a diffusing particle in different situations are examples of scaling laws. In this temporal context we will return to the distinction presented in Chap. 1 between mean field exponents and critical exponents, corresponding here to normal diffusion ($\gamma = 1$) and anomalous diffusion ($\gamma \neq 1$). Studying the origin of anomalous diffusion will enable us to better understand the typical mechanisms leading to critical behaviour.
- Diffusion is a phenomenon which can be considered at many different scales. Through this example we will show the subjective, incomplete and even simplistic nature of descriptions at a given scale. We will see that a better understanding is obtained by a trans-scale, multiscale view, striving to connect the different levels of description. As it happens this is the only possible approach we can take as soon as the phenomena show emergent properties, for example critical properties.

4.1 Diffusion: What We Observe

4.1.1 Thermal Energy

Let us begin with a quick reminder of the "motor" of all diffusive motion, that of thermal energy.

A. Lesne and M. Laguës, *Scale Invariance*, DOI 10.1007/978-3-642-15123-1_4,
© Springer-Verlag Berlin Heidelberg 2012

Statistical mechanics books teach that the temperature T of a gas can be *defined* by the average kinetic energy of the constituent molecules:

$$3kT = m\langle v^2 \rangle, \tag{4.1}$$

where $k = 1.381 \ 10^{-23}$ J/K is the Boltzmann constant and m the mass of a molecule. In the following we show that at equilibrium this kinetic temperature coincides well with the familiar concept of temperature, that which we measure with a thermometer [43, 46]. Strictly speaking, the statement only concerns dilute gases, but the qualitative idea of *thermal energy* remains valid in all fluids. In this way heat is connected to the kinetic energy of molecular movement. Inversely, the molecules of a system maintained at temperature T in thermal equilibrium are spontaneously and continuously moving,[1] with an average velocity of $\sqrt{3kT/m}$, called the *thermal velocity*, which depends only on their mass and of course on T. More generally, in a system at thermal equilibrium, each degree of freedom will show fluctuations called *thermal fluctuations* of average energy $kT/2$. This result is known by the name of the *equipartition of energy theorem* [35, 36]. This statement is actually only valid for degrees of freedom in which the excitation energy (quantum) is inferior to the typical thermal energy kT, however this is in fact the case at normal temperatures for the translational degrees of freedom that are involved in the context of diffusion.

The qualitative idea to take from this discussion is that of thermal fluctuations of amplitude controlled by the temperature. This "thermal energy" is the origin of all microscopic movement in the absence of external fields. It plays an essential role in establishing the equilibrium states of matter; it is responsible for the fluctuations affecting these equilibrium states and the spontaneous transitions produced between states when they coexist (see Chap. 1). Diffusion is a manifestation of thermal energy that is observable at our scale, providing almost direct access to this molecular phenomenon. This point was understood and exploited by Einstein and Perrin at the beginning of the twentieth century, as we will see in the following paragraph.

4.1.2 Brownian Motion

Brownian motion takes its name from the biologist Brown, who in 1827 very accurately studied the continual and erratic movement of a grain of pollen in suspension in water. He not only observed this phenomenon in many fluids including

---

[1]There is no paradox or perpetual motion in this statement. Either we consider a perfectly insulated container, in which case the movement continues simply because the kinetic energy of the molecules is conserved, or alternatively the movements will have the tendency to dampen, for example due to inelastic collisions with the walls of the container if this is at a temperature lower than T. Saying that we maintain the system at temperature T therefore is exactly saying that the molecular motion is maintained by an adequate supply of energy.

in a drop of water trapped within a piece of amber, but he also observed the motion with mineral grains, leading him to definitively reject any explanation based on the living character of the grain. This was not the movement of a living being and the question of its origin left the domain of biology for that of physics. It took almost a century and Einstein's analysis of the phenomenon in 1905, to have an explanation. The theoretical study was then validated by the extremely precise experimental work of Perrin. Their two complementary approaches definitively established that Brownian motion was due to collisions of the grain with water molecules in thermal motion; this explanation provided a strong argument in favour of the atomic theory of matter, which was still new at the time.

Perrin used emulsions of resin in water, which he made monodisperse (that is the grain size is uniform, in this case of the order of a micron) by fractional centrifugation. By observing the movement of the grains initially very close to each other, he began by verifying that convection is not the origin of the displacement of the grains. He then carried out an exhaustive study by varying the size of the grains, their mass, the viscosity of the supporting fluid and the temperature. He found that the movement is all the more active as the temperature is increased, as the fluid viscosity is decreased and as the grain size is reduced. He also verified that movement of grains of equal size is not affected by the nature of the grains or by their mass.

By recording the successive positions of a grain, projected onto a plane, Perrin showed the irregularity of the trajectories, their random nature and the existence of details at all scales: if he increased the time resolution of his recording, the linear sections interpolating two successive positions transformed into segmented lines composed of shorter segments of random orientations (see Fig. 4.1). He concluded from this that the instantaneous velocity of the grains as far as he could deduce from analysing the recorded trajectories, was badly defined since it depended on the scale with which he observed the trajectories. This absence of characteristic scale fitted perfectly with the explanation proposed a little earlier by Einstein: if the movement of the grain did result from kicks it was subjected to from the water molecules, its velocity must then show constant changes of direction and this property must persist if you increase the resolution all the way down to the molecular scale.

Furthermore, analysis of trajectories shows that the average displacement $\langle r(t) - r(0)\rangle$ is zero due to the isotropy of the erratic movement of the grains. It is therefore neither the velocity nor the displacement but a third quantity, the *root mean square displacement* of the grains $R(t) \equiv \langle [r(t) - r(0)]^2 \rangle^{1/2}$, which is the relevant observable for a quantitative description of the movement. By tracing $\log R^2(t)$ as a function of $\log t$, Perrin could show that the grains obeyed, at long times, a statistical law:

$$R^2(t) \sim 2dDt \qquad\qquad (t \to \infty). \qquad\qquad (4.2)$$

We generally talk about *normal diffusion* when the asymptotic behaviour of the root mean square displacement $R(t)$ of the diffusing object is, as here, proportional to $\sqrt{t}$. The constant of proportionality defines the *diffusion coefficient D* of the object, here a grain of resin, in the supporting fluid (conventionally, the dimension of space

d is shown explicitly in the law of diffusion). Once the normal diffusion law is found to be satisfied by the existence of a linear section of slope 1 on the graph of $\log R^2(t)$ as a function of $\log t$, we determine the value of D as being the slope of the linear section of the graph of $R^2(t)$ as a function of $2dt$.

A few years before Perrin's work, Einstein [18] had established a formula which now carries his name:

$$D = \frac{RT}{\mathcal{N}_{Av}} \frac{1}{6\pi r_0 \eta} \qquad \text{(Einstein's formula)}. \qquad (4.3)$$

This formula, which we will present in more detail in Sect. 4.4.2, expresses the diffusion coefficient D as a function of the temperature T, the ideal gas constant $R = 8.314$, the (measurable) radius r_0 of the grains (assumed to be spherical), the dynamic viscosity η of the fluid and Avogadro's number $\mathcal{N}_{Av}$. The quantity D gave Perrin an experimental access to Avogadro's number[2] and achieved what was considered as a direct proof of the existence of atoms [19, 62]. We refer the reader to the original work of Perrin [53] *Les Atomes*, and its translation [53], for all the details of this historic breakthrough.

We finish with a comment on methodology. The direct analysis of the experimental data showed that $R^2(t)$ behaves as $2dDt$ at long times, measured D, and then showed how D varies with the *observable* parameters (mass and size of grain, fluid viscosity, temperature). It is however essential to have an underlying theory, in this case the kinetic theory of fluids, to interpret these results in terms of molecular mechanisms (the collisions of molecules with the grain), and to have an explicit theoretical result, the Einstein formula, to extract from the observations the value of a *microscopic* parameter (i.e. related to the underlying microscopic reality, unobservable at the micrometer experimental scales envisaged here), namely Avagadro's number, which also makes the microscopic image falsifiable.

4.1.3 Self-Similarity of the Trajectories

Brownian motion is an example of a mechanism generating fractal structures, here the typical trajectories of the grains. We have just seen that Perrin, in 1913, had highlighted the property of the trajectories that we today call (statistical) *self-similarity*: if we observe the trajectories with a better spatial resolution, k times finer, they statistically conserve the same appearance, more specifically the same statistical properties, after resetting the scale by a factor k (Fig. 4.1).

This self-similarity comes from the microscopic origin of the movement: a displacement of the grain corresponds to a transient anisotropy in the variation

---

[2]Perrin obtained 6.85×10^{23}, compared to 6.02×10^{23} known today.

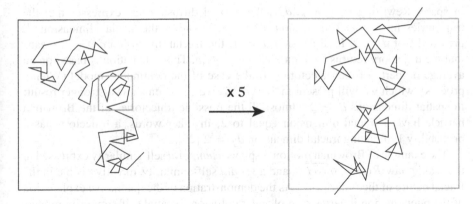

Fig. 4.1 Diagram illustrating the statistical self-similarity of trajectories of Brownian motion: a detail enlarged (here by a factor $k = 5$) shows the same statistical properties as the initial observation. This property is at the origin of the continuous but not differentiable character of the trajectories, typical of fractal curves

of momentum resulting from a very large number of random collisions of water molecules with the grain. When we refine the spatial resolution by a factor of k, we can see the consequences of anisotropies of shorter duration: a segment of length λ after the change of resolution appears as a segmented line made up of segments of length λ/k.

Changing the temporal resolution has the same effect: on decreasing the time step Δt by a factor k^2, less instantaneous fluctuations in the position[3] are accumulated and the root mean square displacement, proportional to the square root of the duration, is reduced by a factor k. The observed motion therefore reflects the statistical velocity distribution of the water molecules and the way in which fluctuations in the velocity of the impacting water molecules will accumulate their effects, at all scales, finally leading to the diffusion law $R(t) \sim \sqrt{2dDt}$ for the grain. We will return in a more precise manner to this statistical law of addition of microscopic fluctuations, which is at the origin of the self-similarity of the trajectories, in Sect. 4.3.1, as well as in Sect. 4.5, where we will present the limits of this law.

The self-similarity nature of the trajectories can be quantified. The scaling law $R(t) \sim \sqrt{2dDt}$ relating the root mean square displacement at time t elapsed since the start can be interpreted differently: t is also the arc length coordinate along the trajectory, in some sense the "mass" of the object while $R(t)$ is its linear extension

---

[3]We have just been talking about the movement of the grain in terms of fluctuations in momentum and now in terms of fluctuations in position: both descriptions are possible and they are obviously linked. We will elaborate on this point in Sect. 4.4.2, once we have introduced the necessary technical tools.

in space. Rewriting $t \sim R^2/2dD$, the normal diffusion law expresses that the trajectories have fractal dimension $d_f = 2$, whatever the spatial dimension d, provided that $d \geq 2$. In dimension $d = 1$, the fractal dimension $d_f = 2$ means that the trajectory returns back on itself very often. This calculation is based on an average property of the trajectories. In the case of the continuous model (Wiener process), which we will present in Sect. 4.3.2, we can even show a stronger result: in spatial dimensions $d \geq 2$, almost all the possible trajectories of the Brownian particle have a fractal dimension equal to 2; in other words, a trajectory has a probability 1 of having fractal dimension $d_f = 2$ [20].

The example of Brownian motion displays a *temporal* self-similarity expressed in the scaling law $R(t) \sim \sqrt{2dDt}$, and a *spatial* self-similarity, directly visible in the fractal nature of the trajectories. As the demonstration of the qualitative explanation of the phenomenon in terms of molecular collisions, spatial self-similarity reflects the self-similarity of the dynamic process – an accumulation of variations in momentum – at the origin of the observed motion. It is a prime example of the close link between the spatial patterns observed (here the trajectories) and the dynamics that generated them. We will encounter many other examples, in particular fractal growth of interfaces in Chap. 8 and strange attractors in Chap. 9.

In the same fashion, this example shows that spatial fluctuations, as we can show with a statistical description, are closely coupled to temporal fluctuations, together reflecting the spatiotemporal organisation of the phenomenon. In particular, the correlation time and correlation length will diverge simultaneously, as observed for example in critical phase transitions (see Chaps. 1 and 3). Sticking within the context of diffusion, in the following paragraph we will present another manifestation of self-similarity in diffusion, as spatial as it is temporal: diffusion fronts.

4.1.4   Diffusion Fronts

The results obtained by Perrin came from an analysis of individual trajectories, however this approach is only easily implemented experimentally for quite large particles, of the order of a micron. Another type of experiment, real or numeric, is the observation of a *diffusion front*. The idea is to start from a situation where the species A whose diffusion we want to study is localised, with for example a step function profile in a direction Ox, the step being of finite size[4] (limited by the walls of the container). At the macroscopic scale, we then observe a broadening and dampening of the profile over time under the influence of diffusion (Fig. 4.2).

---

[4]The diffusion profile describing the evolution of a semi-infinite (towards negative x) step is different; its expression is shown to be $c(x,t) = \mathrm{erfc}(x/\sqrt{4Dt}) = \int_{x/\sqrt{4Dt}}^{\infty} e^{-u^2} du$, also a solution to the diffusion equation $\partial_t c = D \partial_{xx}^2 c$ but verifying *at every instant* the boundary condition $c(x = 0, t) = 1$.

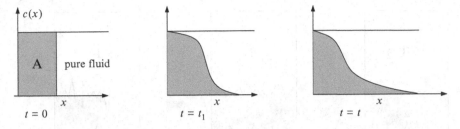

Fig. 4.2 Diffusion front observed at the macroscopic scale. The diffusion coefficient of a species A (molecules, macromolecules, colloids or dust particles) in a given fluid can be measured by observing the evolution of its average concentration $c(x, t)$, starting from an initial step function profile of finite size in the direction Ox (flat profile in the directions Oy and Oz orthogonal to the direction Ox of the concentration gradient)

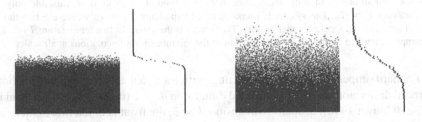

Fig. 4.3 Diffusion front observed at the microscopic scale, here of a numerical simulation in two dimensions (initial profile consists of 150 particles with a height of 75 lines). The interface and the corresponding average profile are shown at two successive time steps, highlighting the development of inhomogeneities. The front looks like a rough interface, with spatial fluctuations at all scales: it is a fractal structure

The spatialtemporal evolution of the average concentration gives the diffusion coefficient D: we will see in Sect. 4.2.1 that at long enough times, the average profile in the direction Ox of the initial gradient behaves as $c(x, t) \sim (1/\sqrt{2dDt}) \exp[-x^2/4Dt]$. The quality of this method of measuring D is limited by density fluctuations, both longitudinal (in the direction of the gradient Ox) and lateral (in the directions Oy and Oz perpendicular to the gradient). The profile, which is regular if we observe it at the macroscopic scale, actually shows a complex microscopic structure reflecting the molecular and random origin of diffusion.

Numerical simulation of the phenomenon in $d = 2$ or three dimensions, shows that the interface marking the boundary of the cloud of particles (i.e. the diffusion front in the strict sense) is extremely irregular (Fig. 4.3). The irregularity of the interface accentuates over time; its surface (or length in the two dimensional case) and its thickness increase indefinitely. Specifically this border has a highly convoluted fractal structure at spatial scales smaller than its thickness (but larger that the particle size).

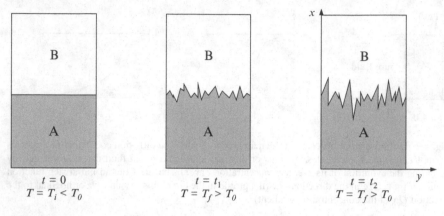

Fig. 4.4 Self-similarity of a diffusion front, here of a fluid A in an fluid B, miscible only at temperatures $T > T_0$. The system is prepared at a temperature $T_i < T_0$ where the two fluids are not miscible and separate under gravity. The system is then brought to a temperature $T_f > T_0$. A complex interface develops: the composition of the mixture shows fluctuations at all scales

For hard (impenetrable) non-interacting particles A, a theoretical model[5] predicts a fractal dimension $d_f = 7/4$ in spatial dimension $d = 2$ (in which case the front is a line) [Gouyet 1996]. In spatial dimension $d = 3$, the front is much more extended along Ox with an advanced fractal region of dimension $d_f \approx 2.5$ and an internal region which is porous but homogeneous in the sense that its fractal dimension is 3.

The fractal structure of diffusion fronts can also be demonstrated experimentally by starting with a situation where the concentration profile has an abrupt step. For example take two fluids A and B that are non miscible at low temperature and miscible at temperatures $T > T_0$. Preparing the system at low temperature the denser liquid, say A, will be underneath B and its concentration profile therefore has a step in the vertical direction. Let us increase the temperature to above T_0 at time $t = 0$. The fluids A and B start to diffuse into each other. We then observe the interface between the two fluids by looking at the system in the direction of the gradient Ox: the surface appears rough with irregularities at all scales (Fig. 4.4). These horizontal spatial fluctuations (i.e. from one point to another along the directions perpendicular to the gradient) develop rapidly[6] and only disappear when the system has relaxed completely [63, 69].

Let us point out that diffusion fronts are *nonequilibrium* structures which, in infinite medium, continue to develop indefinitely. It is only in finite mediums that we observe a relaxation to an equilibrium diffusive state of homogeneous concentration at long times.

---

[5]The concepts used in the microscopic analysis of these fronts, in particular to establish the value of their fractal dimension, directly result from the theory of percolation, which we will present in Chap. 5 [58]. We will meet other examples of the growth of fractal interfaces in Chap. 8.

[6]The only limitation to the size of these fluctuations in the envisaged experimental system is that imposed by gravity.

A more recent measurement method: FRAP

Recently, state of the art fluorescence techniques have led to an ingenious method of studying diffusion of macromolecules, for example proteins inside a cell or polymers in a solvent. The method is known by the name FRAP, acronym for Fluorescence Recovery After Photobleaching [48]. The molecular species whose diffusion is to be studied are labelled with fluorophore (fluorescent molecules).[7] The fluorescent markers must be small enough not to change the diffusion properties of the species under consideration, in practice limiting the method to the study of large macromolecules. The lifetime of the fluorescent emission, excited by a laser of appropriate wavelength, is long, of the order of several hundreds of seconds. The principle of FRAP is to destroy ("bleach") the fluorescence locally at an instant in time t_0 using another laser pulse: the molecules in the irradiated area no longer emit a fluorescent signal. This area is then observed with a confocal microscope. The irradiated area which becomes dark ("bleached") at the instant t_0 progressively recovers its fluorescence, which can only be due to the penetration of fluorescently labelled molecules coming from surrounding areas by diffusion. The fluorescent signal emitted by the bleached region is therefore directly related to the flux of labelled molecules passing through the region. Comparing this signal with predictions made by a model describing the diffusion of the species under consideration gives the diffusion parameters. This method is more reliable than that consisting of imposing a strong gradient in concentration of the studied species and observing how the gradient relaxes by diffusion since such an imposed gradient is actually itself affected by large spatial fluctuations (see Sect. 4.1.4).

Most importantly, FRAP enables an observation of diffusion occurring during the "natural" behaviour of a system such as a living cell.

4.1.5 Diffusion: Exceptional Motion

We started this chapter with the example of Brownian motion. This comes about as an indirect consequence of the thermal motion of molecules in the medium (of size less than a nanometre) exerted on a supramolecular object (typically of order a micron).

---

[7] It is now known how to insert a gene for a naturally fluorescent protein into the gene coding for a given protein, such that the cell makes modified proteins containing a fluorescent domain but without disrupting their biological function. This allows us to follow the localisation and dynamics of these proteins inside a living cell.

A second example of diffusion is that of the mixing of a drop ink deposited in water, directly resulting from the thermal energy of the dye molecules and the water molecules. However mixing is only provided by diffusion at small spatial scales; at larger spatial scale convection may play a role. The comparison of diffusive terms with convective terms defines a dimensionless number called the *Peclet number Pe*, determining the order of magnitude of these "small" scales l at which diffusion dominates.[8] This number is the ratio of a characteristic diffusion time l^2/D to a characteristic convection time l/V:

$$\frac{\text{convection}}{\text{diffusion}} \sim \frac{V/l}{D/l^2} \sim \frac{Vl}{D} \equiv Pe, \tag{4.4}$$

where V is the typical convection speed in the medium (water here) in which the phenomenon is observed. We see that this ratio Pe is small compared with 1 as soon as l is small enough; the upper bound on l is larger the smaller the convection speed V. Therefore we could almost abolish the effects of convection by observing diffusion of a dye in a gel: crosslinking of macromolecules in a gel completely suppresses convection whilst relatively insignificantly slowing down diffusion of small molecules, which "pass through the gaps in the gel mesh".

The two examples of diffusion we have just seen, Brownian motion and mixing of ink in water, lead to very similar (and often confused) observable phenomena. This simply shows that the self-similarity of the phenomenon extends over a wide range of scales. This self-similarity is observed whenever the motion is induced by random kicks from molecules in the supporting fluid, whether the particle being hit is another molecule of comparable size (the example of ink in water) or a much larger particle (a grain of pollen or resin performing Brownian motion): the motion is therefore characterised by the diffusion coefficient D of the particle/fluid pair. Perrin's observations and the expression (4.3) of this coefficient show that the diffusive motion depends on the viscosity η of the medium where the diffusion takes place ($D \sim 1/\eta$), the temperature T ($D \sim T$), the size a of the diffusing particle ($D \sim 1/a$) and to a lesser extent the shape of the particle (unless this shape is very particular, for example a long linear chain). To give an idea of the orders of magnitude: in water, D is typically 10^{-9} m²/s for a small molecule, 10^{-11} m²/s for a macromolecule and 10^{-13} m²/s for a micrometre sized grain.[9]

A point worth noting is that the coefficient D *does not depend on the mass* of the diffusing particle. This independence highlights a fundamental characteristic of diffusion phenomena: *inertial effects are negligible*. It is precisely when we can

[8]For example, heating a room of 20 m³ takes about 10 hours by diffusion (with $D \approx 2 \times 10^4$ m²/s for a gas) compared to 5 min by convection (with $V \approx 1$ cm/s). In this case the Peclet number is of the order of 100, corresponding to a critical length separating the diffusive and convective regimes of the order of a centimetre.

[9]The dynamic viscosity of water is $\eta = 1.14 \ 10^{-3}$ kg/(m·s). For an approximately spherical grain of radius r_0 expressed in microns, Einstein's formula (4.3) therefore gives $D \approx (2/r_0) \ 10^{-13}$ m²/s.

ignore the acceleration term (also called the inertial term) in the equation of motion of the particles that we talk about diffusion.[10]

As the size of the particle increases, the effect of diffusion quickly becomes negligible compared to other mechanisms at work: Brownian motion becomes indiscernible at macroscopic scales. In the case of sedimentation of particles in suspension, an upper bound on the size of the particle appears, which we call the *colloidal limit* and is typically of the order of a micron. Above this size gravity dominates over diffusion and the particles fall to the bottom of the container; below this size Brownian motion is sufficient to maintain the particles in suspension.

In summary, diffusion is a random motion which is statistically isotropic; the average displacement is therefore zero.[11] Diffusion is only the principal cause of motion in situations where inertia is negligible (in practice small objects, less than a micron) and at scales small enough that convection doesn't play a role. The dependence $R(t) \sim \sqrt{2dDt}$ of the root mean square displacement shows that diffusion is only an efficient method of transport at short times (or equivalently, at small length scales) since the apparent velocity $R(t)/t \sim 1/\sqrt{t}$ decreases with increasing time.

4.2 The Diffusion Equation and its Variants

4.2.1 Fick's Law and Diffusion Equation

The simplest description of diffusion and that which is the closest to current observations (see for example Fig. 4.2) is the partial differential equation describing the evolution of the local concentration c of the diffusing species:

$$\partial_t c(\boldsymbol{r}, t) = D\Delta c(\boldsymbol{r}, t), \tag{4.5}$$

where Δ is the Laplacian operator: $\Delta = \nabla^2$, written in Cartesian coordinates as $\Delta = \partial_{xx}^2 + \partial_{yy}^2 + \partial_{zz}^2$. The justification of this macroscopic description seems better when we decompose the above equation, known as the *diffusion equation*, into two

---

[10]We will see in paragraphs Sects. 4.2.3 and 4.5.4 that different mechanisms can lead to similar macroscopic behaviours to that of a cloud of diffusing particles, described by analogue macroscopic equations. The diffusion that we describe in this paragraph is "thermal" diffusion, which originates from the thermal energy of the molecules and during which the diffusing particles do not experience any acceleration.

[11]The slowness and isotropy of diffusion explains why in living organisms faster and orientated "active" transport mechanisms very often take over; for example advective transport in circulatory flow and intracellular transport provided by molecular motor proteins.

coupled equations. The first is a *conservation equation*,[12] simply expressing the conservation of the total number of diffusing particles in the absence of any source or chemical reactions:

$$\partial_t c(r,t) + \nabla . j(r,t) = 0. \tag{4.6}$$

It involves the current density j of the species (j has dimensions of velocity multiplied by concentration). To obtain a closed system this equation needs to be completed by a *constitutive equation* giving the expression for j as a function of concentration. It is this second equation which takes into account the physical mechanism underlying the movement. The phenomenological equation proposed by Fick in 1855 is a linear response, expressing that j is proportional to the concentration gradient ("the flow of a river is greater the steeper the slope"):

$$j = -D\nabla c. \tag{4.7}$$

By using this expression, now known as *Fick's law*, in the particle conservation equation, we obtain the usual diffusion equation $\partial_t c = D\Delta c$. The conservation equation (completed with a source term if necessary) is actually quite simple; it is Fick's law applied to diffusion. At this stage, it was purely empirical and descriptive; it was not until several decades after Fick that a fundamental microscopic explanation for it was given (see paragraph 4.1).

It is worth mentioning that the diffusion equation is formally identical to the heat equation by Fourier (c there is simply replaced by the temperature and D by the thermal conductivity of the medium). Just as for the diffusion equation, the heat equation must be completed by initial conditions $c(r,t = 0)$ and boundary conditions of the spatial domain where the phenomenon is produced, if it is finite, or by the behaviour required at infinity if it is not bounded. These boundary conditions critically influence the solutions.[13] Figure 4.5 illustrates intuitively the spatiotemporal evolution associated with the diffusion equation.[14]

---

[12]To obtain it we write that the variation $\partial_t \int_{\mathcal{V}} c(r,t)d^d r$ of the number of particles in a volume $\mathcal{V}$ can only be due to particles entering or leaving the volume, in other words the flow $\oint_{\delta\mathcal{V}} j.dS$ across the closed surface $\delta\mathcal{V}$ surrounding the volume $\mathcal{V}$ (the surface element dS being oriented along the outward normal to the surface). A (very useful) result of vector analysis transforms this surface integral into a volume integral $\int_{\mathcal{V}} \nabla . j(r,t)d^d r$. Since the volume $\mathcal{V}$ here is arbitrary, we can deduce from this the local form of the conservation equation: $\partial_t c + \nabla . j = 0$.

[13]At least from a mathematical point of view, since these boundary conditions determine the functional space where the solutions occur.

[14]A technical result associated with the diffusion equation (derived from the maximum principle for parabolic equations) is as follows: if $c_1(x,t)$ and $c_2(x,t)$ are two bounded real positive solutions of the diffusion equation $\partial_t c = D\partial_{xx}^2 c$, such that $c_1(x,t = 0) \leq c_2(x,t = 0)$ at all x, then $c_1(x,t) \leq c_2(x,t)$ at all subsequent times [54]. By taking $c_1 \equiv 0$, this result shows that $c(x,t)$ remains positive if it is positive at the initial time (which is satisfied if c describes a concentration). This theorem also shows that $c(x,t)$ remains bounded above by the maximum value $\sup_x c(x,t = 0)$ taken by the initial condition.

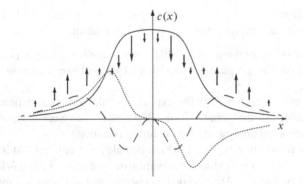

Fig. 4.5 Graphical interpretation of the diffusion equation $\partial_t c = D\partial^2_{xx}c$. A localised initial condition c (*continuous line*), its spatial derivative $\partial_x c$ (*dotted line*) and the term $D\partial^2_{xx}c$ (*dashed line*) – how this term evolves c is shown by the *arrows*

4.2.2   Scale Invariance of the Diffusion Equation

The usual diffusion equation $\partial_t c = D\Delta c$ shows a scale invariance with respect to the transformation $(r \rightarrow \lambda r, \; t \rightarrow \lambda^2 t)$, the verification of which is sufficient to establish the normal diffusion law.[15]

Let us simplify the problem by working in one dimension and consider an initial profile of width l and mass A_0, then we can write $c_0(x) = (A_0/l)\phi_0(x/l)$ where ϕ_0 is of *bounded support* and is normalised to 1. The solution of $\partial_t c = D\partial^2_{xx}c$ is written as the product of the convolution $c(x,t) = [G_t * c_0](x)$ of the initial profile by the Green's function of the diffusion equation (which is the solution for an initial profile $\delta(x)$):

$$G_t(x) = \frac{1}{\sqrt{4\pi Dt}}\, e^{-x^2/4Dt} \qquad \left(\lim_{t\to 0} G_t(x) = \delta(x)\right). \qquad (4.8)$$

Explicitly:

$$c(x,t) = \int_{-\infty}^{\infty} e^{-\frac{1}{2}\left[\frac{x}{\sqrt{2Dt}} - z\frac{l}{\sqrt{2Dt}}\right]^2} \frac{A_0\phi_0(z)}{\sqrt{4\pi Dt}}\, dz \qquad (4.9)$$

ϕ_0 being of bounded support, we can take the limit $l/\sqrt{2Dt}$ tends to 0 in the integrand. Then we obtain a solution $c_\infty(x,t) \equiv G_t(x)$, which is invariant under

---

[15]More explicitly, let us take $T = \lambda^2 t$, $R = \lambda r$ and $C_\lambda(R,T) = c(r,t)$. Then for all $\lambda > 0$, C_λ satisfies the same equation: $\partial_t C_\lambda = D\Delta C_\lambda$.

the transformation $[x \to \lambda x, t \to \lambda^2 t]$, i.e. showing the same scale invariance as the starting equation. Several points are worth highlighting:

- The total mass A_0 moving under diffusion is conserved; this point proves to be essential for the properties stated here (we will see a counter example in Sect. 4.5.4).
- The exponent $\gamma = 1$, which is the exponent of the temporal dependence of the law of diffusion (in the general form $R^2(t) \sim t^\gamma$, see Sect 4.5), obtained by a simple dimensional analysis of the evolution equation.
- The asymptotic solution $c_\infty(x, t)$, which is scale invariant, has lost its memory of the initial condition. Due to the conservation of total mass A_0, if it is localised in a *bounded* interval, at $t = 0$ this could be concentrated at 0 or spread over a width l. The initial width l of the profile just introduces a correction $\mathcal{O}(l/\sqrt{2Dt})$ which is negligible at long times. This explains why this solution $c_\infty(x, t)$ coincides with the Green's function $G_t(x)$.
- This asymptotic scale invariant solution can be obtained directly from the diffusion equation by exploiting a self-similarity argument. We introduce an auxiliary variable $z = x/\sqrt{2Dt}$ and the auxiliary function φ, by the relation (in $d = 1$ dimension):

$$c_\infty(x, t) = \frac{A}{\sqrt{2Dt}} \, \varphi\left(\frac{x}{\sqrt{2Dt}}\right). \qquad (4.10)$$

The factor in front of φ follows from the normalisation of c (probability density). The scale invariance of diffusion means that the phenomenon can be described with one single variable z. It immediately follows that the function φ satisfies:

$$\begin{bmatrix} \varphi'' + z\varphi' + \varphi = 0 \\[2mm] \varphi(\pm\infty) = 0 \\[2mm] \int_{-\infty}^{\infty} \varphi(z)dz = 1 \\[2mm] \varphi \ \text{even.} \end{bmatrix} \qquad (4.11)$$

The problem is now reduced to solving an ordinary differential equation that can be easily integrated:

$$\varphi(z) = \frac{1}{\sqrt{2\pi}} \, e^{-z^2/2} \qquad \text{where} \qquad c_\infty(x, t) = \frac{1}{\sqrt{4\pi Dt}} e^{-x^2/4Dt}. \quad (4.12)$$

The scale invariant diffusive regime takes place at intermediate times, long enough that the transient influence of the initial condition has disappeared ($t \gg l^2/2D$), but before having attained a trivial state of $c \approx 0$ or sensing the effect of confinement due to the finite size L of the domain accessible to the diffusion movement ($t \ll L^2/2D$).

The diffusion equation says nothing about the trajectories of a *single* particle. The scaling behaviour $(\Delta x)^2 \sim \Delta t$ results from this equation driven to a divergence at short times of the "apparent velocity" $\Delta x / \Delta t$. This behaviour simply reveals the existence of a specific phenomenon at the microscopic scale that must be described in a totally different framework, as we will see in Sect. 4.3.

Diffusive coupling

In a macroscopic description of "continuous medium" type, a diffusion term appears in the equation for the evolution of an observable $A(r,t)$ when the coupling between different regions of the system remains local. Let us expand on this point. The most general coupling (in one dimension for simplicity) is written in integral from:

$$\partial_t A(x,t) = \int_{-\infty}^{\infty} K(x-y)A(y,t)dy + \cdots , \qquad (4.13)$$

where $(\ldots)$ represent the terms associated with local interactions that do not concern us here. The kernel $K(x-y)$ describes the weight with which the value A in y contributes to the evolution of its value in x, with $\int_{-\infty}^{\infty} K(x)dx = 0$ (coupling term) and $\int_{-\infty}^{\infty} xK(x)dx = 0$ (by symmetry). By rewriting the integral term $\int_{-\infty}^{\infty} K(y)A(x-y,t)dy$ and replacing A by its Taylor expansion in x (assuming that the kernel K decays fast enough), the first non zero term is

$$\partial_t A(x,t) = K_2 \partial_{xx}^2 A(x,t) + \cdots \quad \text{with} \quad K_2 = \frac{1}{2}\int_{-\infty}^{\infty} x^2 K(x)dx. \qquad (4.14)$$

If the coupling is short ranged (i.e. if the kernel $K(x)$ is peaked around 0 with a small enough width), this term is dominant and we can neglect terms involving derivatives of A of higher order and we therefore talk about *diffusive coupling*.

Here we see an intuitive interpretation appearing of the presence of such a term proportional to $\partial_{xx}^2 A$ not just in the equation describing the diffusion of a population of particles (A being then the local instantaneous concentration of particles) but also, for example, in the equation describing the evolution of magnetisation (which is then A) of a system of spins (Landau's theory[16]). In certain situations, the next term in the expansion, $K_4 \partial_{xxxx}^4 A$ where $K_4 = \int_{-\infty}^{\infty} x^4 K(x)dx/4!$ must be taken into account. This is the case in systems where an activator coupling is superimposed at short range and an inhibitor coupling at longer distances (or the other way round), which we meet for example in reaction-diffusion systems or in neural networks [49,50]. We will also see, in Chap. 8, that terms of this type appear in the modelling of certain growth mechanisms.

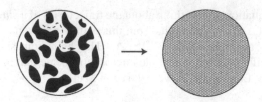

Fig. 4.6 *Left*: the real porous medium: the intervening space (*white*) is full of fluid in which diffusion takes place, described by the equation $\partial_t c_0 = D\Delta c_0$ completed by the boundary conditions, reflecting the fact that the particles, of local concentration $c_0(\mathbf{r}, t)$, cannot penetrate the solid regions (*black*), where the concentration is therefore $c_0 = 0$. *Right*: the effective homogeneous medium: we describe the local state by a concentration $c = \langle c_0 \rangle$, obtained by a spatial average over a representative volume. The diffusion is therefore governed by the equation $\partial_t c = D_{\text{eff}}\Delta c$. The effective diffusion coefficient D_{eff}, proportional to D, takes into account in an average way how the diffusion slows down due to the reduction of space actually accessible. Here the diffusion remains normal

4.2.3 Diffusion in a Porous Medium

A question of practical and fundamental interest is the diffusive behaviour observed when the fluid in which diffusion is occurring is confined in a porous medium.

If the structure formed by the pores (spaces accessible to the diffusion) is fractal ($d_f < 3$), the exponent of the diffusion law is affected: $R(t) \sim t^{\gamma/2}$ with $\gamma < 1$ (diffusion is slower). This is an example of anomalous diffusion, which we will study in more detail in Sect 4.5.

Here we will consider the opposite situation, in which the interstitial space still has a fractal dimension 3. In particular, the volume fraction is finite and defined by a coefficient $\alpha = \mathscr{V}_{\text{pores}}/\mathscr{V}_{\text{total}}$ called the *porosity* of the medium. The concentration c_0 of diffusing particles, which is non zero in regions occupied by the fluid, changes abruptly to zero in the solid regions inaccessible to diffusion. The challenge is then to solve the diffusion equation $\partial_t c_0 = D\Delta c_0$ in the region occupied by the fluid, taking into account the boundary conditions prescribed by the very complicated and irregular geometry of the accessible volume $\mathscr{V}_{\text{pores}}$ and its surface $\mathscr{S}_{\text{pores}}$.

The trick is to bypass this difficulty by noting that the interesting quantity, for example from an experimental point of view, is a concentration $c = \langle c_0 \rangle$ obtained by averaging the real concentration of diffusing particles in a representative volume which includes the solid regions (in which $c_0 \equiv 0$) and the pores in which diffusion takes place (see Fig. 4.6). This volume must be chosen to be small enough to conserve a significant spatial dependence but larger than the typical pore size, in such a way as to obtain a function that is regular throughout space. The crucial step in the analysis will be to establish the effective equation that c obeys, by

---

[16]The Landau free energy $F(m)$ contains the description of the dynamics of relaxation of the system towards equilibrium, by the equation $\partial_t m = -\delta F/\delta m$; the term $(\nabla m)^2$ in $F(m)$ gives a term Δm in the evolution equation (see Sect. 3.4.3).

performing the average at the level of the conservation equation and Fick's law. Spatially averaging Fick's law $j = -D\nabla c_0$ involves a relation $\langle j \rangle = -D\langle \nabla c_0 \rangle = -D\nabla\langle c_0 \rangle + \mathscr{I}$ where $\mathscr{I}$ is a surface integral describing the additional contribution to $\langle \nabla c_0 \rangle$ coming from the boundaries of zone accessible to the liquid (surface $\mathscr{S}_{\text{pores}}$) [50]. This integral appears as a mean field adding its influence to that of the linear response term $-D\nabla\langle c_0 \rangle$ (response term at the scale where the medium appears homogeneous). We show that it is written $\mathscr{I} = (1 - \kappa)D\nabla\langle c_0 \rangle$ with $\kappa = 1$ if the medium is homogeneous (and $\kappa < 1$ if the medium is porous). In this way we obtain a Fick's law $\langle j \rangle = -\kappa D\nabla\langle c_0 \rangle$ for the average quantities, where $\kappa D \equiv D_{\text{eff}} < D$ is the effective diffusion coefficient of the porous medium.

This mean field approach is based on the fact that the physics, which is very complicated at the microscopic scale (here the scale of the pores), simplifies at a larger scale. This approach was developed in many contexts, from diffusion in porous rocks [38] to diffusion in living tissues of complex structure, for example the brain [50]. Under the name of *homogenisation*,[17] it was formalised and proved mathematically by various theorems establishing the validity of the effective averages and their properties [5, 28]

So the result is remarkably simple: at the scale in which the medium appears homogeneous (but of porosity, also called relative density $\alpha < 1$), diffusion taking place within it obeys an effective diffusion equation:

$$\partial_t c = D_{\text{eff}}\Delta c \qquad \text{with} \qquad D_{\text{eff}} \equiv \frac{D}{\lambda^2}. \qquad (4.15)$$

The effective diffusion coefficient D_{eff} is proportional to the "bare" diffusion coefficient D. This is generally written in the form $D_{\text{eff}} = D/\lambda^2$ where λ, called the *tortuosity*, is a geometric parameter if the medium. This parameter λ can be calibrated by observations or calculations based on a local model of the porous medium and the perturbation it induces on a random walk. Theoretical arguments [1] suggest a relation $\lambda^2 \sim \alpha^{-\beta}$, where $1/2 < \beta < 2/3$, but a local analysis of the medium, for example numerical, remains the most efficient way of determining D_{eff}, knowing a priori that this coefficient has meaning from the above analysis.

The procedure of homogenisation we have just described is a mean field theory. As such, it fails when the phenomenon becomes critical, here when the interstitial space becomes fractal, due to the existence of pores of all sizes. In this case the diffusion is therefore anomalous and its exponent $\gamma < 1$ deviates from the value $\gamma = 1$ obtained in "mean field".

---

[17]Other procedures of homogenisation are met in the context of diffusion, e.g. when D is a rapidly oscillating spatially periodic function (i.e. of small wavelength). We use a mean field approximation consisting of neglecting correlations between the function D and the function j. Averaging Fick's law here involves the approximation $\langle j/D \rangle \approx \langle (1/D) \rangle \langle j \rangle$, leading to the expression $D_{\text{eff}} = \langle (1/D) \rangle^{-1}$ for the effective diffusion coefficient of the medium, $\langle \rangle$ designating here a spatial average over the period of D.

Other "diffusion equations in porous media".

The term "diffusion in porous media" covers more generally various phenomena and equations. In particular we should distinguish two classes of problems.

The first concerns diffusion of substances in a porous substrate entirely filled with fluid. As for Brownian motion, this diffusion results from thermal motion of the fluid molecules. The porosity of the medium superimposes an effect of *quenched inhomogeneity* on the Brownian diffusion taking place inside the pores. In this situation, the confinement of the diffusing particles is weak enough not to destroy normal diffusion; the particular geometry of the substrate is taken into account simply by a renormalisation[18] of the diffusion coefficient D.

This interstitial diffusion can deviate from normal diffusion if the porous substrate has a fractal structure. In such a case the existence of pores of all sizes makes the procedure of homogenisation invalid (because the procedure rests on the hypothesis that the medium is homogeneous at a mesoscopic scale in order to perform the average defining $c = \langle c_0 \rangle$). It is therefore the exponent of the diffusion law and not just the diffusion coefficient D that must be modified (*anomalous diffusion*).

The second class of problems concerns the diffusive displacement of a liquid in a porous rock that is initially dry. It behaves as a hydrodynamic phenomenon and the movement is described at a scale much larger than that of the fluid molecules. Here there is a real "diffusion force" at work (hydrodynamic pressure) and not just the entropic (purely statistical) force underlying Fick's law. This diffusion force operates through Darcy's law producing a current proportional to the pressure gradient.

There could be a barrier to penetration of the liquid, when the pores are very small or linked to each other by narrow channels. In this case we use *percolation* models, such as those presented in Chap. 5 [Gouyet 1992].

Even when this effect of obstruction is negligible, another effect introducing a *dynamic asymmetry* will lead to a modification of the diffusion equation: when a pore empties a thin film of liquid remains on the sides of the pore (Fig. 4.7).

---

[18]The term "renormalisation" is used here in the sense of "redefinition of a parameter to include in an effective manner influences that we do not want to describe explicitly" (e.g. microscopic details or correlations). This procedure, introduced more than a century ago in the context of hydrodynamics, is not unrelated to the renormalisation methods presented in Chap. 3: it constitutes the elementary step. It is sufficient in non critical situations such as here where a separation of scales makes a homogenisation procedure possible.

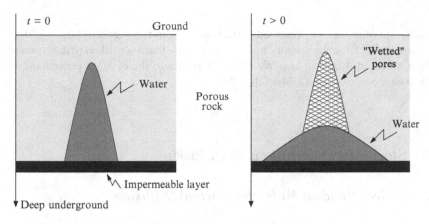

Fig. 4.7 Diffusion in a porous rock substrate limited by a deep horizontal impermeable layer. The initial situation is one of a localised surplus of liquid (with mass A_0 and horizontal linear extension l). The evolution of this profile due to diffusion of liquid in the porous rock is affected by the retention of a film of liquid in the pores that empty leading to a decrease in the mass of moving liquid. This causes a memory of the initial condition and of the elapsed time that affect the diffusion law (see also Sect. 4.5.4)

The conservation equation must be modified to take into account this retention of water or other liquids by making the diffusion coefficient dependent on the sense of variation of local concentration. If $\partial_t c > 0$, corresponding to an influx of liquid, we take $D = D_0$ and if $\partial_t c < 0$, corresponding to a local drying we take $D = D_0(1 + \epsilon)$. We end up with an equation, called the *porous medium equation*, of the form:

$$\partial_t c = D(\partial_t c)\,\Delta c^n,\qquad\qquad (4.16)$$

where the exponent n depends on the nature of the porous medium, in particular on its deformability [4, 25]. It is a typical model of out of equilibrium spatiotemporal phenomena, such as those we will meet in Chap. 8. In the example shown in Fig. 4.7 $n = 2$ and we therefore talk of non linear diffusion. The case of $n = 1$, corresponding to an elastoplastic medium and called *Barenblatt's equation*, is more interesting to us. Comparing it with the usual diffusion equation identifies the consequences of non conservation of moving mass, in particular the initial mass will be one of the parameters controlling the evolution. The unusual form of the diffusion coefficient, depending at each point and each moment in time on the trend with which the local density evolves, destroys the self-similarity of normal diffusion. The solutions show a behaviour called "anomalous diffusion", which we will present in Sect. 4.5.4. The full treatment of this second example of diffusion in porous media, very different from that we presented at the beginning of this section, is done in the framework of a scaling theory by using a renormalisation method to determine the observed diffusion law. We refer the reader to the detailed presentation shown in the work by Goldenfeld [25].

4.3 Stochastic Descriptions of Diffusion

4.3.1 Ideal Random Walk and Normal Diffusion

The basic microscopic model to describe diffusion of a particle is that of a *random walk*, called *ideal* or *Brownian* to differentiate it from models with bias, broad distributions of steps or correlations, which we will look at later (Sect. 4.5). In the simplest version, the particle makes jumps of length a over duration τ, in a direction chosen at random (from a distribution with a constant probability) and independently from the previous step. Numerical implementation simplifies the model further by restricting the steps to edges on a square ($d = 2$) or cubic ($d = 3$) lattice of lattice spacing a. The position at time $t = n\tau$ is therefore given by the random vector:

$$X(n\tau) = \sum_{i=1}^{n} a_i, \tag{4.17}$$

where the vectors $(a_i)_i$ are identically distributed, independent random vectors of length a and zero average (isotropic distribution). It follows immediately that the average displacement of the particle is zero:

$$\langle X(n\tau) \rangle = 0. \tag{4.18}$$

Due to the *independence of successive steps*, the variances simply add, which leads to the diffusion law describing the root mean square displacement $R(t)$ of the particle:

$$R^2(n\tau) \equiv \langle X^2(n\tau) \rangle = na^2 \equiv 2dDt. \tag{4.19}$$

The diffusion coefficient D, describing the average behaviour at long times is, in this way, related to the microscopic nature of the movement by knowing the length a and duration τ of the elementary steps in the model under consideration:

$$D \equiv \frac{a^2}{2d\tau} \qquad \text{(in } d \text{ dimensions)}. \tag{4.20}$$

We can generalise the model without affecting the asymptotic result by allowing a certain dispersion in the length and duration of the steps. In this case diffusion remains normal and we have another expression $D = a^2/2d\tau$, where τ is now the average duration and a^2 is now the variance of the step length.[19] We will see in Sect. 4.5 that it is essential that a and τ are finite in order for diffusion to remain normal. We can also relax the condition of statistical independence of the steps $(a_i)_i$ of the random walk: it can be easily shown that *finite* range correlations do not affect the normal character of diffusion.

Within these models, the concept of normal diffusion extends to situations where the root mean square displacement $R(t) = \langle X^2(t) \rangle^{1/2}$, i.e. the standard deviation of the instantaneous position, behaves *asymptotically* as $\sqrt{t}$. The scaling law $R(t) \sim \sqrt{2dDt}$ is therefore only valid in the limit $t \to \infty$, which in practice means for sufficiently long times. From this we see that the diffusion coefficient is not only an average, but also an asymptotic characteristic of diffusion:

$$D = \lim_{t \to \infty} \frac{R^2(t)}{2dt}. \tag{4.21}$$

In the presence of an external field, the relevant model becomes a *biased* random walk. In one dimension (to simplify the analysis), the probability of a step to the right becomes $p = (1 + \epsilon)/2$ where $0 < |\epsilon| \leq 1$; steps to the right are preferred if $\epsilon > 0$. This results in a drift:

$$\langle X(t) \rangle = (2p - 1)\,at/\tau = \epsilon at/\tau. \tag{4.22}$$

However the variance continues to follow a normal diffusion law:

$$\langle X^2(t) \rangle - \langle X(t) \rangle^2 = p(1 - p)\,ta^2/\tau = 2Dt(1 - \epsilon^2). \tag{4.23}$$

If $\epsilon \ll 1$, at short times ($t \ll t^*(\epsilon) = \tau/\epsilon$) we observe a diffusive motion, although at long times ($t \gg t^*(\epsilon)$) we observe a drift corresponding to a deterministic motion of uniform velocity $v(\epsilon) = \epsilon a/\tau$. The resulting observed behaviour is therefore a superposition of two scaling laws with different exponents, the relative weights of which vary with the (temporal) scale of observation. So we observe a *crossover* in the region around $t = t^*(\epsilon)$, which is longer the weaker the bias.

Diffusion of the particle could also be described by its *probability density*[20] $P(r,t)$, such that $P(r,t)d^d r$ is the probability that the particle is found at time t in a volume $d^d r$ around r. The density $P(r,t)$ (the integral of which is normalised

---

[19]If we consider diffusion of a molecule amongst other molecules (e.g. ink in water, Sect. 4.1.5), this formula is still valid if we take the mean free path l (distance travelled by the molecule between two collisions) for a and the mean free path time ($\tau = l/v_{th}$ where v_{th} is the thermal velocity of the molecule) for τ, giving $2dD = lv_{th}$.

[20]We write here r for the argument of the distribution $P(.,t)$, to distinguish the random variable $X(t)$ from the values r that it can take.

to 1) is then simply the law of probability of the random variable $X(t)$. Since the observable $X(t)$ is a sum of identical statistically independent random variables, which are centred on zero (in the case of no bias) and have finite variance a^2, a general result of probability theory, the central limit theorem,[21] tells us that the distribution $P(r,t)$ is asymptotically Gaussian, centred on zero and with variance $2dDt$:

$$P(r,t) \sim \left(\frac{1}{4\pi Dt}\right)^{d/2} e^{-r^2/4Dt} \qquad (t \to \infty). \qquad (4.24)$$

Effect of non critical correlations

Let $(a_i)_i$ be a series of centred random variables which are statistically stationary in the sense that the joint distributions are invariant under translation of indices. In particular, the correlation function $\langle a_i.a_j \rangle$ only depends on $|j-i|$ and we will therefore write it as $C(j-i)$. Explicitly calculating the variance of $\sum_{i=1}^{n} a_i$, shows that if the correlations decay fast enough that the sum $\sum_{n=-\infty}^{\infty} |C(n)|$ is finite, then $X(t = n\tau) = \sum_{i=1}^{n} a_i$ follows a normal diffusion law, with

$$D = \frac{1}{2d\tau} \sum_{n=-\infty}^{\infty} C(n). \qquad (4.25)$$

Even though there are correlations between successive steps, the behaviour at long times is that of an ideal random walk: $R(t) \sim \sqrt{2dDt}$, where the expression of D shows that it acts as an effective diffusion coefficient taking into account the effect of correlations. We recover the formula $D = a^2/2d\tau$ if the steps are independent (and we therefore have $C(n) = 0$ as soon as $|n| \geq 1$). If the correlations are positive, diffusion is accelerated ($D > a^2/2d\tau$). If on the other hand the correlations are negative, the particle tends to return on itself and diffusion is slowed down ($D < a^2/2d\tau$), however the exponent $1/2$ of the diffusion law is not changed.[22]

---

[21] We will expound this, along with its generalisations and implications in Sect. 4.5.2.

[22] Note that the criterion ensuring the persistence of the normal character of the diffusion involves a sum of absolute values, whereas it is the sum $S = \sum_{n=-\infty}^{\infty} C(n)$ of values with their signs which comes into the expression for the effective diffusion coefficient (4.25). $\sqrt{S}$ is interpreted as the effective step size of an ideal random walk which is asymptotically equivalent. $n_{corr} = \sum_{n=-\infty}^{\infty} |C(n)|/2C(0)$ is interpreted as the number of steps over which the correlations are appreciable (with $n_{corr} = n_0$ if $C(n) = C(0)e^{-|n|/n_0}$). It is therefore important to take the sum of *absolute* values because the sign of the correlations must not affect the estimation of their range in a compensatory way.

4.3.2   Mathematical Modelling: The Wiener Process

By definition, the random walks we have discussed above depend explicitly on the scale at which we describe the movement (time step $\Delta t = \tau$) and the specific details of the movement at each scale (step length and orientation distributions for example). However their long time behaviours, characterised by a normal diffusion law, are very similar, even identical if the diffusion coefficients are the equal. It makes sense to try to unify random walks leading to the same diffusion law in a continuous time description, valid at all scales. It is soon clear that a new mathematical object is necessary to reproduce the particular properties of diffusive motion, knowing the random nature of the trajectories and the fact that they are continuous but not differentiable.

This continuous time model, known by the name of the *Wiener process* and denoted $W_D(t)$, is entirely defined[23] by the following properties (expressed in dimension $d = 1$) [41, 68, 70]:

1. It is statistically stationary.
2. Its trajectories are continuous.
3. $W(0) = 0$.
4. If $t_1 < t_2 \le t_3 < t_4$ then $W_D(t_4) - W_D(t_3)$ and $W_D(t_2) - W_D(t_1)$ are statistically independent
5. $W_D(t) - W_D(s)$ is Gaussian, centred and of variance $2D|t - s|$.

We extend this process to dimension d by considering that the d components of motion (along each coordinate axis) are independent Wiener processes. In practice, we only need to remember the absence of temporal correlations and the Gaussian distribution of this process:

$$P_D(r, t) = \left(\frac{1}{4\pi Dt} \right)^{d/2} e^{-r^2/4Dt}. \tag{4.26}$$

We should note that this distribution is equal, for all values of t, to the asymptotic distribution of the random walks considered in the previous paragraph. As such, it is exactly scale invariant:

$$\text{for all } k > 0, \qquad k^d \, P_D(r, t) = P_D(kr, k^2 t). \tag{4.27}$$

A mathematical result ensures that the fractal dimension of almost all Wiener process trajectories is $d_f = 2$ (for dimension $d \ge 2$) [20].

---

[23]Incidentally it is sufficient to assume it is statistically stationary and that $W_D(t = 1)$ has variance $2D$ to obtain the general expression for the variance. The coherence of these hypotheses is given by the fact that a sum of independent Gaussian variables is a Gaussian variable with variance the sum of variances. A theorem, the general version of which is due to Kolmogorov ensures that this defines a unique process.

The Wiener process is an idealisation: real diffusion actually has a natural cutoff length, the mean free path of the diffusing particle (between two collisions), but this scale is so small that the continuous approximation is very good. The non rectifiable nature of the trajectories must therefore be seen as an artifact of this idealisation and not as damaging the physical meaning.

The Wiener process seems to be a universal basis, common to all discrete descriptions of the same dimension d and the same asymptotic diffusion law. It also seems to be the continuous limit of discrete diffusion models (random walks). We show in the following paragraph that the renormalisation ideas presented in Chap. 3 establish that the Wiener process and Brownian random walks effectively belong to the same universality class, that of normal diffusion. These ideas will also prescribe the transition procedure generating the Wiener process at the continuous limit. Here the situation is simple enough that we can say without further calculation that the step length a and duration τ should be *jointly* taken to zero, with $a^2/2d\tau = cte = D$ (in d dimensions).

Renormalisation of a random walk

The asymptotic equivalence of ideal random walks and the Wiener process is a result that can be obtained using a renormalisation approach. It consists of showing that random walks can be divided into universality classes of the sort that the diffusion law is asymptotically the same in each class.

The first step is to build a renormalisation transformation $\mathscr{R}_{k,K}$, with two real positive parameters k and K, acting on the probability density $p(r,t)$ according to[24]:

$$(\mathscr{R}_{k,K}p)(r,t) = k^d\ p(kr, Kt) \qquad \text{(for dimension } d\text{)}. \qquad (4.28)$$

If p is defined on a lattice of parameter a, with a temporal discretisation of step τ, we see that $\mathscr{R}_{k,K}p$ will be defined on a finer spatiotemporal mesh, of parameters $(a/k, \tau/K)$. In which case the fixed points p^* of $\mathscr{R}_{k,K}$, that is to say the probability densities satisfying $\mathscr{R}_{k,K}p^* = p^*$, will be necessarily continuous processes. The root mean square displacement $R(p,t)$ (the notation makes it explicit that $R(t)$ is a function of p) satisfies:

$$R[\mathscr{R}_{k,K}p,t] = k^{-1}\ R(p, Kt). \qquad (4.29)$$

---

[24]We note that the renormalisation can be just as well defined on the characteristic functions [40].

We confirm that the Wiener process W_D, of distribution P_D, is a fixed point of the transformations $\mathcal{R}_{k,k^2}$, which is another way to show its self-similarity.[25] We have seen in Chap. 3 the role and interpretation of fixed points of renormalisation transformations as typical representatives of universality classes. In the present case, taking $p_{a,\tau}$ as the probability distribution of a random walk with parameters $\Delta x = a$ and $\Delta t = \tau$, we have the following convergence property towards the fixed point W_D under the action of the renormalisation:

$$\lim_{k\to\infty} \mathcal{R}_{k,k^2}\, p_{a,\tau} = \lim_{n\to\infty} \mathcal{R}^n_{k_0,k_0^2}\, p_{a,\tau} = P_D \quad \text{where} \quad D = \frac{a^2}{2d\tau}. \quad (4.30)$$

The intermediate equality is obtained by noting that iterating the renormalisation modifies its scale parameters: $\mathcal{R}^n_{k_0,k_0^2} = \mathcal{R}_{k_0^n,k_0^{2n}}$. It proves the asymptotic equivalence of discrete random walks and Wiener processes; the Wiener process W_D is seen as the typical representative of normal diffusion processes with diffusion coefficient D. In this example we see that renormalisation is a constructive and demonstrative general method to correctly implement transition at the continuous limit.

The renormalisation approach may seem an unnecessary sophistication for the question of the asymptotic equivalence of Brownian random walks and Wiener processes, which can be shown directly. However, the same method is the only method that can answer this question for more complex random walks (the implementation is obviously more technical). For example it can show that a finite memory (the distribution of a step depends on the realisation of the k previous steps, $k < \infty$) and that a "weak disorder" (random transition probabilities, varying at each time step) does not destroy normal diffusion [9].

Note that the renormalisation just presented only applies to Markovian processes.[26] We will see in Sect. 6.3.5 that other methods must be developed in the context of self avoiding random walks with infinite memory.

---

[25] We can follow up the study with the analysis of other transformations $\mathcal{R}_{k,K}$. Fractal Brownian motions with exponent H, which we introduced in Sect. 4.5.5, are fixed points of the transformation $\mathcal{R}_{k,k^{1/H}}$, whereas Levy flights with exponent α (Sect. 4.5.2) is a fixed point of $\mathcal{R}_{k,k^\alpha}$. The self-similarity of these processes therefore translates into invariance by usable renormalisation to test the robustness of the associated diffusion laws with respect to various perturbations to the underlying laws of motion (perturbations to the short time dynamics).

[26] A Markovian process is a stochastic process $(X_t)_t$ without memory, in the sense that knowledge of an instantaneous state $X_{t_0} = x_0$ is sufficient to predict subsequent behaviour for all $t > t_0$, without it being necessary to know all the evolution history at times $t < t_0$, or even part of this history. This property is explained by the conditional probabilities: $\text{Prob}(X_t = x | X_{t_0} = x_0, X_{t_1} = x_1, \ldots, X_{t_n} = x_n) = \text{Prob}(X_t = x | X_{t_0} = x_0)$ for all integer $n \geq 1$ and all times $t > t_0 > t_1 > \cdots > t_n$.

4.4   From One Scale to the Next

Over the preceding sections we have seen several descriptions of the diffusion of
particles (ink molecules, grains of resin, etc.) in a supporting fluid, for example,
water. These correspond to experiments performed at different scales and therefore
subscribe to different theoretical frameworks, each one adapted to its scale. Before
going further, let us summarise the results.

- **In the experimental context of Brownian motion**, the diffusion law
  $R(t) \sim \sqrt{2Ddt}$ is obtained from statistical analysis of individual trajectories
  (Sect. 4.1.2). These trajectories are self-similar: they are fractal structures with
  dimension $d_f = 2$ (Sect. 4.1.3).
- **We can also observe the evolution in time**[27] **under the influence of diffusion** of a population of particles with an initial concentration profile that is
  localised (a finite length step, Fig. 4.2, or point like: $c(x,0) = \delta(x)$). At
  the macroscopic scale, at long times we obtain an even profile $c_\infty(x,t) =
  (4\pi Dt)^{-1/2} \exp[-x^2/4Dt]$. At the microscopic scale, we see a fractal interface
  developing (Sect. 4.1.4).
- **The phenomenological law** $j = -D\nabla c$ proposed by Fick leads to a description
  of the macroscopic evolution of any concentration profile by the diffusion
  equation $\partial_t c = D\Delta c$ (Sect. 4.2.1). This equation is invariant under the scaling
  transformation $(r \to kr, \ t \to k^2 t)$ for all real $k > 0$ (Sect. 4.2.2).
- **Several microscopic models have been proposed** to describe normal diffusion
  of a particle: random walks in discrete time (Sect. 4.3.1) and the Wiener process
  in continuous time (Sect. 4.3.2).

   We will see in this Sect. 4.4 that a remarkable coherence of these seemingly
disparate results can be obtained by showing that they can be deduced from each
other with the aid of simplifications or approximations justified by the change in
spatiotemporal scale of the descriptions. In particular, we show that it is the *same
diffusion coefficient D* that appears in these formulae.

4.4.1   How the Different Descriptions Are Related

The different descriptions of the phenomenon of diffusion can be arranged by
increasing spatial scales. We distinguish three levels of description.

- **The first level**, the lowest, corresponds to the *deterministic* and *reversible*
  description that we can make (at least formally) at the *molecular* scale. We
  consider the equations of motion of all the molecules within the framework of

---

[27]A new interpretation of the diffusion law $R(t) \sim \sqrt{2Dt}$ is possible: if we imagine sitting at a
point x_0, the concentration $c(x_0, t)$ increases with time t provided $x_0 > R(t) = \sqrt{2Dt}$; whilst if
the characteristic size $R(t)$ of the profile is longer than x_0 so we are sitting inside the profile, the
concentration decreases with time.

classical dynamics: the system is described by a Hamiltonian and the associated evolution equation. That is the evolution of the probability distribution of the presence of molecules in the phase space of all the degrees of freedom of the molecules. This evolution equation is known by the name of the *Liouville equation* and is equivalent to the set of equations of the molecular dynamics. The interactions between molecules are described by a specific potential, but since this is short ranged we generally absorb these interactions into the elastic collisions; the potential enters only in an effective manner in the collisional cross section of the molecules. Taking into account the exponential amplification of any noise at each collision and the large number of particles (and therefore of collisions), the resulting evolution has an extremely short correlation time and becomes rapidly unpredictable in the long term. We therefore talk of *molecular chaos* (see Fig. 4.8).

At larger scales the evolution of the population of molecules appears totally random and *only its statistical properties can be described*. Theoretical analysis of the deterministic model shown in Fig. 4.8 (one of the versions of a model known as a *Lorentz gas*) can be implemented using tools from chaos theory (see Sect. 9.2). It shows that the resulting asymptotic movement follows a normal diffusion law in an appropriate range of concentrations of obstacles. We therefore know how to relate the diffusion coefficient to other statistical characteristics of the microscopic dynamics.

One of the implications of this result is to reconcile determinism and stochasticity. When modelling a phenomenon, these two concepts are partly subjective because they depend on the scale that is regarded as the elementary scale and the scale at which the behaviour is to be predicted by the model. Diffusion is a case in point and we will see alternate deterministic and stochastic descriptions as we increase the minimum scale of the model.

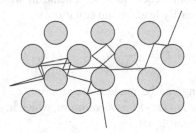

Fig. 4.8 Cartoon of microscopic model (*Lorentz gas*) explaining the concept of molecular chaos underlying the random nature of diffusion. Here a particle moves in a network of fixed diffusers, with whom it undergoes elastic collisions: its velocity, well defined between collisions, remains the same as its thermal velocity. Despite the perfectly deterministic nature of this model, the trajectory is unpredictable at long times because each collision doubles the angular uncertainty in the initial condition, due to inevitable noise present in the system. In practice, the phenomenon observed at long times and large spatial scales has all the characteristics of a stochastic motion. We show that it follows a normal diffusion law, as long as the obstacles are not be too numerous (trapped) nor too sparse (rectilinear movement, in a straight line). These properties are still observed when the particle moves within a cloud of other particles that are also moving

- **Due to the unpredictability** of molecular motion which we have just talked about, it is simpler and above all more practical to adopt a *stochastic* model that is equally valid at the scale of a single molecule within a population and at the scale of a grain suspended in a population of solvent molecules. We can adopt a discrete model (random walk) or a continuous model (stochastic process). Showing the equivalence of these two descriptions requires taking the limit carefully. The step length a and duration τ of the random walk must be simultaneously taken to zero, with $a^2/2d\tau = D$. We have already seen that it is very productive to use the method of renormalisation (Sect. 4.3.2).

Note that, unlike the molecular *dynamic* description, this stochastic description is only kinematic in the sense that the basic laws of motion are given. For example, in situations more complicated than simple diffusion, additional physical mechanisms (e.g. trapping by adsorption, external fields) and constraints (e.g. walls, excluded volume) playing a role in the diffusion are simply taken into account in an effective manner in the choice of probability laws governing the particle. The validity of the *choice* can be established by comparing the predictions of the stochastic model with the observed behaviours or by rooting it in the dynamic description at a molecular level as described in the previous point.

The stochastic description can also be placed at a slightly larger spatial scale, forgetting the individual trajectories in order to describe the evolution of the probability distribution. Let us illustrate this point with the simplest example, that of a non biased random walk in one dimension. During the time step Δt, the particle has a probability of $1/2$ of taking a step Δx to the right and a probability of $1/2$ of taking a step Δx to the left, $(\Delta x)^2 = 2D\Delta t$, which is written:

$$P(x, t + \Delta t) = \frac{1}{2} \, P(x - \Delta x, t) + \frac{1}{2} \, P(x + \Delta x, t). \qquad (4.31)$$

An expansion up to second order, valid at a scale in which Δx and Δt are the infinitesimal quantities, easily leads to the equation:

$$\partial_t P = \frac{\Delta x^2}{2\Delta t} \, \partial_{xx} P = D \, \partial_{xx} P. \qquad (4.32)$$

This type of reasoning can be made more rigorous and put in a more general framework leading to the *Fokker–Planck equation*, which in the example considered is equivalent to the diffusion equation.

We refer the reader to [31, 39, 56, 64] for a more complete presentation of this derivation, its conditions of validity and implicit approximations.

Kinetic theory

It is worth mentioning that a statistical description of a population of N particles that is more complete than that of Fokker–Planck, is formulated in terms of the joint probability densities $f_1(r, v, t)$, $f_2(r_1, v_1, r_2, v_2, t)$, $\ldots f_N(r_1, v_1, \ldots, r_N, v_N, t)$ with $2d$ degrees of freedom[28] (r, v) of the different particles. For example, we define $f_2(r_1, v_1, r_2, v_2, t) dr_1 dv_1 dr_2 dv_2$ as being the probability of finding, at time t, two particles at (approximately) r_1 and r_2 with velocity (approximately) v_1 and v_2 respectively. This approach is known by the name of *kinetic theory*. Introduced by Boltzmann, it was essentially developed for gases, i.e. media dilute enough that the molecular interactions do not play a dominant role. The approach is rooted directly in the deterministic molecular description. From the Liouville equation for f_N we can derive a hierarchy of equations describing the evolution of the different distributions $(f_j)_{j \geq 1}$, called the *BBGKY hierarchy* (from the names Born, Bogoliubov, Green, Kirkwood and Yvon). It is an infinite system of coupled equations and we talk of hierarchy because the evolution equation of f_j involves the higher order functions $(f_k)_{k>j}$. Various relations are introduced to close the set of equations to obtain a closed finite system of (approximate) equations. In a dilute gas, we recall that the *Boltzmann approximation* neglects the correlations between two particles before collision, written $f_2(r_1, v_1, r_2, v_2, t) \approx f_1(r_1, v_1, t) f_1(r_2, v_2, t)$. The justification of this is molecular chaos, illustrated in Fig. 4.8, and the fast decorrelation of associated motion [23, 24]. Using this approximation, the first equation in the hierarchy, describing the evolution of f_1, becomes a closed equation: the *Boltzmann equation*. In the presence of an external field of force a (field on the unit of mass), this equation is written:

$$\partial_t f_1 + v.\nabla_r f_1 + a.\nabla_v f_1 = \mathscr{I}(f_1), \qquad (4.33)$$

where $\mathscr{I}$ describes the contribution of the collisions (*the collision integral*); which is expressed (exactly) as a function of f_2 but the Boltzmann approximation transforms it into a quadratic functional of f_1 (integral in velocity space). This kinetic approach, explicitly taking into account the molecular velocities, allows us to describe phenomena of relaxation towards thermal equilibrium. After integrating over the velocities, we find[29] the Fokker–Planck

---

[28]Unlike the velocity of a Brownian particle, which depends on the scale at which we observe it, the molecular velocities are well defined degrees of freedom (the molecules being here depicted as point particles).

[29]The procedure, known as the *Chapman–Enskog method* consists of a perturbative expansion of the Boltzmann equation about the equilibrium state, in which the velocity distribution is Maxwellian ($\phi_0(v) \sim \exp[-mv^2/2kT]$ up to a normalisation constant); here we therefore plug an expression of the form $f_1(r, v, t) = P(r, t)\phi_0(v)[1 + \text{higher orders}]$ into the Boltzmann equation. After integrating over v, we obtain, to lowest order, the Fokker–Planck equation mentioned above for the unknown function $P(r, t)$.

equation in a more justified and better microscopically based way. Kinetic theory is, in this way, a route leading from Newtonian molecular dynamics to the Fokker–Planck equation. More generally, it is a theory positioned "between" the Hamiltonian deterministic microscopic description and the continuous spatiotemporal field description (hydrodynamic theory in the case of fluids, Sect. 9.5), which enables us to derive the field description from the microscopic equations and, if necessary, complete it with information about the fluctuations and correlations of the molecular velocities.

This summary only scratches the surface of kinetic theory; we suggest [11, 17, 55] for a more substantial presentation.

- **Finally, at our macroscopic scale**, we adopt a global spatiotemporal description of a population of particles. The observable is then the concentration $c(r, t)$ such that $c(r, t)d^d r$ is the number of particles in the volume $d^d r$, considered as small at the scale of description, but large enough at the microscopic scale to contain many particles (hypothesis called *continuous media*). If N is large enough and the particles are sufficiently non-interacting, we can identify, up to a normalisation factor, this concentration c with the probability distribution $P(r, t)$ of a random walk with equation $\partial_t c = D \Delta c$, according to the Fokker–Planck equation. The argument invoked in this identification is the law of large numbers. We therefore see the connection between the microscopic description in terms of trajectories and the macroscopic description in terms of concentrations; it shows that *the diffusion coefficients involved in the diffusion law on one hand and the diffusion equation on the other hand are identical*. One single coefficient, the diffusion coefficient D of a particle in a carrier fluid, describes the basic essentials of diffusive motion, whatever the scale of observation. Note that at this macroscopic scale, we recover a *deterministic* evolution equation, but it is *irreversible*, unlike the molecular equations (see Sect. 4.4.3).

The diffusion equation $\partial c = D \Delta c$, the microscopic foundation of which we have just shown, can also be introduced, or completed, in a phenomenological manner, as it was done historically by Fick's law. We can therefore now better interpret Fick's law $j = -D \nabla c$: the current j is a probability current. The origin of the apparent "force" driving the particles in the direction opposite to the concentration gradient towards less densely populated regions is therefore *only statistical*; the effect only occurs in the presence of a large number of particles. There is no real force exerting at the molecular level and there is no "diffusion force" that exists. The term *entropic force* is sometimes used to indicate this statistical effect perceived at the macroscopic scale. We will show in Sect. 4.4.3 that this microscopic and statistical origin of diffusive motion is also reflected in the irreversibility of the diffusion equation.

Law of large numbers

This law is the most famous *statistical law*. It states that the arithmetic average of a sequence $(\chi_i)_i$ of statistically independent random variables, identically distributed and with a finite variance, converges to its statistical average, i.e. to a number: $\lim_{N \to \infty} N^{-1} \sum_{i=1}^{N} \chi_i = \langle \chi \rangle$. It justifies the often observed fact that a large number of microscopic random events can produce a deterministic and reproducible macroscopic behaviour. For example, the average of a sequence of tosses of a coin, heads (1) or tails (0), tends to $1/2$ when the number of throws tends to infinity. This law in particular enables the identification of the frequency of an event $X \in [x, x + \Delta x]$ with its probability $P(x)\Delta x$. Analogously, here we apply this to the random variable $\chi_i(x, t)$ taking the value 1 if the particle i is found between x and $x + \Delta x$ at time t; we therefore have $\text{Prob}[\chi_i(x, t) = 1] = P(x, t)\Delta x$ and $\sum_{i=1}^{N} \chi_i(x, t) = \mathscr{N}_{Av} \, c(x, t)\Delta x$ if we measure the concentration c in number of moles per unit volume. The central limit theorem then ensures that the deviation between $\mathscr{N}_{Av} \, c(x, t)$ and $N P(x, t)$ goes as $\sqrt{N}$.

4.4.2   Einstein Formula and the Fluctuation-Dissipation Theorem

At the microscopic scale, the effect of microscopic collisions, which is reflected in the diffusion coefficient D, also results in the viscosity of the medium and in the friction coefficient Γ of a moving particle: when the particle is moving with velocity v, the viscous friction force exerted on it is Γv. We can view this force as the "coherent" effect of molecular collisions on the *moving* particle producing work, while the diffusion describes the "incoherent" contribution. This common origin between diffusion and viscosity is expressed in the *Einstein formula*, already seen in Sect. 4.1.2 [18]. The most general form, also known by the name of the *fluctuation-dissipation theorem*, of which the Einstein relation is a particular case, is written[30]:

$$D = \frac{kT}{\Gamma},\tag{4.34}$$

where D is expressed in m^2/s and Γ in $kg\ s^{-1}$. We should highlight that Γ, like D, is a characteristic of the particle-medium pair in question: Γ is in fact related

---

[30]This relation can hold in certain situations where the diffusion is anomalous. In this case the diffusion coefficient $D(t) \equiv R^2(t)/2dt$ and the mobility $1/\Gamma$ depend on time [51].

to the dynamic viscosity[31] η of the fluid environment by *Stokes formula*, written as $\Gamma = 6\pi r_0 \eta$ for a spherical particle of radius r_0. This formula leads to the more explicit, but also more specific, form of the Einstein formula:

$$D = \frac{kT}{6\pi r_0 \eta}. \tag{4.35}$$

We can express the Boltzmann constant k as a function of the ideal gas constant $R = 8,314$ (known experimentally well before the first studies of Brownian motion) and the Avagadro number $\mathcal{N}_{Av}$. This gives the Einstein formula in its original form:

$$D = \frac{RT}{6\pi r_0 \eta \, \mathcal{N}_{Av}} \tag{4.36}$$

used by Perrin to determine the value of the Avogadro number $\mathcal{N}_{Av}$. The quantity $1/\Gamma$, called the *mobility* of the particle, is a linear response coefficient[32]: if the particle is submitted to an external force F, its velocity takes the value F/Γ. We can therefore write the Einstein formula as:

$$\text{response} \equiv v = \frac{DF}{kT} \equiv D \, \frac{\text{excitation}}{kT}. \tag{4.37}$$

This formula shows the linear response relation seen in Chap. 1 (for example Sect. 1.1.7). In addition, D occurs as a coefficient describing the spontaneous velocity fluctuations. In the case of diffusion of a "labelled" (at least mentally) fluid molecule in a medium of identical molecules, we can demonstrate[33] the *Green–Kubo formula* relating D to the temporal autocorrelation function of the molecular velocities (in d dimensions):

---

[31] η is an *effective* quantity appearing in the deterministic macroscopic description of hydrodynamics. It is measured in units of poise, equal to $0.1 \text{kg}/(\text{m s}) = 0.1 \text{Pa s}$ (for example η is 1 millipoise for water at $20°\text{C}$). It is in this context, for macroscopic spheres, that Stokes showed "his" formula; it therefore does not apply to atoms, nor in dilute media where the "continuous media" approximation underlying hydrodynamics no longer applies. More precisely, η is introduced in a phenomenological manner by a law similar to Fick's law describing the transport of the moving quantity: $J_{ij} = -\eta \, \nabla_i v_j$ where v is the (macroscopic) velocity field of the fluid and J is the tensor describing the current density of the moving quantity, generated in the fluid in response to the velocity gradient (J_{ij} is a force per unit surface, in other words a pressure). Nevertheless, the origin of viscosity ultimately lies in collisions of fluid molecules with each other and with the moving object. All the analysis presented in this paragraph relating to the diffusion coefficient can therefore be applied to the viscosity η.

[32] The equation of motion of a particle, of mass m, is written $m\dot{v} = F - \Gamma v$. In the stationary regime this becomes $v = F/\Gamma$.

[33] The approach used starts from the Liouville equation (molecular dynamics); it can be generalised to a number of other transport phenomena [17, 35].

$$D = \frac{1}{d} \int_0^\infty \langle v_0 . v_t \rangle \, dt \qquad \text{(Green–Kubo formula)}. \qquad (4.38)$$

The mobility $1/\Gamma$ can be seen as a "susceptibility" since we can write, within the framework of linear response, that $\partial v_i / \partial F_j = \delta_{ij}/\Gamma$, where v is the velocity of the particle subjected to an applied force F and δ_{ij} is the Kronecker delta symbol (zero unless $i = j$, when it takes the value 1). The Green–Kubo formula, rewritten $\int_0^\infty \langle v_0 . v_t \rangle \, dt = d \, kT \, (1/\Gamma)$, is therefore exactly analogous to the relationship between the magnetisation correlation function (quantifying the fluctuations) and the magnetic susceptibility χ in a system of spins (of n components):

$$\int_0^\infty \langle M(0) . M(r) \rangle \, d^d r = n \, kT \, \chi. \qquad (4.39)$$

These two relations are particular cases of the fluctuation-dissipation theorem, which stipulates that at *thermal equilibrium*, the response function G (linear response to an external field) and the correlation function C of fluctuations (in the absence of an external field) are proportional [36]. In other words, as long as we stay in the linear response regime, that is close to equilibrium, the system reacts to an external perturbation in the same way as it reacts to its spontaneous internal perturbations. The most general and most commonly found statement of this theorem expresses the functions C et G in terms of Fourier components. By denoting the external perturbation as kTf (in general time dependent) and the conjugate variable as q (in the sense where $kTfq$ is an energy) with which we measure the response of the system, we can show[34]:

$$\widehat{C}(\omega) = \frac{2kT}{\omega} \, \text{Im}[\widehat{G}(\omega)] \quad \text{with} \quad \begin{cases} \langle \hat{q}(\omega) \rangle = \widehat{G}(\omega) \hat{f}(\omega) \\ \langle \hat{q}(\omega) \hat{q}(\omega') \rangle = 2\pi \delta(\omega + \omega') \widehat{C}(\omega). \end{cases} \qquad (4.40)$$

where the equalities to the right are the expressions in conjugate space of $C(t) = \langle q(t) q(0) \rangle$ and $\langle q(t) \rangle = \int_0^\infty G(t - s) f(s) ds$ defining $C(t)$ and $G(t)$.

In conclusion, we will regroup the different expressions for the diffusion coefficient that we have seen (in d dimensions), from the microscopic descriptions to the macroscopic and phenomenological descriptions:

(1) D related to other statistical (chaotic) properties of deterministic microscopic dynamics.
(2) $D = (1/d) \int_0^\infty \langle v(0) . v(t) \rangle dt$ (Green–Kubo formula).
(3) $D = a^2 / 2d\tau$ (random walk with step length a and duration τ).
(4) D parameterising the Wiener process W_D.

---

[34]This theorem, shown to be true for equilibrium systems, has been recently generalised: it remains partially valid in certain stationary *nonequilibrium* states, in which fluxes flow across the system, such as we observe in dissipative systems [57].

(5) $D = kT/\Gamma$ (Einstein formula).

(6) $j = -D\nabla c$ (Fick's law).

(7) $D = \lim_{t\to\infty} R^2(t)/2dt$ (diffusion law for a particle).

(8) $P_\infty(r,t) = (4\pi Dt)^{-d/2} \exp[-r^2/4Dt]$ (asymptotic probability distribution in the case of an isotropic diffusion in d dimensions).

(1) and (2) come from statistical analysis of deterministic microscopic dynamics, in the framework of ergodic theory and kinetic theory respectively; (3) and (4) follow from a purely stochastic description; (6) is phenomenological and (5) expresses a coherence between the descriptions at microscopic and macroscopic scales. Finally, the asymptotic relationships (7) and (8), which are directly comparable to observation, can be obtained from each of the different descriptions, showing the equivalence of the different ways of defining the diffusion coefficient D.

Theory of Brownian motion

We started this chapter with Perrin's *experimental* results of Brownian motion of colloids (grains of resin) (Sect. 4.1.2). We then gave a *kinetic* description of the random trajectories of particles by a stochastic process, the Wiener process (Sect. 4.3). Here we will describe more precisely the *dynamics* of the grains and its microscopic roots in the thermal motion of water molecules. We will find the formulae we have already seen, for example the Einstein formula, but here from a point of view which allows us to *prove* them. The objective is also to include Brownian motion in the rigorous framework of stochastic integration, giving new calculation rules to use, for example to solve an evolution equation involving Brownian motion or more generally thermal noise [21, 68].

The evolution equation of a grain

The starting point of the theoretical analysis is the physical explanation of Brownian motion: it boils down to saying that the irregular and random trajectories of the grain are entirely determined by water molecules in thermal motion colliding with the grain. The idea is then to decompose the force exerted on the grain into two contributions:

- A *deterministic* component Γu where Γ is the friction coefficient of the grain in the fluid environment.
- A *random* component called a *Langevin force*, b, which has a zero mean by construction.

As we have already highlighted in Sect. 4.4.2 when introducing Γ, these two components have the same origin, namely the collisions of the fluid molecules with the grain. The first describes the resulting deterministic contribution and the second describes the fluctuating part. The evolution equation for the velocity, u, of a grain of mass m, is then written:

$$m\dot{u} = -\Gamma u + b. \tag{4.41}$$

The heart of the analysis will be to quantify the characteristics of this "noise" b, and then integrate (4.41) to deduce the (observable) statistical properties of the trajectory $r(t)$.

Different time scales of Brownian motion
We bring to light three characteristic time scales:

1. The correlation time τ_0 of the random force b; it is related to the time between collisions of water molecules (mean free time), characterising both the relaxation time of the medium after the grain has passed and its autocorrelation. According to the molecular chaos hypothesis (illustrated in Fig. 4.8) it is very short, of the order of a picosecond. Whenever we are at a time scale larger than τ_0, we can assume that $b(t)$ is not correlated to the instantaneous velocities $u(s)$ and has no autocorrelation, which can be written for two components i and j at instants of time t and s: $\langle b_i(t) b_j(s) \rangle \approx A^2 \delta_{ij} \delta(t-s)$ where the constant A is, at this stage, still to be determined;
2. The viscous relaxation time $\tau_m = m/\Gamma$ where m is the mass of the grain. This is the time after which all memory of the initial conditions (u_0, r_0) has been lost, as we can see by formally integrating (4.41):

$$u(t) = u_0 \, e^{-\Gamma t/m} + \frac{1}{m} \int_0^t e^{-\Gamma(t-s)/m} \, b(s) \, ds, \tag{4.42}$$

where u_0 is the well determined initial velocity of the grain. τ_m is of the order of a nanosecond for a grain of the order of a micron diffusing in water;
3. The observation time τ_{obs}, in general longer than a millisecond, at which we describe the diffusive movement of the grain.

Thermal noise: white noise
In situations in which we normally study Brownian motion, we have $\tau_0 \ll \tau_{obs}$ and it is therefore legitimate to replace the random force b by a term $A\zeta$ satisfying (exactly) $\langle \zeta_i(t) \zeta_j(s) \rangle = \delta_{ij} \delta(t-s)$. This noise ζ is called *white* because its intensity is uniformly distributed among different Fourier modes, just as the energy of white light is uniformly distributed among the different monochromatic wavelengths: $\langle \hat{\zeta}_i(\omega) \hat{\zeta}_j(\omega') \rangle = 2\pi \delta_{ij} \delta(\omega + \omega')$.

Saying that the noise is white is therefore equivalent to saying that it is isotropic, stationary and not correlated in time. This condition entirely defines the noise if we assume in addition that the noise is Gaussian and centred, since a Gaussian process is entirely prescribed by its average and its correlation function. We can then use (4.42). It first of all follows that

$\langle \boldsymbol{u}(t) \rangle = \boldsymbol{u}_0 \, e^{-\Gamma t/m}$: the average velocity of the grain is asymptotically zero and in practice zero as long as $t \gg \tau_m$. We then express the velocity correlations of the grain:

$$\langle u_i(t) u_j(s) \rangle - \langle u_i(t) \rangle \langle u_j(s) \rangle = \frac{A^2}{2m\Gamma} \, \delta_{ij} \, \left(e^{-\Gamma |t-s|/m} - e^{-\Gamma(t+s)/m} \right).$$
(4.43)

For $t = s \gg \tau_m$, we obtain the kinetic energy $m \langle u^2/2 \rangle = d \, A^2/4\Gamma$, which is identified with the thermal energy $d \, kT/2$ (equipartition of energy theorem for d translational degrees of freedom) showing that $A = \sqrt{2kT\Gamma}$. The random component of Brownian motion, which we call the *thermal noise* is therefore entirely determined by:

$$\boldsymbol{b} = \sqrt{2kT\Gamma} \, \boldsymbol{\zeta} \quad \text{with} \quad \begin{cases} \langle \zeta_i(t) \rangle = 0, \, i = 1 \ldots d \\ \langle \zeta_i(t) \zeta_j(s) \rangle = \delta_{ij} \delta(t-s) \end{cases} \quad \text{(white noise)}.$$
(4.44)

Solution in the overdamped regime

In the limit $m \to 0$, the expression for the velocity correlations of the grain simplifies to $\langle u_i(t) u_j(s) \rangle = \sqrt{2kT/\Gamma} \, \delta_{ij} \, \delta(t-s)$, in a way which is coherent with the equation $\boldsymbol{u} = \Gamma^{-1} A \boldsymbol{\zeta} = \sqrt{2kT/\Gamma} \, \boldsymbol{\zeta}$ obtained in this limit. When the time scale at which we describe the motion is long compared to τ_m (in particular, the grain must not be too heavy), we can simplify the analysis by neglecting the inertial term $m\dot{\boldsymbol{u}}$ in (4.41). This is referred to as the "overdamped regime".[35] Integrating the evolution equation $\boldsymbol{r} = \sqrt{2kT/\Gamma} \boldsymbol{\zeta}$ gives $\boldsymbol{r}(t) - \boldsymbol{r}_0 = \sqrt{2kT/\Gamma} \int_0^t \boldsymbol{\zeta}(s) ds$. So $< [\boldsymbol{r}(t) - \boldsymbol{r}_0]^2 > \sim 2dkTt/\Gamma$; by comparing with the diffusion law $< (\boldsymbol{r}(t) - \boldsymbol{r}_0)^2 > \sim 2dDt$, we obtain the fluctuation-dissipation relation $\Gamma D = kT$. This relation reflects the fact that the thermal noise term (Langevin force) and the friction coefficient have the same origin: collisions with molecules in thermal motion. They describe the influence of the same phenomenon at two different scales. It is easy to check that $\int_0^t \boldsymbol{\zeta}(s) ds = \boldsymbol{W}_{1/2}$ (normalised Wiener process); so we find that $\boldsymbol{r}(t) - \boldsymbol{r}_0 = \sqrt{2D} \, \boldsymbol{W}_{1/2} = \boldsymbol{W}_D$. The description of Brownian motion as a Wiener process relies on the equipartition of energy between translational

---

[35] Equation (4.42) shows that a white noise $\boldsymbol{b}(t)$ in the equation for $\boldsymbol{u}(t)$ gives a correlated noise $\boldsymbol{B}(t)$ in the equation for $\boldsymbol{r}(t)$: $\langle B_i(t) B_j(s) \rangle \sim cte.\delta_{ij} \, e^{|t-s|/\tau}$ to leading order. If $\tau_m \ll \tau_{obs}$, we can neglect the autocorrelation of the noise $\boldsymbol{B}(t)$. This is also the approximation underlying the hypothesis of the overdamped regime; in other words, it amounts to eliminating from the description those variables that rapidly relax to zero.

degrees of freedom and it involves two approximations: firstly, the noise term is assumed to be "white", non correlated, justified by the fact that the time scales are large compared to the correlation time of molecular collisions: $\tau_{obs} \gg \tau_0$; secondly, we neglect the inertial effects (overdamped regime), which is justified by the small mass of envisaged particles: $\tau_{obs} \gg \tau_m$ [67].

Starting from this description of the grain, it is finally possible to derive three Fokker–Planck equations, describing the respective evolution in velocity $\{u\}$ space, in phase $\{u, r\}$ space and in real space (position $\{r\}$ space) [12].[36] The latter takes the form of the usual diffusion law[37]: $\partial P(r, t) = D\Delta P(r, t)$ [46].

4.4.3 Irreversibility of Diffusion

Let us finish this section by underlining a fundamental aspect of diffusion: its irreversibility. This is even visible in the form of the macroscopic diffusion equation $\partial_t c = D\Delta c$, which is not invariant under time reversal: Changing t to $-t$ changes D to $-D$ but a diffusion equation with a negative diffusion coefficient has no physical sense since its solutions are unstable under the smallest of perturbations. This point can be understood without calculations: in Fig. 4.5, the vertical arrows could be drawn with their directions reversed whilst remaining proportional to the second derivative of the profile. In other words, such an equation describes an amplification of the smallest inhomogeneities, the amplification being larger the more localised the inhomogeneity. So, when we try to go back in time following the diffusion equation, the smallest hitch in the profile $c(x, t)$ results in a divergence over a time $\tau \sim l^2/2D$ if l is the characteristic scale of the defect. This instability of the dynamics with respect to the smallest noise reflects the physical impossibility of evolving a diffusion profile in reverse: to do so we would need to perfectly control the tiniest source of noise.

The microscopic description of Brownian motion gives a different formulation to the same explanation. To reverse the diffusive motion of a cloud of particles (molecules or grains) and return to an initial concentration profile, it would need to be possible, at a given time t_0, to reverse *exactly* the velocities of *all* the particles in the system. The reversal must be applied to, not just the velocities of the particles we

---

[36]We sometimes talk about a *Brownian particle* when the motion is described by its position $r(t)$ and a *Rayleigh particle* when the motion is described more accurately by its velocity $v(t)$ [64].

[37]More generally, the equation describing overdamped Brownian motion of a particle in a force field $F(r)$ is written: $\dot{r} = F(r, t)/\Gamma + \sqrt{2kT/\Gamma}\zeta$. The associated probability flux, involved in the Fokker–Planck equation $\partial_t P = -\nabla.J$, is therefore $J = FP/\Gamma - kT\nabla P/\Gamma$ [60].

observe diffusing but also, those of the fluid molecules in which the diffusion takes place, whose thermal motion, transmitted to the diffusing particles over countless collisions, is itself the origin of the diffusive motion of the diffusing particles. In addition, this reversal must be an exact reversal of the sign of the velocity $v \rightarrow -v$ without the slightest fluctuation in direction and without any time delay. Even if such a reversal was possible, it would, in all probability, lead to an evolution which rapidly deviated from the "projection backwards in time" of the evolution observed between 0 and t_0, due to the amplification of the effect of noise (inevitably present in the system) in the individual trajectories (Fig. 4.8).

This explanation of the irreversibility we observe for all diffusive motion anticipates the concept of chaos, which we will present in Sect. 9.2 [22, 23]. It can also be stated within the framework of statistical mechanics.[38] In the example of ink diffusing in water, the fact that the reverse evolution (ink reconcentrating into a droplet) is never observed is explained by the fact that the volume $\mathscr{V}_i$ in phase space (of $6N$ dimensions if there are N particles) occupied by the initial state is infinitely smaller than the volume $\mathscr{V}_f$ of the set of diluted states. Using the reversibility of molecular motion itself, we can show that

$$\frac{\text{Prob}(f \rightarrow i)}{\text{Prob}(i \rightarrow f)} = \frac{\mathscr{V}_f}{\mathscr{V}_i} \ll 1, \tag{4.45}$$

where $\text{Prob}(i \rightarrow f)$ denotes the probability of observing an evolution from the initial state to a diluted state within $\mathscr{V}_f$ and $\text{Prob}(f \rightarrow i)$ that of an evolution in the reverse direction, to a configuration within $\mathscr{V}_f$, to return to the initial state. The evolution of one of these diluted states to a state where the ink is concentrated is not impossible, it is just improbable.[39]

Diffusion illustrates in an exemplary way the "degradation" of molecular evolution equations, deterministic and reversible when written at the molecular scale, into irreversible and random processes at larger scales. This irreversibility is, in this way, an example of an *emergent* phenomenon, that is a phenomenon that appears

---

[38]This is not surprising if we remember the dynamic root of statistical mechanics; Boltzmann's ergodic hypothesis replaces the temporal description by a static picture in which time has disappeared (Sect. 9.3.1).

[39]This explanation, already presented in Boltzmann's work is now known as the microscopic foundation of the second law of thermodynamics for closed systems [6, 37].

To avoid any confusion, we remark that the H-theorem of Boltzmann is a different result: it follows from an approximation consisting in neglecting pairwise correlations (hypothesis of factorisation of joint distributions of several particles), justified by the randomising nature of microscopic dynamics, which we call the hypothesis of molecular chaos (Fig. 4.8, Sect. 9.3.1). This result is "subjective", concerning the irreversibility of our model and it is a bit of a jump to relate it to an "objective" irreversibility of the phenomenon under consideration.

Whilst we are on the subject, it is worth mentioning that the explanation of irreversibility is more subtle and still under discussion in the case of open systems far from equilibrium, in the context of chaos theory. We suggest the works, [10, 17, 45], for a profound discussion of these questions; a few foundational texts are presented in [2].

only at macroscopic scales, due to the large number of coupled degrees of freedom that play a role at these scales, and that cannot be predicted from the underlying mechanisms.

4.5 Anomalous Diffusion

4.5.1 Possible Origins of Anomalies

We have obtained in Sect. 4.3.1 an expression $D = a^2/2d\tau$, giving the diffusion coefficient D as a function of the average length a and average duration τ of steps of a random walk describing the motion of diffusion. This expression and the arguments used to reach it immediately make apparent situations in which the movement does not follow a normal diffusion law:

1. The first example is a situation in which *the average duration of steps diverges*: $\tau = \infty$. This is the case whenever a trapping mechanism can keep the particle at a given site for arbitrarily long times.
2. The second example is a situation in which *the step length has an infinite variance* $a^2 = \infty$. This is the case whenever a mechanism generating long straight excursions is superimposed on the thermal diffusion.
3. A third situation is that in which *the correlation time diverges*. Whenever we can no longer claim that the steps $a(s)$ and $a(t + s)$ are statistically independent for all times t longer than a certain fixed time, t_0, in other words whenever the correlations between steps have an infinite temporal range, the exponent of the diffusion law is different. A typical example is that of self-avoiding random walks used to model the spatial conformation of linear polymers. The fact that they cannot cross themselves, entails a permanent memory of their previous route (see Chap. 6). Negative correlations of infinite range can also exist, in which case diffusion is slowed down. On the other hand, we have shown in Sect. 4.3.1 that *finite* range correlations do not affect the normal character of diffusion.
4. A final example is a situation where the random walk does not take place in Euclidian space (of d dimensions) but is confined *in a medium of fractal geometry*, of dimension $d_f < d$.

We talk of *anomalous diffusion* when the deviation from an ideal random walk is as far as to change the exponent of the diffusion law to $R(t) \sim t^{\gamma/2}$, with $\gamma \neq 1$ and $0 \leq \gamma \leq 2$ (other values of γ are incompatible with the stationary state of the statistics[40]). We talk of *superdiffusion* if $\gamma > 1$. It is understood intuitively,

---

[40]Note that this constraint $\gamma < 2$ only comes into play if the motion has finite variance and we can define a finite root mean square displacement ($R(t) < \infty$). The constraint disappears when the constitutive steps have infinite variance (Lévy flights) and the diffusion law is expressed by a convergence of the renormalised sum to the Lévy law $\mathscr{L}_\alpha$ (see p.150) with $\alpha < 1$.

and we will show in the following paragraphs, that such a motion is found when a random walk is performed with large deterministic steps (case 2 above) or has positive correlations of infinite range, associated for example with an excluded volume constraint (case 3). The motion is said to be *subdiffusive* if $\gamma < 1$. We expect to observe such a motion in the case where diffusion is slowed down by trapping phenomena (case 1), negative correlations (case 3) or a fractal geometry of the supporting medium (case 4). Typical models of anomalous diffusion replace the Brownian walk and the Wiener process associated with normal diffusion. These are, on the one hand, Lévy flights and, on the other hand, fractal Brownian motions, and will be presented in the two following subsections.

4.5.2   Lévy Flights

Starting from a Brownian random walk, a model showing an anomalous diffusion law is obtained by keeping the statistical independence and isotropy of successive steps $(a_i)_i$, but taking a distribution $P(a)$ for their length which has infinite variance, without a characteristic scale, which we call a *heavy tailed* (broad) distribution. The typical example is that for which this distribution asymptotically takes the form of a power law $P(a) \sim a^{-(1+\alpha)}$ for large a, with $0 < \alpha \le 2$ (which has an infinite variance). The random walk obtained with such a distribution for the constituent steps is called a *Lévy flight*.[41] The generalised central limit theorem, which we will state precisely below, means that Lévy flights asymptotically follow an anomalous diffusion law $R(t) \sim t^{1/\alpha}$. In addition, the asymptotic form of their density probability is written $P(r,t) \sim t^{-1/\alpha} \mathscr{L}_\alpha (rt^{-1/\alpha})$ where $\mathscr{L}_\alpha$ is a particular distribution called a *Lévy distribution* (Fig. 4.9).

Central limit theorem

Let $(a_i)_i$ be a set of random, independent, identically distributed, real variables. We then construct $X_N = \sum_{i=1}^{N} a_i$. If the variance $\mathrm{Var}(a_i) = \sigma^2$ is finite, the average $m = \langle a \rangle$ is also finite and the central limit theorem states that $[X_N - Nm]/\sqrt{\sigma^2 N}$ tends to a centred Gaussian distribution with variance 1 (called the normal distribution). This theorem is called the *central limit theorem* because it rigorously states a principle which is omnipresent in nature: phenomena which are essentially random can show a regular deterministic collective behaviour. We refer to this as a *statistical law*. Critical

---

[41]If we take for the distribution of steps a law $P(a) \sim e^{-a/\sigma}$, with a finite characteristic scale σ, we obtain a *Rayleigh flight* which follows a normal diffusion law. Going from a Rayleigh flight to a Lévy flight, we find again the passage from a power law to an exponential law, typically associated with the emergence of critical behaviour.

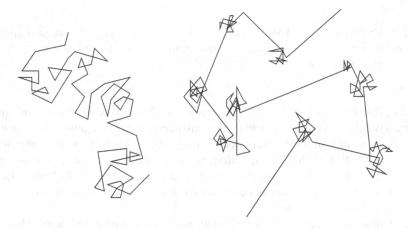

Fig. 4.9 Diagram illustrating the difference between an ideal random walk (*left*) and a Lévy flight (*right*), in which the probability to make long straight excursions is not negligible. This difference is reflected quantitatively in the exponent $\gamma > 1$ of the diffusion law $R(t) \sim t^{\gamma/2}$ associated with a Lévy flight compared to the exponent $\gamma = 1$ of the normal diffusion law observed in, for example, Brownian motion

phenomena are precisely those in which this principle fails, either because of a divergence of the range of correlations, or due to broad distributions (i.e. with infinite variance) of constituent events. We will consider the second case here.

Suppose that the asymptotic behaviour of the distribution $P(a)$ of random variables a_i is a power law: $P(a) \sim |a|^{-(1+\alpha)}$ for $|a| \to \infty$ with $\alpha > 0$ (so that P remains normalisable). The moment $\langle a^q \rangle$ diverges when $q \geq \alpha$. We distinguish the following three cases:

- If $\alpha > 2$, we recover the normal behaviour: the average m and variance σ^2 are finite, $X_N - Nm \sim \sqrt{\sigma^2 N}$ and the usual theorem applies.
- If $1 < \alpha < 2$, the variance diverges but the average m is still finite, and $X_N - Nm \sim N^{1/\alpha}$. We can state a generalised limit theorem: $(X_N - Nm)N^{-1/\alpha}$ converging to the Lévy distribution $\mathscr{L}_\alpha$.
- If $0 < \alpha < 1$, m is not defined (or diverges) and the dominant behaviour is $X_N \sim N^{1/\alpha}$ (note that $1/\alpha > 1$). The generalisation of the limit theorem is then expressed as: $N^{-1/\alpha} X_N$ converging to the Lévy distribution $\mathscr{L}_\alpha$.

If we give back the steps their duration τ, we must replace N by t, m by m/τ and σ^2 by σ^2/τ^2 in the formulae above.

The Lévy distribution $\mathscr{L}_\alpha$, appearing here as an asymptotic distribution, has itself a power law behaviour: $\mathscr{L}_\alpha(x) \sim |x|^{-(1+\alpha)}$ for $|x| \to \infty$. It can be shown[42] that the largest value $x_{max}(t)$ taken by the set $(a_i)_i$ between 0 and t behaves as $t^{1/\alpha}$, that is to say as their sum: *the rare events dominate the collective behaviour.*

Lévy distributions $(\mathcal{L}_\alpha)_\alpha$ replace Gaussian distributions when the central limit theorem is extended to cases in which the constituent random variables have a power law behaviour and an infinite variance [42]. Correspondingly, they are stable with respect to addition of the corresponding random variables (as for all Gaussians): if X_1 and X_2 follow Lévy distributions $\mathcal{L}_\alpha$, then $X_1 + X_2$ also follows a Lévy distribution $\mathcal{L}_\alpha$. This stability is necessary for it to take its place as an asymptotic distribution in a limit theorem.

Lévy distributions are explicitly defined, not by their probability density but, by their characteristic function ϕ_α (Fourier transform of $\mathcal{L}_\alpha$). This has a particularly simple form, which by limiting ourselves to the case of centred symmetric distributions is given by:

$$\phi_\alpha(u) \equiv \langle e^{iuX_\alpha} \rangle = e^{-A|u|^\alpha}, \qquad (4.46)$$

where X_α is the underlying random variable. It can be seen that the asymptotic form of the corresponding probability density $\mathcal{L}_\alpha$ is indeed a power law, proportional to $|x|^{-(1+\alpha)}$ for large values of x (but $\mathcal{L}_\alpha$ is continuous at $x = 0$). From this expression we easily find the "stability" already mentioned of Lévy distributions, also called *stable distributions* for this reason. In mathematical terms this amounts to stating the stability by a convolution of the parametric forms of the associated probability distributions, equivalent to the stability by a simple product of the characteristic functions. The density is only explicit in the case where $\alpha = 1$, thereby giving the *Cauchy distribution* (it is also known as the *Lorentz distribution*), parameterised by its width λ:

$$\mathcal{L}_1(x) = \frac{\lambda}{\pi^2\lambda^2 + x^2}. \qquad (4.47)$$

An important property of Lévy flights worth highlighting is their self-similarity. It is in fact fairly predictable given the status of the limiting distributions of Lévy flights. These are involved both to describe the distribution of constitutive steps (as typically broad distributions, behaving as a power law) and to describe the long time behaviour of the process obtained by adding such steps in a statistically independent manner, just like the Gaussian distribution describes the asymptotic behaviour of a Brownian walk in the case of normal diffusion. However the self-similarity of Lévy flights is different from the self-similarity of Brownian walks. For the case of Lévy flights, by denoting $X_\alpha(t)$ the process at time t, we obtain (maintaining the independence of successive steps):

---

[42] $x_{max}(t)$ is such that $t \int_{x_{max}(t)}^{\infty} \mathcal{L}_\alpha(x)dx = 1$: we have at most one value greater than $x_{max}(t)$ between 0 and t. By replacing $\mathcal{L}_\alpha(x)$ by its power law behaviour, we obtain $x_{max}(t) \sim t^{1/\alpha}$ [8].

$$\phi(t, u) \equiv \langle e^{iu.(X_\alpha(t)-X_\alpha(0))} \rangle = \phi(1, u)^t \sim e^{-At|u|^\alpha}. \qquad (4.48)$$

The scale invariance is therefore written:

$$\phi(t, u) = \phi(kt, k^{-1/\alpha}u) \qquad (4.49)$$

that is to say that the process is invariant with respect to the combined transformation $(t \to kt, u \to k^{-1/\alpha}u)$ or, returning to real space, with respect to the transformation $(t \to kt, r \to k^{1/\alpha}r)$. In this way we find that the diffusion law of a Lévy flight is $R(t) \sim t^{1/\alpha}$.

The model of Lévy flights has the advantage of entering into the well explored mathematical framework of the generalised central limit theorem, but it has the drawback of attributing the same duration to all jumps, whatever their length. This, sometimes not very realistic, ingredient can be rectified by introducing a finite displacement velocity v for the particle such that the duration of a step of length a will be a/v. We keep a broad distribution, in a power law, for the step length. This new model is called a *Lévy walk*. The diffusion law will be different from that of a Lévy flight of the same distribution $P(a)$, however it remains anomalous. For example we show that if $P(a) \sim a^{-(1+\alpha)}$, then $R^2(t) \sim t^{3-\alpha}$, that is a value $\gamma = 3 - \alpha > 1$. More complex models, better taking into account particular experimental situations, are obtained by taking a velocity $v(a)$ which is dependent on the length of the step [33, 59].

Chaotic transport and anomalous diffusion

One interesting question is to study the movement of (non charged) particles transported by a fluid. The motion behaves a priori as convection and not diffusion. However when the fluid is driven by a non trivial motion, for example if it obeys a Hamiltonian dynamics with a phase space containing chaotic regions (Sect 9.2), the only possible description is made in terms of statistical properties of the motion. The concepts are therefore those used to characterise diffusive motion resulting from a random walk forgetting that the trajectory results from a deterministic dynamical system.

In this context, anomalous diffusion is often observed, typically Lévy flights or walks. The trajectory remains trapped in a region where the dynamics is slow or localised, then makes a long "straight" excursion and then the motion is localised again. The exponent of the anomalous diffusion law will depend on the distribution of trapping times and flight times [33, 59, 61]. This example occurs in various concrete problems such as the dispersion of pollutants in the sea or the atmosphere, which we want to control, or at least predict.

4.5.3   Fractal Brownian Motions

We have already mentioned that the central limit theorem, and the associated normal diffusion law, no longer apply in the presence of long range correlations. One necessary condition to see the appearance of anomalous diffusion is therefore that $\sum_t |C(t)|$ diverges. A typical model of an associated anomalous random walk is provided by the family of *fractal Brownian motions*. They are defined by the characteristic function of their growth:

$$\phi(t, u) \equiv \langle e^{iu.(X_{t+s}-X_s)} \rangle = e^{-Au^2|t|^{2H}}, \qquad (4.50)$$

where $H \leq 1$ is called the *Hurst exponent* [32]. Like the Wiener process that they generalise, fractal Brownian motions show a self-similarity property, in which their exponents (different from those of Brownian motion) reflect their anomalous behaviour:

$$\phi(kt, k^{-H}u) = \phi(t, u). \qquad (4.51)$$

By using that $R^2(t) = -\partial^2\phi/\partial u^2(t, u = 0)$, we immediately deduce from this scale invariance the diffusion law obeyed by these processes:

$$R(t) \sim t^H \qquad \text{or} \quad \gamma = 2H. \qquad (4.52)$$

The autocorrelation function of the growth:

$$\langle [X_{t+s} - X_s].[X_s - X_0] \rangle = A[\,|t+s|^{2H} - |s|^{2H} - |t|^{2H}\,] \qquad (4.53)$$

shows that the growth is positively correlated if $H > 1/2$ (persistent hyperdiffusive motion) and negatively correlated if $H < 1/2$ (subdiffusive motion). This power law form of correlations reflects the divergence of the correlation time, typical of dynamic critical phenomena.

4.5.4   Examples

Examples are encountered in many domains. We find them throughout this book: subdiffusion in a cluster at critical percolation (Chap. 5); superdiffusion of self avoiding random walks, the trajectories of which model spatial conformations of (long) linear polymers (Chap. 6); the relaxation of a spin glass can be described as a random walk in a landscape of complex topography, the energy landscape of the spin glass; electrical transport properties in amorphous disordered or quasiperiodic materials can also be described in terms of anomalous diffusion [7]. For example, the subdiffusive motion of electrons in quasi-crystals[43] leading to these materials having insulating properties even though they are metal alloys.

Another example is that of Lévy flights observed in giant micelles [52]. Micelles are continually rearranging: they break and after diffusing in the solvent the fragments recombine differently. The phenomenon can be studied by labelling the surfactant molecules making up the micelles with fluorescent markers allowing individual molecules to be followed. By observing in this way the motion of a surfactant molecule of the system, alternating phases are seen of the molecule diffusing within a micelle and jumping from one micelle to another leading to a superdiffusion.

It has been observed that albatross migration changed, from Brownian motion if prey was abundant, to hyperdiffusive Lévy flights when fish was scarce. The same phenomenon is observed in certain bees following the abundance of pollen sources [66]. It is also observed that the motion of an individual within a herd shows a hyperdiffusive behaviour as the herd begins collective motion [29].

As we have detailed in Sect. 4.5.2, transport by a fluid driven by chaotic motion and superdiffusion phenomena observed in turbulent motion are well described by Lévy walks.

Fractal Brownian motion has been used to model, among other things, flooding of the river Nile and stock market prices [47].

A final example, which we will describe in detail below, is that of diffusion of a liquid in a porous rock that is initially dry. In contrast to the examples we have just cited, this is not treated just in a stochastic "microscopic" framework, but at a macroscopic level, by a modified diffusion equation to take into account a temporal asymmetry fundamental to the motion, underlying the anomalous nature of the observed diffusion.

Anomalous diffusion in a porous medium

The final example introduced in Sect. 4.2.3, is that of diffusion of a liquid in a porous rock that is initially dry. The asymmetry between filling and drying of pores, the latter being only partial, invalidates the usual diffusion equation. The phenomenon can be taken into account by taking for D a discontinuous function of the rate of change of local and instantaneous concentration $\partial_t c$:
$\partial_t c = D(\partial_t c) \, \Delta c$. D just needs to change value with the sign of $\partial_t c$, taking a smaller value during the filling phase than during the drying phase, D_0 and $D_0(1 + \epsilon)$ respectively. Due to the fact that a film of liquid remains trapped inside the pores after the liquid has passed though (see Fig. 4.7), the total mass of moving liquid decreases. This non conservation of mass gives the system a "memory" and, contrary to the case of normal diffusion, we cannot eliminate reference to the initial condition (A_0, l) where l is the extension of the domain

---

[43]Quasi-crystals are assemblies of metallic atoms that have a symmetry of order 5, for example Al, Fe and Cu, and are thus incompatible with crystalline order and lead to quasiperiodic structures [65].

in which the initial mass A_0 is localised. Whilst it is possible to let l tend to zero to determine the universal asymptotic properties of normal diffusion (see Sect. 4.2.2), here A_0 and l must be changed simultaneously. The spreading out of liquid from a new fictitious initial extension $l' < l$, to the original extension l is accompanied by a loss in mass of moving material and must therefore be associated with a new value of the mass A_0' with extension l' to continue to describe the same phenomenon. Letting l tend to zero must therefore be accompanied by a change in the parameter A_0. The renormalisation methods presented in Chap. 3 were specially designed to solve this type of difficulty. Renormalisation determines the "right way" to let l tend to zero whilst jointly transforming ("renormalising") the other system parameters in order to leave the physics at work in the observed phenomenon unchanged. Using this, the non trivial asymptotic regime emerges, which is not at first sight evident in the initial diffusion equation. In this way a condition $A_0 l^{2\alpha(\epsilon)} = const.$ is explicitly determined, completing the procedure of taking the limit $l \to 0$. A diffusion law with exponent $\gamma = 1 - \alpha(\epsilon)$ then follows [25]. Here the anomalous diffusion reflects the aging of the system. In a way, the system "measures the time passed" through the evolution of the mass. It is this extra temporal dependence that is responsible for the anomalous diffusion. The equation is of the form (in one dimension):

$$c(x,t) \sim \frac{1}{(2Dt)^{1/2+\alpha(\epsilon)}}\, g\left(\frac{x}{\sqrt{2Dt}}, \epsilon \right) \qquad (4.54)$$

implying the same scaling variable $z = X/\sqrt{2Dt}$ as normal diffusion. It is the anomalous prefactor $1/(2Dt)^{1/2+\alpha(\epsilon)}$ that leads to an exponent $\gamma = 1 - \alpha(\epsilon) < 1$ in the diffusion law. As in critical phenomena, dimensional analysis alone does not determine the anomalous exponent (unlike the case of normal diffusion); and more complete analysis by renormalisation is required.[44]

To conclude let us note that here anomalous diffusion appears at the level of deterministic description by a diffusion equation without it being necessary to introduce stochastic fluctuations explicitly.

---

[44] We can clarify this point by distinguishing [3]:

– a normal similarity $U = f(X_1, X_2)$ where X_1 and X_2 are dimensionless variables and U is the dimensionless function that we wish to determine; here the function f is a *regular* function of its two arguments;

– an anomalous similarity $U = X_2^{\alpha}\, g(X_1/X_2^{\alpha})$ where g is a regular function; an anomalous exponent α appears here that could not emerge from a simple dimensional analysis of the starting equation. The example presented here illustrates this second case.

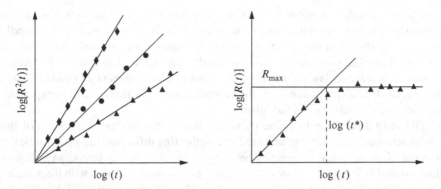

Fig. 4.10 Typical appearance of a graph representing $\log R^2(t)$ as a function of $\log t$, $R(t)$ being the root mean square displacement. *Left*: normal diffusion (*circles*) and anomalous diffusion, subdiffusive (*triangles*) or hyperdiffusive (*ovals*); symbols represent experimental points, *straight lines* a linear interpolation found by the method of least squares. The slope of the lines gives an estimate of the exponent γ of the diffusion law $R^2(t) \sim t^\gamma$. *Right*: a situation in which at long times ($t > t^*$) the root mean square displacement shows a saturation at a value R_{max}, corresponding to a confinement of the particle in a region of linear extension proportional to R_{max}; the system achieves a homogeneous diffusive equilibrium state (in this way the link between the linear extension L of the confinement region and the saturation value R_{max} can be calculated). The motion can remain diffusive at short times, as seen on the graph by the linear part for $t < t^*$, of slope equal to the exponent γ of the diffusion law

We will end this subsection with a warning for analysing experimental data. Starting from a recording of particle trajectories (in the general sense: molecules, cells or grains) performing diffusive motion, it is always possible to calculate a root mean square displacement $R(t)$ and to trace $\log R^2(t)$ as a function of $\log t$. However it is more subtle to interpret the graph obtained. For example, diffusion between obstacles or traps typically leads to a subdiffusive motion ($\gamma < 1$) whereas diffusion in confined media leads to a saturation of the root mean square displacement (Fig. 4.10). However it is difficult to distinguish, only by looking at the graph, a diffusion between very dense obstacles inducing trapping and confined diffusion inside a volume enclosed by walls (for example membranes in a biological context). This is all the more true when more complex mechanisms come into play, for example diffusion between dense but fluctuating obstacles and diffusion in a confined space but with mobile walls. It is therefore essential to use a model of diffusion to analyse the observations, interpret them in terms of microscopic mechanisms and to extract quantitative information.

4.5.5 Spectral Dimension

Random walks are an illustration of the connection existing between static geometric properties of a fractal structure and the underlying dynamic properties.

A more complex question, but one of particular interest is that of diffusion in a pre-existing fractal structure. The properties of diffusive motion will be profoundly changed by the particular geometry of the accessible regions and paths. Nevertheless, since this geometry is self-similar, the asymptotic behaviour will remain described by a power law $R(t) \sim t^{\gamma/2}$. In addition to the exponent γ of this diffusion law and the fractal dimension d_f of the substrate, we will introduce a third exponent, the *spectral dimension* d_s of the substrate.

The most direct way to define d_s is to consider the modes of vibration of the fractal structure, here serving as a medium supporting diffusion. Let $\rho(\omega)d\omega$ be the number of modes with frequencies between ω and $d\omega$. For an Euclidian lattice of dimension d (for example a crystalline solid), the number of modes with frequencies less than ω varies as ω^d where $\rho(\omega) \sim \omega^{d-1}$. This spectral density will be strongly affected by the fractal nature of the medium under consideration. We therefore define d_s by the relation $\rho(\omega) \sim \omega^{d_s-1}$. For example, we can calculate explicitly the spectral dimension of a Sierpinski gasket: $d_s = 2\log 3/\log 5 = 1.364....$ In general it can be shown that $d_s \leq d_f \leq d$ [Gouyet 1992].

Let us now consider a random walk on a fractal lattice of fractal dimension d_f and spectral dimension d_s. Since $R(t) \sim t^{\gamma/2}$ is the only characteristic scale in the problem, the probability of a particle being present has the scaling behaviour:

$$P(r,t) \sim R(t)^{-d_f} f\left(\frac{r}{R(t)}\right), \tag{4.55}$$

where f is a some scaling function, at this stage unknown (besides, it does not concern us). The factor $R(t)^{-d_f}$ comes from ensuring P is normalised, the domain of integration being here reduced to the fractal medium. In addition, it can be shown [7] that the probability of first return to the origin (first passage) behaves as $t^{-d_s/2}$. Comparing this result with the behaviour of $P(r,0)$, considering $R(t) \sim t^{\gamma/2}$, we find:

$$\gamma = \frac{d_s}{d_f}. \tag{4.56}$$

The fractal dimension of a typical trajectory remains equal to $2/\gamma$ (so equal to $2d_f/d_s$). We always have $d_s \leq d_f \leq d$, hence $\gamma \leq 1$, even with $\gamma < 1$ if the structure is actually fractal (in which case $d_s < d_f < d$). Diffusion on a fractal structure is therefore subdiffusive and the dimension $2/\gamma$ of trajectories is always greater then 2. These properties quantitatively reflect the fact that diffusion is hampered and slowed down by the restraining geometry where it takes place. Observations of diffusion of a labelled particle (a "tracer") on a structure will therefore quantitatively reveal these properties.

The spectral dimension d_s thereby describes the geometry of the supporting medium *from a dynamic point of view*. A fractal structure with many dead ends and a well connected fractal structure (obtained for example by adding to the first structure some connections, of zero total mass, to transform dead ends into passageways) can have the same fractal dimension. However they would have very different spectral dimensions, far smaller in the first case. This point is illustrated, for example, in the solution–gel transition of polymers: d_s considerably increases the moment the chains are crosslinked (leading to the formation of a gel) whereas d_f does not vary much.

4.6 Driven Diffusion and Processes Far from Equilibrium

Diffusion processes have served as a test bed for statistical physics at and near to equilibrium (linear response theorem). We will see that they also play a central role in certain recent developments in statistical physics far from equilibrium. A model example is that of *"forced" or "driven" diffusion*, in which the system, composed of impenetrable particles in thermal motion, is maintained far from equilibrium by imposed boundary conditions. Studying this example will elucidate the theoretical analysis and properties of all systems that are placed between two different reservoirs imposing different values on their associated intensive variables (for example, temperature, density or electrical potential) and by thus maintaining a flux (of heat, particles or charges respectively) through the system.[45]

4.6.1 A Typical One Dimensional Model

The most basic one dimensional model we will present is that of *displacement with exclusion*, in which particles diffuse under the constraint that two particles cannot occupy the same site, which corresponds to a short range repulsive interaction between particles. In one dimension, this interaction creates strong geometrical constraints on the relative positions of particles. As a result, their movements are correlated with each other and we therefore expect radically different results from those observed in the case of non-interacting particles. This model has two variants; a "symmetric simple exclusion process" (SSEP) in which the diffusion is symmetric or an "asymmetric simple exclusion process" (ASEP) in which the diffusion is biased in a given direction [13]. The particular case of a totally asymmetric process, "totally asymmetric exclusion process" (TASEP), in which all the jumps are in a single direction, is analytically tractable and so shows in an indisputable way the fundamental differences existing between systems at equilibrium and systems maintained far from equilibrium, through which a flux flows. Let us immediately give the three essential differences: (i) boundary conditions will control

---

[45]Let us underline straight away that non trivial nonequilibrium properties only appear if a complex internal dynamics, coupling microscopic degrees of freedom of the system, add up to destabilise the boundary conditions. This coupling is ensured here by the exclusion constraint; it can also follow from chaotic microscopic dynamics [10].

the stationary state of the system and the phase transitions it can show; (ii) the free energy is only defined globally, without being us able to identify a local functional; (iii) fluctuations of macroscopic quantities around their mean values are no longer Gaussian.

4.6.2 Density Profile and Phase Diagram

The system consists of a one dimensional chain of L sites on which particles move.[46] Its discrete nature makes this system particularly well adapted to numerical studies. The particles make their attempted moves independently of each other and without memory, over time intervals distributed following a Poisson law: over a time interval dt, each particle attempts a jump with probability dt/τ. Following the exclusion principle, a displacement is only accepted if the new site is free. At the two extremities the system is connected to two reservoirs of particles of densities ρ_A and ρ_B fixing the boundary conditions: $\rho(0) = \rho_A$ and $\rho(L+1) = \rho_B$ (Fig. 4.11). The exclusion principle imposes $\rho \leq 1$, so ρ_A and ρ_B are varied in the interval $[0, 1]$. The left hand boundary condition (reservoir of particles with density ρ_A) attempts to inject particles at $x = 0$ with rate $\rho_A dt$, injection being actually realised when the entry site $x = 1$ is empty. In the same way the right hand boundary condition (reservoir of particles with density ρ_B) extracts particles on the exit site ($x = L$) with a rate $(1 - \rho_B)dt$. The number of particles present in the system is not fixed a priori but is one of the observables to be determined, along with the density $\rho(x)$ and the current $j(x)$. The control parameters are the densities ρ_A and ρ_B of the reservoirs controlling the flux in and out of the system respectively. The resulting dynamics is written by a master equation, that is by the evolution for the probability

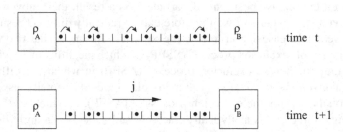

Fig. 4.11 Totally asymmetric process with exclusion: particles move by elementary jumps only towards the right and only if the new site is unoccupied. The system is maintained nonequilibrium by the densities ρ_A and ρ_B imposed at the extremities being different from each other ($\rho_A \neq \rho_B$), inducing a uniform current of particles j in the stationary state

---

[46]These particles can be fictitious and correspond to local excitations (for example quanta of vibration called phonons).

distribution of the presence of N particles in terms of the transition probabilities from one configuration of particles on the chain to another.

Symmetric processes: The simplest situation is that in which the diffusion of particles is symmetric: each particle has the same probability $p = 1/2$ of attempting a jump to its neighbouring site to the right or to the left, the jump being accepted if the envisaged new site is empty. It can then be shown that a stationary state is established in which the density profile is linear:

$$\rho(x) = \rho_A + (\rho_B - \rho_A)x/L. \tag{4.57}$$

On balance, the current j of particles crossing through the system is zero, The difference from equilibrium systems concerns the fluctuations which are no longer Gaussian [16].

Asymmetric processes: A richer scenario is that in which one of the directions is preferred: the particle attempts a jump to the right with probability $p \neq 1/2$ and a jump to the left with probability $1 - p$. This is the case in which a constant uniform field is applied to the system. Particle movement is then ruled by two effects:

- On the one hand a stochastic diffusive motion *statistically* leads, by an entropic effect, to a current of particles towards regions of low concentration, here towards the right if $\rho_B < \rho_A$.
- On the other hand a deterministic movement in which each particle *individually* feels a force of amplitude proportional to $2p - 1$, pushing it in one of the two directions. If $\rho_B < \rho_A$, this force cooperates with the diffusive motion if $p > 1/2$ or opposes it if $p < 1/2$.

A concrete example is that of the diffusion of ions in a transmembrane ion channel with a potential difference δV between its extremities across the membrane (electrodiffusion[47]) [26].

Totally asymmetric processes: The extreme case of $p = 1$ corresponds to the case in which a field, or the nature of the problem itself, forces the particles to jump from one site only to its neighbouring site to the right, provided it is unoccupied. This is a simple model of road traffic; it is also used to account for the movement of ribosomes (molecular machines that assemble amino acids into proteins) along a messenger RNA.[48]

---

[47]The macroscopic description of this problem involves the sum of the current $-D\boldsymbol{\nabla} n$ given by Fick's law (D being the diffusion coefficient of ions and n the ion concentration) and the current $-zen(D/k_BT)\boldsymbol{\nabla} V$ given by Ohm's law (z being the valance of ions, V the electric potential and the Einstein relation has been used according to which the mobility of an ion is equal to D/k_BT). These two terms are multiplied by ze and by Avogadro's number to obtain a charged current density. The interest of the microscopic model presented in this section is to go beyond this macroscopic mean field description. This thus allows us to evaluate the role of fluctuations and correlations, as well as the fine structure of the density profile.

[48]It was in this context that the model was introduced [44].

Fig. 4.12 Phase diagram of the totally asymmetric process (see text)

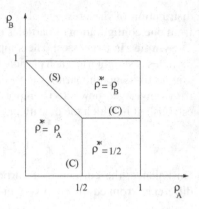

For this totally asymmetric model, we know how to calculate the configurational probability distribution in the stationary state [14]. The density in the middle of the system converges to a stationary value ρ^*, as a function of ρ_A and ρ_B, enabling a phase diagram to be drawn. It can also be shown that the spatially uniform current in the stationary state is a function of the density in the centre: $j^* = j(\rho^*) = \rho^*(1 - \rho^*)$. We obtain the phase diagram shown in Fig. 4.12 [34]. Three phases are observed: a low density phase $\rho^* = \rho_A$ if $\rho_A < 1/2$ and $\rho_A < 1 - \rho_B$, a high density phase $\rho^* = \rho_B$ if $\rho_B > 1/2$ and $\rho_B > 1 - \rho_A$ and a phase $\rho^* = 1/2$. The line S is a first order transition line since ρ^* has a discontinuity $\rho_B - \rho_A = 1 - 2\rho_A > 0$. Suppose we start at time $t = 0$ with a step function density profile: $\rho(x) = \rho_A$ if $x < x_0$ and $\rho(x) = \rho_B$ if $x > x_0$; the current in the left hand side is $\rho_A(1 - \rho_A)$ whereas it is $\rho_B(1 - \rho_B)$ in the right hand side. This singularity gives rise to a *shock wave*. In other words, to reconcile the two currents, the position of the step will move at velocity $v = 1 - \rho_A - \rho_B$ (as long as it remains far from the boundaries). On the line S, we have $v = 0$ and the step is the nonequilibrium analogue of the coexistence of two phases $\rho^* = \rho_A$ and $\rho^* = \rho_B$. On the other hand, ρ^* remains continuous on crossing the lines C. In the zone enclosed by these lines C, we have $\rho^* = 1/2$ and the current is maximal: $j^* = 1/4$. This can be understood intuitively: on average one site in two is occupied; each particle (in this average, of course very approximate, picture) therefore always sees an empty site in front of it and follows a free, unconstrained, motion, as if it never saw the other particles and was alone on the chain. Throughout this zone we see a *spontaneous critical behaviour*: the system must organise itself to reconcile the boundary condition $\rho = \rho_A$ and the value at the centre $\rho^* = 1/2$. To do so, it develops *boundary layers*, which are scale invariant in the sense that their density profiles follow power laws (in the stationary state): to the left, $\rho(x) - \rho^* \sim x^{-1/2}$ and to the right, $\rho(x) - \rho^* \sim (L - x)^{-1/2}$. Spatial correlations in these boundary layers also have a power law behaviour [30].

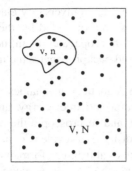
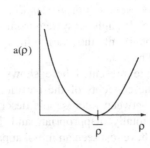

Fig. 4.13 Analysis of large deviations. The large deviation function $a(\rho)$ describes the probability of local density fluctuations in a sample of volume v: $\mathrm{Prob}(v/n = \rho) = \exp[-va(\rho)]$. This probability is maximum when ρ is equal to the average density $\bar{\rho} = V/N$

4.6.3   Large Deviations

A very interesting quantity for the theoretical description of nonequilibrium systems is the *large deviation function* (Fig. 4.13). Let us consider a volume V containing N particles so the average density is $\bar{\rho} = N/V$. If we now imagine a sample element of volume $v \ll V$, it will contain a random number n of particles. The large deviation function $a(\rho)$ describes the local density fluctuations $\rho = n/v$ about $\bar{\rho}$:

$$\mathrm{Prob}\left[\frac{n}{v} = \rho\right] = e^{-va(\rho)}. \tag{4.58}$$

The interesting point is that in the case of equilibrium this function, which is minimum at $\bar{\rho}$, coincides with the free energy of the system; it is therefore involved in the fluctuation-dissipation theorem, which is written:

$$\langle(n - \langle n\rangle)^2\rangle = \frac{v}{a''(\bar{\rho})} \qquad \text{(at equilibrium).} \tag{4.59}$$

So the large deviation function is a potential candidate to extend the concept of free energy to nonequilibrium systems [15, 16].

4.6.4   Crucial Difference from Equilibrium Systems

The concluding message to remember is that nonequilibrium stationary states, i.e. driven flows, differ profoundly from equilibrium states. We have just seen that they show phase transitions in one dimension. Peierls' argument no longer applies because the interfaces between phases are susceptible to being moved. In addition, fluctuations are generally not Gaussian, the fluctuation-dissipation theorem (Sect. 4.4.2) does not apply a priori and new universality classes are to be determined. Nonequilibrium stationary states are controlled by the boundary conditions; flux through the system from left to right propagates the influence of the left hand boundary into the heart of the system and backpropagates the influence of the right hand boundary.

The example presented here shows that a statistical mechanics formalism, founded on the concepts of the partition function and free energy, and forming a connection between microscopic descriptions and macroscopic thermodynamic properties, is totally inappropriate and does not have an evident nonequilibrium analogue. We have just seen an initial approach at establishing a formalism valid far from equilibrium, which relies on the large deviation function. A second approach will be presented in Sect. 9.3, in which the relevant macroscopic quantities are defined and relationships between them are established for particular dynamical systems (hyperbolic dynamical systems, whose invariant measures belong to a particular class called SRB). The relationships obtained are then assumed to be more general than the case (chaotic hypothesis) in which the were rigorously established [17]. The role of models, such as this one, that are analytically solvable or easy to simulate is therefore very important because by using them we can test the concepts introduced and the conjectured relationships.

4.7   Conclusion

Diffusion is a directly observable manifestation of thermal motion. Thermal motion is an inevitable ingredient of all natural phenomena. It introduces a random component (we speak of *thermal noise* or *intrinsic stochasticity*) which averages and only appears at larger scales implicitly in the values of effective parameters, *apart from* in the neighbourhood of critical points (encountered in Chap. 1), dynamic singularities (bifurcations) which we will describe in Chap. 9, or when the system has certain geometries (fractal structures).

The scaling law $R(t) \sim t^{\gamma/2}$ describing the diffusive behaviour only emerges at long times; the limit $t \to \infty$ is the exact analogue of the thermodynamic limit $N \to \infty$ in which singularities and scaling laws associated with phase transitions appear. An analogy with the concepts presented in Chap. 1 may follow: the normal diffusion law can be thought of as a "mean field description", in which one single effective parameter, the diffusion coefficient D, is enough to describe the phenomenon. The exponent takes its mean field value $\gamma = 1$ and the distribution of fluctuations is Gaussian. The mean field approach fails if the underlying microscopic events follow a broad distribution (Lévy flights) or if they have long range temporal correlations. We then observe a "critical exponent" $\gamma \neq 1$. Anomalous diffusion can therefore be seen as the temporal equivalent of the static critical phenomena discussed in Chap. 1. It is critical in the sense that it reflects a "statistical catastrophe", that is to say a dramatic effect of fluctuations, either because they are amplified by correlations with diverging range, or because the local fluctuations are themselves anomalous, distributed with a broad distribution (of infinite variance).

References

1. G.E. Archie, The electrical resistivity log as an aid in determining some reservoir characteristics. Trans. Am. Inst. Min. Metall. Petrol. Eng. Inc. **146**, 54 (1942)
2. A. Barberousse, *La mécanique statistique: De Clausius à Gibbs* (Belin, Paris, 2002)
3. G.I. Barenblatt, *Similarity, self-similarity and intermediate asymptotics*, Consultants Bureau, New York, 1979; revised edition Cambridge University Press, 1996
4. G.I. Barenblatt, *Dimensional Analysis* (Gordon and Breach, New York, 1987)
5. A. Bensoussan, J.L. Lions, G. Papanicolaou, *Asymptotic Analysis for Periodic Structures* (North Holland, Amsterdam, 1978)
6. L. Boltzmann, *Lectures on Gas Theory* (New edition Dover Publications Inc., 1995)
7. J.P. Bouchaud, A. Georges, Anomalous diffusion in disordered media: statistical mechanics, models and physical applications. Phys. Rep. **195**, 127 (1990)
8. J.P. Bouchaud, M. Potters, *Theory of financial risks* (Cambridge University Press, 2003)
9. J. Bricmont, A. Kupiainen, Renormalization group for diffusion in a random medium. Phys. Rev. Lett. **66**, 1689 (1991)
10. P. Castiglione, M. Falcioni, A. Lesne, A. Vulpiani, *Chaos and coarse-graining in statistical mechanics* (Cambridge University Press, Cambridge, 2008)
11. C. Cercignani, *Mathematical Methods in Kinetic Theory*, 2nd edn. (Plenum Press, New York, 1990)
12. S. Chandrasekhar, Stochastic problems in physics and astronomy. Rev. Mod. Phys. **15**, 1 (1943)
13. Derrida B., An exactly soluble non-equilibrium system: the asymmetric simple exclusion process. *Phys. Rep.* **301**, 65 (1998)
14. B. Derrida, M.R. Evans, V. Hakim, V. Pasquier, Exact solution of a 1D asymmetric exclusion model using a matrix formulation. J. Phys. A. **26**, 1493 (1993)
15. B. Derrida, J.L. Lebowitz, Exact large deviation function in the asymmetric exclusion process. *Phys. Rev. Lett.* **80**, 209 (1998)
16. B. Derrida, J.L. Lebowitz, E.R. Speer, Free energy functional for nonequilibrium systems: an exactly solvable case. Phys. Rev. Lett. **87**, 150601 (2001)
17. J.R. Dorfman, *An Introduction to Chaos in Non Equilibrium Statistical Mechanics* (Cambridge University Press, 1999)
18. A. Einstein, On the movement of small particles suspended in a stationary liquid demanded by the molecular-kinetic theory of heat. *Ann. Physik.* **17** (1905)
19. A. Einstein, *Investigations on the Theory of Brownian Motion* (Dover, London, 1926, New edition 1956)
20. K. Falconer, *Fractal Geometry* (Wiley, New York, 1990)
21. C.W. Gardiner, *Handbook of Stochastic Methods*, Collection Springer Series in Synergetics, vol. 13 (Springer, Berlin, 1985)
22. P. Gaspard, *Chaos, Scattering and Statistical Mechanics* (Cambridge University Press, 1998)
23. P. Gaspard, Microscopic chaos and chemical reactions. *Physica A* **263**, 315 (1999)
24. P. Gaspard, M.E. Briggs, M.K. Francis, J.V. Sengers, R.W. Gammon, J.R. Dorfman, R.V. Calabrese, Experimental evidence for microscopic chaos. Nature **394**, 865 (1998)
25. N. Goldenfeld, *Lectures on phase transitions and the renormalization-group* (Addison-Wesley, Reading Mass., 1992)
26. D.E. Goldman, Potential, impedance and rectification in membranes. J. Gen. Physiol. **27**, 37 (1943)
27. J.F. Gouyet, *Physics and fractal structures* (Springer, New York, 1996)

28. W.G. Gray, P.C.Y. Lee, On the theorems for local volume averaging of multiphase systems. Int. J. Multiphase Flow **3**, 333 (1977)
29. G. Grégoire, H. Chaté, Y. Tu, Active and passive particles: Modeling beads in a bacterial bath, Phys. Rev. E **64**, 011902 (2001)
30. J.S. Hager, J. Krug, V. Popkov, G.M. et Schtz, Minimal current phase and universal boundary layers in driven diffusive systems. Phys. Rev. E **63**, 056110 (2001)
31. J. Honerkamp, *Statistical physics*, 2nd edn. (Springer, Heidelberg, 2002)
32. H.E. Hurst, Long-term storage capacity of reservoirs. Trans. Am. Soc. Civ. Eng. **116**, 770 (1951)
33. J. Klafter, M.F. Shlesinger, G. Zumofen, Beyond Brownian motion. Physics Today **49**, 33 (1996)
34. A.B. Kolomeisky, G.M. Schütz, E.B. Kolomeisky, J.P. Straley, Phase diagram of one-dimensional driven lattice gases with open boundaries. J. Phys. A. **31**, 6911 (1998)
35. R. Kubo, M. Toda, N. Hatsuhime, *Statistical Physics II. Non Equilibrium Statistical Mechanics*, chapter 1 (Springer, Berlin, 1991)
36. R. Kubo, The fluctuation-dissipation theorem. Rep. Prog. Phys. **29**, 255 (1966)
37. J.L. Lebowitz, Boltzmann's entropy and time's arrow. Physics Today **46**, 32 (1993)
38. F.K. Lehner, On the validity of Fick's law for transient diffusion through a porous medium. Chem. Eng. Sci. **34**, 821 (1979)
39. H. Lemarchand, C. Vidal, *La réaction créatrice: dynamique des systèmes chimiques* (Hermann, Paris, 1988)
40. A. Lesne, *Renormalization Methods* (Wiley, 1998)
41. P. Lévy, *Processus stochastiques et mouvement Brownien*, (Gauthiers-Villars, Paris, 1948, new edition 1992 Editions J. Gabay, Paris)
42. P. Lévy , *Théorie de l'addition des variables aléatoires* (Gauthiers-Villars, Paris, 1954, new edition 1992 J. Gabay, Paris)
43. S.K. Ma, *Modern Theory of Critical Phenomena* (Benjamin, Reading Mass., 1976)
44. J.T. Mac Donalds, J.H. Gibbs, A.C. Pipkin, Kinetics of biopolymerization of nucleic acid templates. Biopolymers **6**, 1 (1968)
45. R.M. Mackey, *Time's Arrow: Origins of Thermodynamic Behaviour* (Springer, 1992)
46. D.A. Mac Quarrie, *Statistical Mechanics* (Harper & Row, New York, 1973)
47. B. Mandelbrot, *The Fractal Geometry of Nature* (Freeman, San Francisco, 1982)
48. T. Misteli, Protein dynamics: implication for nuclear architecture and gene expression. Science **291**, 843 (2001)
49. J.D. Murray, *Mathematical Biology*, 3rd edn. (Springer, 2002)
50. C. Nicholson, Diffusion and related transport mechanisms in brain tissue. Rep. Prog. Physics **64**, 815 (2001)
51. G. Oshanin, O. Benichou, S.F. Burlatsky, M. Moreau, in *Instabilities and non-equilibrium structures IX*, ed. by E. Tirapegui and O. Descalzi. Biased tracer diffusion in hard-core lattice gases: some notes on the validity of the Einstein relation (Kluwer, Dordrecht, 2003)
52. A. Ott, J.P. Bouchaud, D. Langevin, W. Urbach, Anomalous diffusion in "living polymers": a genuine Lévy flight?, Phys. Rev. Lett. **65**, 2201 (1990)
53. J. Perrin, *Atoms*, (Ox Bow Press, Woodbridge, 1990) Translation of the French original published in 1913 in Editions Felix Alcan
54. M.H. Protter, H.F. Weinberger, *Maximum Principles in Differential Equations* (Prentice Hall, Englewood Cliffs, 1967)
55. P. Résibois, M. De Leener, *Classical Kinetic Theory of Fluids* (Wiley, New York, 1977)
56. H. Risken, *The Fokker–Planck Equation* (Springer, Berlin, 1984)
57. D. Ruelle, General linear response formula in statistical mechanics, and the fluctuation-dissipation theorem far from equilibrium. Physics Letters A **245**, 220 (1998)
58. B. Sapoval, M. Rosso, J.F. Gouyet, The fractal nature of a diffusion front and relation to percolation. J. Physique Lett. **46**, L149 (1985)
59. M. Shlesinger, J. Klafter, G. Zumofen, Above, below and beyond Brownian motion. Am. J. Phys. **67**, 1253 (1999)

60. M. Smoluchowski, Über den Begriff des zufalls und den Ursprung der Wahrscheinlichkeitsgesetre in der Physik. Naturwissenschaften **17**, 253 (1918)
61. T.H. Solomon, E.R. Weeks, H. Swinney, Observation of anomalous diffusion and Lévy flights in a two-dimensional rotating flow. Phys. Rev. Lett. **71**, 3975 (1993)
62. J. Stachel, *Einstein's Miraculous Year: Five Papers that Changed the Face of Physics* (Princeton University Press, Princeton, NJ, 1998)
63. A. Vailati, M. Giglio, Giant fluctuations in a free diffusion process. Nature **390**, 262 (1997)
64. N.G. Van Kampen, *Stochastic Processes in Physics and Chemistry* (North Holland, Amsterdam, 1981)
65. J. Villain, A. Pimpinelli, *Physics of Crystal Growth* (Cambridge University Press, 1998)
66. G.M. Viswanathan, S.V. Buldyrev, S. Havlin, M.G. da Luz, E.P. Raposo, H.E. Stanley, Optimizing the success of random searches. Nature **401**, 911 (1999)
67. M.C. Wang, G.E. Uhlenbeck. Rev. Mod. Phys. **17**, 323 (1945)
68. N. Wax (ed.), *Selected Papers on Noise and Stochastic Processes* (Dover, New York, 1954)
69. D.A. Weitz, Diffusion in a Different Direction. Nature **390**, 233 (1997)
70. N. Wiener, *Collected Works*, ed. by P. Massani (MIT Press, Cambridge Mass., 1976)

Chapter 5
The Percolation Transition

5.1 Introduction

What physical concept can unite the flow of liquid through ground coffee, electrical conduction in a conductor–insulator mixture, target collapse on shooting, the birth of a continent, polymer gelification and the spread of epidemics or forest fires? One question is common to all these processes: how is "something" produced at large scales by contributions at small scales? In all these situations, a quantity (liquid, electric charges, target fracture, dry land, molecular crosslinks, disease or fire) may or may not propagate from one element to its neighbour. As in the previous chapters, we are interested in asymptotic properties resulting at large scale, that is to say in a system that is large compared with the size of the individual elements. The analogue of temperature T here is the inverse of the relative population density p of elements (pores, conducting regions, impacts, etc) which varies between 0 and 1 for the maximum population density. $p \sim 0$ corresponds to a large amount of disorder for a dilute population and $p \sim 1$ corresponds to a large amount of order established. In these situations we can ask ourselves the same questions as for changes of states of matter in thermal systems:

- Is there a clear cut transition and a critical population density p_c?
- How does the transition take place?
- Is there a critical behaviour?
- Can we propose a *mean field* description'?
- What are the limits of its validity?
- Do universality classes exist and if so what are they?

The answer to the first question is that there does exist a clear transition. In 1941 the physicist Flory [8] applied a phase transition type description to polymer gelification, and then in 1957 Broadbent and Hammersley [5] introduced the name *percolation transition*. An almost perfect analogy can be established between phase transitions and the percolation transition, even though this may appear surprising. For phase transitions, order is established in the material due to a

A. Lesne and M. Laguës, *Scale Invariance*, DOI 10.1007/978-3-642-15123-1_5,
© Springer-Verlag Berlin Heidelberg 2012

Table 5.1 Examples of large scale processes generated by accumulation of contributions at small scales

Small scale element	Large scale process
Pores in ground coffee	Flow of liquid
Conducting regions	Electric current
Impact of a bullet	Collapse of the target
Emergence of an island as sea level drops	Formation of a continent
Interactions between polymers	Gel: "molecule" of macroscopic dimensions
Contamination of an individual	Epidemic
Tree catching fire	Burning forest

competition between microscopic interactions and thermal motion, whereas nothing of the sort exists in percolation scenarios. For percolation there is no interaction between elements and no competition with thermal motion, in fact percolation is a purely topological effect. Distances do not count, the only useful quantity for the percolation transition is the number of neighbours of a site and not the exact form of the lattice. It is precisely these topological analogies at the critical point, such as the divergence of the characteristic length and scale invariance, that mean we can use the same methods to describe the two cases. This is what gives the percolation transition the same universal nature as phase transitions. The same critical behavior is indeed observed in very different physical situations (Table 5.1).

5.1.1   An Initial Mean Field Approach

A simple example of a percolation transition is given by an Ising model where each site is randomly occupied by a spin or not. How does the ferromagnetic–paramagnetic transition depend on the population density p of spins? This complex problem involving two generalised fields, temperature and population density p, (see the end of this section and reference [12] for a rigorous approach) can be described by the mean field approach, which we used for the Ising model in Sect. 1.1, (1.18). By simply replacing the average number of neighbours q of an occupied site by pq, the average magnetisation can be evaluated by the implicit equation:

$$m = \tanh\left(\frac{pqJ}{kT}m\right).$$

Near the transition, the magnetisation is weak and $\tanh(x)$ can be replaced by x. This leads to a transition temperature:

$$T_c = p\frac{qJ}{k} = pT_{MF}$$

proportional to p, where T_{MF} is the critical temperature predicted by the mean field approach for an Ising lattice in which all the sites are occupied. At nonzero temperature, an order, that is to say a path allowing step by step propagation of spin orientation, is established within this approximation when $p > 0$. This is obviously wrong because when the system is very diluted the occupied sites are far from one another and cannot interact to establish long range order, whatever the temperature. In the next section we will see Cayley–Bethe's description, which is a less "primitive" mean field approach. But first let us see why the mean field on a randomly occupied Ising lattice is not an appropriate description.

5.1.2 A Path Through the Network: The Infinite Cluster, Its Mass and Branches

Experimentally and by numerical simulations the existence of a critical population density has been seen, above which a path connecting the opposite sides of the system is established (see Fig. 5.1). Even though filling is random, the population density threshold p_c and the "critical" behaviours are perfectly reproducible when the number N of elements in the system is large.

We call the group of elements that are connected to each other a *cluster*. The propagating quantity (flow for a liquid) only propagates *within each cluster* and not between clusters. Above the threshold, the size of at least one *cluster* diverges and is therefore called *infinite*. In two dimensions it is easy to show that there is only one infinite cluster because of the topological impossibility of two or more to exist, however when we are exactly at the threshold, this result remains true in three dimensions. This is justified intuitively by the significant branching of an infinite cluster: even though topologically possible in three dimensions, the existence of two infinite clusters with no points of contact between them is improbable, with strictly zero probability in an infinitely sized system. The number of sites in the infinite cluster, i.e. its *mass*, grows very rapidly above the threshold (see Fig. 5.2), because many isolated clusters suddenly aggregate. Paradoxically, the liquid flow or conductivity increase more slowly above the threshold. This is due to the very branched structure of the infinite cluster at the percolation threshold (Fig. 5.3). Only the *infinite cluster backbone* is useful for the propagation or flow. (To improve the readability of this figure, the population of finite clusters containing only a few

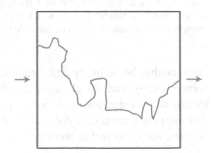

Fig. 5.1 A critical population density threshold p_c exists above which at least one continuous path is established between two opposite sides of the system

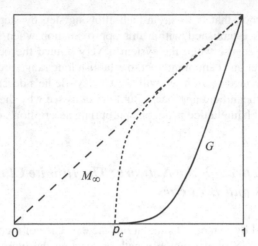

Fig. 5.2 Above the threshold p_c, the mass M_∞ of the infinite cluster (*dotted line*) increases rapidly with the population density p. The *dashed line* shows the M_∞ for a randomly filled Ising lattice described within the mean field approximation. The *solid line* show the electrical conductance G in an insulator–conductor mixture which is zero below p_c and increases much more slowly than M_∞ above p_c. This difference is due to the fact that many "dead ends" exist in the infinite cluster, which are not useful for conducting the electric current

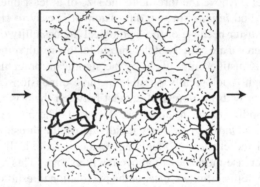

Fig. 5.3 Just above the percolation threshold, the structure of the infinite cluster is very branched. We can distinguish two principle regions: the "backbone", in *thick lines*, made up of the set of paths connecting one side to the other, and the side branches, in *thin lines*, which lead to dead ends. These cul de sacs represent the large majority of the mass of the cluster. The backbone itself incorporates "red bonds" in *grey*, which are the passages the fluid is obliged to pass through and loops in *black*, which offer several possible paths

elements has been highly under represented for readability of the graph.) It can be shown simply that the electrical conductance of the percolating cluster, assumed to consist of elements of electrical resistance, is directly related to the effective diffusion coefficient of a diffuser in the infinite cluster. This physical situation has kept the name *the ant in the labyrinth* proposed by De Gennes [9]. Physicists later have used other insect names to characterise various situations (termites, parasites, etc.)

The particularly large branching of the infinite cluster *at the percolation threshold* can be seen in a simple computer experiment of the lifetime of a forest fire [16] as we will see in the following.

5.1.3 Lifetime of a Forest Fire as a Measure of the Percolation Threshold

We will consider the following computer game: a fire is lit along one side of a square representing a section of forest. The fire propagates from one tree to its nearest neighbours, if they exist, in a given time. A tree that has burnt no longer propagates the fire. We evaluate the lifetime of the fire, the average time required for its extinction, which is strongly affected by the density of trees. The average is taken over a large number of randomly propagated fires lit on the same system. It is observed that:

- If the population of trees is sparse ($p \ll p_c$), the initial bonfires only consume the trees in each isolated cluster in contact with the face set alight. The fire does not propagate, its short average lifetime corresponds to the time required to burn the largest of these isolated clusters.
- If the population of trees is very dense ($p \gg p_c$) with only a few depopulated regions the fire propagates continuously from one side to the other. The average lifetime of the fire corresponds to a step by step "simple route" from one side to the other.
- If the population density corresponds to the percolation threshold ($p \sim p_c$), the fire propagates along all the branches, bends, twists and turns of the infinite cluster. In this case the lifetime of the fire diverges. The divergence of the lifetime of the fire is an effective way to illustrate the percolation threshold (see Fig. 5.4).

It turns out that the value of the percolation threshold ($p_c = 0.59275$ in the case of a square lattice) is very reproducible for large systems. The forest size required to obtain this reproducibility is 10 to 100 million elements, as we will show later in Sect. 5.3.2. It can be easily shown that if the next nearest neighbours are included (leftmost curve in Fig. 5.4), the percolation threshold is complementary to the value for nearest neighbours: $p_c^{(\text{next nearest})} = 1 - 0.59275 = 0.40725$. On a triangular lattice, the percolation threshold is strictly equal to $1/2$. So, we find that, just like the critical temperatures, the percolation threshold depends on the details of the microscopic model: it does not have a universal character.

Table 5.2 shows the exact or numerical values for the percolation thresholds of different regular lattices in two and three dimensions. Depending on the lattice, the threshold value varies by a factor of 1.4 in two dimensions and more than a factor of 2 in three dimensions. These ranges of values become considerably narrower if

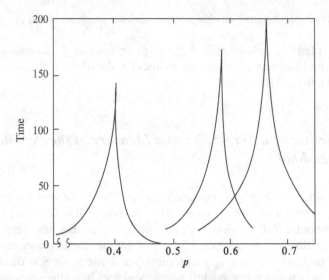

Fig. 5.4 Variation in average lifetime of a forest fire on a square lattice, as a function of the initial population density of trees. The *middle curve* corresponds to the case where the fire propagates to nearest neighbour trees and the curve to the left to propagation to nearest and next nearest neighbours. The *right hand curve* is obtained by assuming that two of its nearest neighbouring trees must be burning for a tree to catch alight (after [16])

Table 5.2 Values of site percolation thresholds, for regular lattices in two and three dimensions (2D and 3D respectively) (See text for the difference from bond percolation.)

Dimensions	Lattice	Percolation threshold (fraction of occupied sites)	Density of a lattice filled by spheres	Percolation threshold (volume fraction)
2D	Square	0.5927603	$\pi/4$	0.4655
	Triangular	0.5	$\pi\sqrt{3}/6$	0.4534
	Honeycomb	0.697043	$\pi\sqrt{3}/9$	0.4214
3D	Simple cubic	0.311608	$\pi/6$	0.1632
	Body-centered cubic	0.2459615	$\pi\sqrt{3}/8$	0.1666
	Face-centered cubic	0.1992365	$\pi\sqrt{2}/6$	0.1475
	Diamond cubic	0.4286	$\pi\sqrt{3}/16$	0.1458

they are evaluated as a proportion of the volume occupied by discs or spheres at each site. By correcting for the density of the equivalent perfectly filled lattice in this way, we obtain about 0.45 in two dimensions and 0.15 in three dimensions. In practice, the "elements" are rarely in circular or spherical form. In three dimensions, their flattening or elongation leads to large variations in the percolation threshold value, typically between 0.05 and 0.15 volume fraction.

5.1.4 Site Percolation and Bond Percolation

Percolation is characterised by the establishment of a path in a large system. A detailed discussion of the effect of the finite size of the system is presented in Sect. 5.3.2. At the scale of the elements, different models are used to consider whether a path is established between two neighbouring points.

- In the case of *site percolation*, which we considered in the previous subsection, a step in the path is established by the presence of two elements on neighbouring lattice sites. The parameter p is then the probability of site occupation.

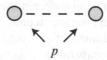

- In the case of *bond percolation*, a step in the path is established by the presence of a bond between two neighbouring points. The population density p is then the probability of bond occupation.

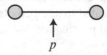

- In the case of *site-bond percolation* introduced to describe polymer crosslinking transitions (*sol-gel* transitions), x is the fraction of occupied sites whereas p is the probability that a bond between two occupied sites is present. This general framework contains the two other cases: $p = 1$ corresponds to site percolation, whereas $x = 1$ corresponds to bond percolation.

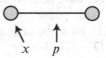

It is interesting to distinguish these different situations in order to apply them to different physical systems. However, they are members of the same universality class: site percolation and bond percolation show the same critical behaviour.

5.1.5 Correlated, Driven and Stirred Percolations

In the previous cases, the elements were assumed to be assigned *randomly*, independent of each other. However, there are many percolation type physical situations in which these conditions are not fulfilled [6].

One family of scenarios, which we described in Chap. 1, correspond to thermal order–disorder transitions in which neighbouring elements have an interaction energy. For example, the Ising model can be considered as a *correlated percolation* problem. The up spins ↑ represent populated sites, down spins ↓ the empty sites and establishment of long range order corresponds to the establishment of a large scale percolating path. We can also imagine a general case of site-bond correlated percolation, characterised by three quantities (three *fields*): temperature T, bond population density p and fraction of occupied sites x. This scenario offers a rich variety of complicated transitions [12].

In the other families introduced below, the element, sites or bonds, are always assumed to be deposited independently of each other. An important family is that of *driven percolations* in which the bonds are "diodes" which let current through only in one direction. Studies have also been made on nonlinear bonds characterised by a *critical current* beyond which they are resistant and on bonds which can switch form a conducting to an insulating state, representing reasonably well what happens in a random network of Josephson superconducting junctions [15].

Stirred percolation is the interesting situation in which clusters are stirred by a fluid, their lifetimes being comparable or shorter than the average passage time of the diffuser (*ant*) in a cluster [14] This model is suitable for electric conductivity in an emulsion or microemulsion.

By this list of a few examples of percolations, we hope to have convinced the reader of the richness of this concept used today in all branches of science. This chapter will be mainly dedicated to site percolation, the universal results of which being also applicable to bond percolation and site-bond percolation.

5.2 Statistics of Clusters and the Ferromagnetic–Paramagnetic Transition

Theoretical percolation studies have been strongly stimulated by the demonstration of their rigorous analogy with phase transitions, for example ferromagnetic–paramagnetic transitions [11]. Near the critical point, the analogue of the relative temperature difference $t = (T - T_c)/T_c$ is $(p_c - p)$. We define the quantity $n_s(p)$ as the number of clusters containing s sites, relative to the total number of lattice sites. Studying the percolation transition amounts to describing the number of clusters as a function of s and p. All the quantities that we will introduce can be deduced from this. The probability of an *occupied site* in a cluster containing s sites is $s\,n_s(p)$. The average size $S(p)$ of a finite cluster is then directly given by

$$S(p) = \frac{\sum_s s^2 n_s(p)}{\sum_s s\, n_s(p)}. \tag{5.1}$$

The order parameter of the percolation transition is the probability $P(p)$ that a site belongs to an infinite cluster:

- $P(p) = 0$ if $p < p_c$.
- $P(p)$ is finite if $p > p_c$.

Finally, the condition that an occupied site is in either an infinite cluster or a finite cluster is expressed by:

$$p = P(p) + \sum_{s} s\, n_s(p). \tag{5.2}$$

5.2.1   Correlations, Characteristic Lengths and Coherence Length

Another important quantity in characterising percolation systems is the pair correlation function, that is the probability $g(p, r)$ that two sites separated by a distance r belong to the same cluster. In the following paragraph we will see how this function can be described in the same way as the correlation function of a phase transition (Chaps. 1 and 4):

$$g(r, p) = \frac{e^{-r/\xi(p)}}{r^{d-2+\eta}}. \tag{5.3}$$

The correlation length $\xi(p)$ can be defined in a percolation system as well as several other characteristic lengths, including the following two:

- *The radius of gyration of finite clusters*, which is also the average connectivity length[1] of a finite cluster. Finite clusters exists on both sides of the percolation threshold and their size diverges as p tends to p_c.
- *The homogeneity cutoff scale of the infinite cluster* is the scale below which, if $p > p_c$, the infinite cluster is fractal and above which it is homogeneous.[2]

It can be shown that all these quantities have the same divergence as a function of $(p - p_c)$:

$$\xi(p) = \xi_0 \,|\, p - p_c |^{-\nu}. \tag{5.4}$$

When we refer to the coherence length $\xi(p)$, this equally represents any one of these characteristic lengths. The universality of the exponent ν is well verified experimentally. However, the prefactors ξ_0 depend on the exact definition of $\xi(p)$ as well as the sign of $(p - p_c)$.

5.2.2   Analogy with Phase Transitions

It can be shown [11] that each of the quantities used to describe the percolation transition is the analogue of a magnetic quantity and that it shows a power law behaviour near the threshold p_c (see Table 5.3). This analogy rests on the scale

---

[1]By the *average connectivity length* of a given cluster, we mean the average distance separating the cluster elements.

[2]That is to say it has classical geometric properties (non fractal), for instance the number of elements contained in a region of linear extension L of the infinite cluster varies as L^d where d is the dimension of space (see also Sect. 3.1.1).

Table 5.3 Analogy between ferromagnetic–paramagnetic and percolation transitions

Quantity	Magnetic transition	Percolation transition	Dependence
Order parameter	$m(t)$	$P(p)$	$\lvert p - p_c \rvert^{\beta}$
Free energy	$f(t)$	Σn_s	$\lvert p - p_c \rvert^{2-\alpha}$
Susceptibility	$\chi(t)$	$S(p)$	$\lvert p - p_c \rvert^{-\gamma}$
Correlation function	$G(r,t)$	$g(r,p)$	$\exp(-r/\xi(p))/r^{d-2+\eta}$
Coherence length	$\xi(t)$	$\xi(p)$	$\lvert p - p_c \rvert^{-\nu}$

invariance of percolation when the population density is equal to the critical population density. The scaling hypothesis (Sect. 3.2.2) is taken here, in other words all the critical behaviours stem from the divergence of the coherence length. Physically, this hypothesis results in the idea of a *dominant cluster*, as we will see in more details in the next subsection.

5.2.3   Hypothesis of the Dominant Cluster and Scaling Relations

The main hypothesis of percolation theory relies on two points. The first is that a *dominant cluster* exists whose characteristic size corresponds to a coherence length $\xi(p)$. We assume that it has a mass s_ξ which obeys a power law:

$$s_\xi(p) = \lvert p - p_c \rvert^{-1/\sigma}. \tag{5.5}$$

We also assume that the number of clusters $n_s(p)$ can be expressed in the following way:

$$n_s(p) = n_s(p_c)\phi\left(\frac{s}{s_\xi}\right) = s^{-\tau}\phi\left(s\lvert p - p_c \rvert^{1/\sigma}\right). \tag{5.6}$$

The power law hypothesis $n_s(p_c) \sim s^{-\tau}$ is well verified experimentally (see Fig. 5.5). All the critical exponents of the percolation transition can then be expressed in terms of the two exponents σ and τ. For example the free energy $f(p)$ is:

$$f(p) = \sum_s n_s(p) = \sum_s s^{-\tau}\phi\left(s\lvert p - p_c \rvert^{1/\sigma}\right) \tag{5.7}$$

or, by choosing a constant value for the argument of ϕ:

$$f(p) \sim \lvert p - p_c \rvert^{\frac{\tau-1}{\sigma}}. \tag{5.8}$$

The expression for α can be extracted by expressing that the exponent characterising f is $2 - \alpha$:

Fig. 5.5 Number of clusters at the percolation threshold determined by numerical simulation on a square lattice of 95,000 × 95,000 (after [16])

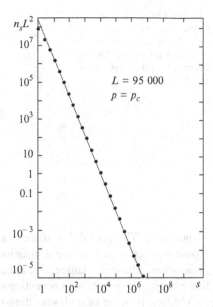

$$\alpha = 2 - \frac{\tau - 1}{\sigma}. \tag{5.9}$$

In the same way we can obtain the values of β and γ:

$$\beta = \frac{\tau - 2}{\sigma} \qquad \gamma = \frac{3 - \tau}{\sigma}. \tag{5.10}$$

This leads us to recover the Rushbrooke scaling relation (Sect. 3.2.1):

$$\alpha + 2\beta + \gamma = 2 \tag{5.11}$$

as well as the other scaling relations. Numerical tests that have been carried out seem to confirm the universality, i.e. that the exponents depend only on the dimension of space. A mean field approach enables these different exponents to be evaluated in a way that is valid for the case of infinite dimension.

5.2.4   Bethe's Mean Field Calculation on Cayley's Tree

Rigorous description of the mechanism of the appearance of a percolating path encounters many difficulties due to the complex structure of the infinite cluster (Fig. 5.3). This structure is the subject of the following section. One of its particularities is the existence of *loops* which are very difficult to take into account exactly. Intuitively the relative weight of these loops decreases as the spatial dimension d

Fig. 5.6 Bethe lattice or
Cayley tree in the case of
$z = 3$ branches emerging
from each point. Here the tree
is limited to three generations
of branches

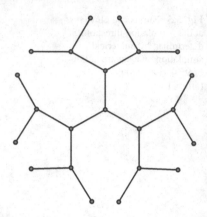

is increased. The probability of two random walks, starting from the same point, crossing is equal to 1 in one dimension, high in two dimensions and less in three dimensions. The Bethe lattice, also called Cayley tree (Fig. 5.6), completely neglects these loops. The percolation transition on this tree is exactly solvable.

The tree is made of nodes sending out z bonds to their neighbours. If we leave from a "central" site and develop the branches up to generation R, the total number of sites is $1 + z + z^2 + \cdots + z^R = (z^{R+1} - 1)/(z - 1)$. The Cayley tree is an object whose mass (total number of sites) varies exponentially as a function of its radius. Since the mass of an ordinary object varies as its radius to the power d, the Cayley tree can be considered as an object of *infinite dimension*. Suppose now that the sites are occupied with a probability p. When we start from an occupied site, at each new node there are $z - 1$ branches leaving it (other than the one we arrive on). The average number of occupied sites that allow the path to continue to the extremity of its branches is therefore $p(z - 1)$. The percolation threshold corresponds to the value of p for which this number is 1, in this way statistically ensuring large scale connectivity. The fact that there are no loops excludes the possibility that the path can come back on itself:

$$p_c = \frac{1}{z - 1}. \tag{5.12}$$

The probability $P(p)$ that a typical site is part of the infinite cluster can be calculated exactly. Let Q be the probability that a site is not connected to the infinite cluster by a given branch. There are no loops in the Cayley tree and the two extremities of a branch lead to completely unconnected universes. One of these universes might or might not be the infinite cluster and this is independently the case for each branch leaving an occupied site. The probability that an origin site is occupied but not connected to the infinite cluster is $p - P(p)$. This probability is also equal to pQ^z, signifying that none of the z branches leaving this site are connected to the infinite cluster. In this way P and Q are related:

$$P = p(1 - Q^z) \quad \text{i.e} \quad Q = \left(1 - \frac{P(p)}{p}\right)^{1/z}. \tag{5.13}$$

Table 5.4 Critical exponents obtained by the Bethe mean field approach

Exponent	τ	ν	α	β	γ	σ
Bethe lattice	5/2	1/2	−1	1	1	1/2

In addition, the probability Q_1 that the corresponding nearest neighbour site is occupied and that none of the $z - 1$ branches leading out of it are connected to the infinite cluster is $p\,Q^{z-1}$. The probability Q is the sum of Q_1 and the probability $1 - p$ that the nearest neighbour site is empty:

$$Q = 1 - p + Q_1 = 1 - p + pQ^{z-1} \qquad (5.14)$$

from which we have an equation to calculate $P(p)$. The expression is complex if z is greater than 3, but whatever z, it can be shown directly that $P(p)$ is linear in $(p - p_c)$, near the threshold (the reader may like to try this as an exercise). The critical exponent of the order parameter is $\beta = 1$ in this mean field approximation. From the same type of argument we can obtain the average cluster size $S(p)$ and show that it is proportional to $1/(p - p_c)$.

The number $n_s(p)$ of clusters of size s can be calculated by evaluating the surface t of empty sites that surround a cluster of size s. For an isolated site $t = z$, then each additional occupied site adds $z - 2$ empty sites to t. The value of t is therefore a linear function of s, namely $t = (z - 2)s + 2$, whilst the value of $n_s(p)$ is:

$$n_s(p) = n_s(p_c)\,p^s(1 - p)^{(z-2)s+2}. \qquad (5.15)$$

which results in an exponential decay of $n_s(p)$ as a function of s. This is particular to the Bethe lattice: for small sized clusters, in small spatial dimensions, the hypothesis of the dominant cluster leads to a power law. After all is said and done, this description does not describe percolation better than the equivalent mean field approaches in phase transitions. On the other hand, renormalisation, associated with numerical simulation techniques, is very effective in the case of percolation.

Table 5.4 summarises the values of the exponents deduced by the Bethe approach. A theoretical renormalisation approach can show that these values are exact in spaces of six or more dimensions. As for phase transitions, there exists a critical dimension, but its value is $d_c = 6$ and not 4.

5.3 Renormalisation of Percolation Models

Here we present just a few examples of renormalisation (see Sect. 3.4.1), however the reader is invited to try this technique themselves on different situations. As we will mention below, it is possible to obtain good quality results at a much lower calculation price that in the case of phase transitions.

5.3.1   Examples of Renormalisation in Two Dimensional Space

5.3.1.1   Site Percolation on a Triangular Lattice

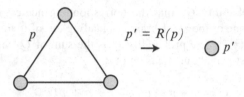

Here we adopt the same approach as for phase transitions. We choose a finite lattice, perform decimation, calculate the renormalisation function, search of a fixed point and calculate the exponents. The simplest finite lattice that we can imagine is a triangular lattice (see diagram). We transform the triangle into a super site by decimation. By using the majority rule, the super site will be considered occupied only if at least two sites of the triangle are occupied:

$$p' = p^3 + 3(1 - p)p^2. \tag{5.16}$$

This relationship leads to two trivial (stable) fixed points, $p^* = 0$ and $p^* = 1$, and an unstable fixed point which corresponds to the transition $p_c = 1/2$. This value is exact even though its evaluation on a small finite lattice is a priori very approximate. Its simple and symmetric value is a reason favouring this result, as we will see later.

The calculation of critical exponents naturally follows the procedure described in Sect. 3.3.3. Remember that the exponent ν corresponding to the coherence length is:

$$\nu = \frac{\log(b)}{\log(\lambda_1)}, \tag{5.17}$$

where b is the linear scale factor and λ_1 is the eigenvalue with the largest absolute value. The scale factor b is obtained by expressing that the number of sites in the "super site" (here 3) is b^2. Here there is only one "coupling coefficient" p. In this case, the eigenvalue reduces to the derivative of the renormalisation relation at the critical point:

$$\lambda_1 = R'(p_c) = 6p_c(1 - p_c) = 3/2. \tag{5.18}$$

From which the value of ν is given by:

$$\nu = \frac{\log(\sqrt{3})}{\log(3/2)} = 1.35. \tag{5.19}$$

Compared to the exact value $\nu = 4/3$, the value obtained by this calculation is an excellent approximation. However this agreement is misleading, as we can see by

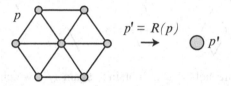

renormalising a slightly larger finite lattice consisting of 7 sites (see diagram). In this case the renormalisation relation is:

$$p' = p^7 + 7(1-p)p^6 + 21(1-p)^2 p^5 + 35(1-p)^3 p^4. \qquad (5.20)$$

Here again, the unstable fixed point corresponding to the transition is $p_c = 1/2$. The eigenvalue this time is $35/16$, which leads to $\nu = 1.243$, from (5.17) (assuming $b = \sqrt{7}$). It could be disappointing that this value is substantially further from the exact values than the previous calculation. This shows that the convergence of renormalisation steps must be carefully checked.

5.3.1.2 Bond Percolation on a Triangular Lattice

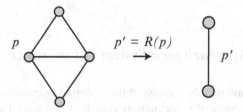

Since bond percolation and site percolation belong to the same universality class, we should get the same exponents, for example $\nu = 4/3$. On the other hand, the percolation threshold is different (see Table 5.2). Here, for bond percolation on a triangular lattice, we expect $p_c = 0.34730$. We will again use a very simple finite lattice, shown in the diagram, in which the probability p is that of the presence of a bond.

The renormalisation relation is obtained in this case by adding the probabilities that a path exists between the upper and lower site in the finite lattice. This probability includes all the situations where 0 or 1 bonds are missing, plus 8 situations out of 10 where 2 bonds are missing and 2 situations in which three bonds are missing:

$$p' = p^5 + 5(1-p)p^4 + 8(1-p)^2 p^3 + 2(1-p)^3 p^2. \qquad (5.21)$$

$$\Diamond \quad + 5 \quad \triangledown \quad + 8 \quad \triangle \quad + 2 \quad \langle$$

The percolation threshold that we obtain is again $1/2$, which is an overestimate by about 50% in this case of bond percolation. Calculating the exponent ν directly gives the value $R'(p_c) = 13/8$. We could be tempted to take $\sqrt{5}$ as the scale factor (reduction in bonds by a factor of 5), which leads to a value of $\nu = 1.66$. If, on the other hand, we return to the definition of b as the reduction in bond length necessary to superimpose the decimated lattice on the initial lattice, we obtain $b = \sqrt{3}$ and $\nu = 1.13$. How should we decide which to take?

In practice, the five bonds of the super site do not have the same status. While the central bond is fully taken into account in the super site, the other four are shared between two neighbouring super sites in the decimated lattice. The *effective* number of bonds contained in the super site is therefore $1 + 4 \cdot \dfrac{1}{2} = 3$, which also leads to $b = \sqrt{3}$:

$$\nu = \frac{\log(\sqrt{3})}{\log(13/8)} = 1.13. \tag{5.22}$$

We leave the reader to extend these approaches to other types of network (square or honeycomb).

5.3.2   Scaling Approach on a System of Finite Size

Let us emphasise again that the success of these simple calculations is misleading in terms of the effectiveness of renormalisation of finite lattices. The example of site percolation on a triangular lattice, treated above, shows that a calculation with 7 sites leads to a less good result than a calculation with 3 sites! It is tempting to pursue the method with larger and larger finite lattices, leading to larger and larger scale factors k, calculating the values of $\nu(k)$ each time and then to extrapolate to an infinite value of k. This programme has been pursued by various methods. It is difficult to use the analytical method used above for finite lattices with large numbers of sites. To go further with this approach, a new idea needs to be developed: the scaling approach on a system of finite size, which we quickly referred to in Sect. 3.5.4.

We are confronted with an apparent contradiction: the scaling approaches are based on scale invariance, which itself assumes that the system is infinite. This problem is dispelled by the fact that although we consider finite systems we assume they are very large compared to the size of the elements. However, the hypothesis underlying scaling indicates that the essential length we must compare the system size to is that of the coherence length. Observing the percolation transition in a large number of identical systems of linear size L (see Fig. 5.7), the percolation threshold is seen to vary from one system to another according to the random filling of sites.

Fig. 5.7 The masses of percolating clusters measured on a finite lattice in two different cases of random filling where the percolation thresholds are p_{c_1} and p_{c_2} respectively, compared to the mass of the percolating cluster in an infinite system

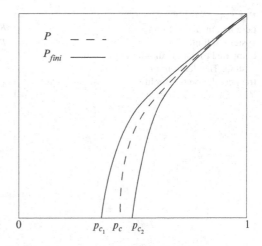

We want to find the distribution of the value of the threshold as a function of L. Let $R(p, L)$ be the probability that a system of size L percolates (i.e. a percolating path between two opposite sides of the system is established) when the population density is p. From the scaling hypothesis, we expect that the behaviour of this probability will be directly linked to the ratio between L and ξ.

- If $L > \xi$, the system behaves as if it was infinite.
- If $L < \xi$, then the boundaries of the system truncate the percolation process.

We can therefore assume that:

$$R(p, L) = \Phi(L/\xi) = \Phi\left[(p - p_c)^\nu L\right]. \qquad (5.23)$$

The distribution $f(p = p_{c_{\text{finite}}}, L)$ of the threshold $p_{c_{\text{finite}}}$ is given by the derivative $\frac{\partial R}{\partial p}$ of $R(p, L)$:

$$f(p = p_{c_{\text{finite}}}, L) = \frac{\partial R}{\partial p} = L^{1/\nu} \Phi'\left[(p - p_c)^\nu L\right]. \qquad (5.24)$$

At the percolation threshold, the argument of the function Φ is in the region of 1. The distribution of the threshold $f(p, L)$ shows a maximum proportional to $L^{1/\nu}$. The integral of this function is 1 and its width Δ is thus proportional to $L^{-1/\nu}$ (see Fig. 5.8). By measuring the width of the distribution, we can therefore deduce a value of ν:

$$y(L) = 1/\nu(L) = -\frac{\log(\Delta)}{\log(L)} + \frac{\text{Cte}}{\log(L)}. \qquad (5.25)$$

By extrapolation, this method determined the exact values of ν [7] as Fig. 5.9 shows.

Fig. 5.8 Probability of
percolation, $R(p, L)$, in a
system of finite size
compared to that of an infinite
system. The distribution of
the percolation threshold
$f(p, L)$ is the derivative $\frac{\partial R}{\partial p}$

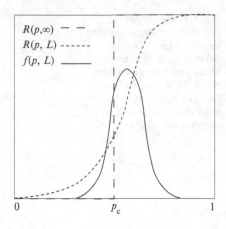

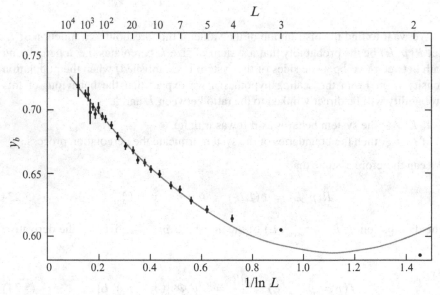

Fig. 5.9 Variation of the width of the distribution of the threshold for triangular lattices of up
to 100 million sites. In agreement with (5.25), the tangent leads to $\nu = 1/y = 4/3$, which is
presumed to be exact for an infinite lattice

5.3.3 Finite Size Exponents

We will show that we expect a form analogous to (5.24) for all the critical quantities
as a function of the size L of the system. The scaling hypothesis (see Chap. 3)
predicts in a general way that a quantity Q near the percolation threshold behaves as:

$$Q(p, L) = (p - p_c)^x q \left(\frac{L}{\xi} \right), \tag{5.26}$$

where the critical exponent corresponding to Q for an infinite system (i.e. for $L \gg \xi$) is x. We can also write the relation:

$$Q(\xi, L) \simeq \xi^{-x/\nu} q\left(\frac{L}{\xi}\right), \qquad (5.27)$$

where $\xi(p) = \xi_0(p - p_c)^{-\nu}$ is the coherence length as it would be in an infinite system. In a finite system where $L \ll \xi$, i.e. in the immediate vicinity of p_c, the coherence cannot physically extend beyond L. It is therefore legitimate to replace ξ by L in (5.27), leading to $Q(p_c, L) \simeq L^{-x/\nu}$. In summary, in a finite system:

- When $L \gg \xi$, i.e. $(p - p_c) \gg \left(\frac{L}{\xi_0}\right)^{-1/\nu}$

$$Q(p, \infty) \simeq \xi^{-x/\nu}. \qquad (5.28)$$

- When $L \ll \xi$, i.e. $(p - p_c) \ll \left(\frac{L}{\xi_0}\right)^{-1/\nu}$

$$Q(p_c, L) \simeq L^{-x/\nu}. \qquad (5.29)$$

Therefore two scaling approaches exist, which are formally symmetric in p or L, depending on whether we are interested in very large systems or systems with a finite size at p_c. The corresponding exponents are directly related to each other: x for the scaling in p corresponds to x/ν for the scaling in L. When we describe transport properties in Sect. 5.5, we will see that for conductivity, the scaling in L, and its universal nature, applies in a vast region, much larger than the region where universality is observed for the scaling in p.

To conclude this subsection, we must note that this approach to systems of finite size is not specific to percolation. It can be applied to all scale invariant situations where a coherence length ξ diverges around a critical value of a control parameter, or field, which could be p but also the temperature T or all other control parameters. Finite size scaling is one of the methods to evaluate the critical exponents precisely. Table 5.5 draws together the results, presumed to be exact, for the main critical exponents of percolation.

5.4 Structure of the Infinite Cluster at the Percolation Threshold

There is always a *largest cluster* for which p is greater or less than the threshold. We are interested in the mass M_L of this largest cluster as a function of the size L of a finite lattice. It can be easily shown [16] that:

Table 5.5 Values of principle exponents of percolation. In two dimensions the values are presumed to be rational

Exponent	σ	τ	α	β	γ	ν
Property	$p - p_c$ dependence of number of clusters	s dependences of cluster	Total number of clusters $\sum n_s$	Weight of P infinite cluster	Average S size of finite clusters	Coherence length ξ
Two dimensions	36/91	187/91	$-2/3$	5/36	43/18	4/3
Three dimensions	0.45	2.18	-0.62	0.41	1.80	0.877
Bethe lattice $d \geq 6$	1/2	5/2	-1	1	1	1/2

- If $p < p_c$, the mass of the largest cluster M_L is proportional to $\log(L)$.
- If $p \gg p_c$, the mass of the largest cluster M_L is proportional to L^d as expected for a full lattice.

What happens at exactly $p = p_c$? At the percolation threshold, the percolating cluster can be characterised by a fractal dimension d_F (see Chap. 2), which describes its spatial mass density.

5.4.1 Fractal Dimension of the Infinite Cluster

The mass M_L of the percolating cluster, which is also the largest cluster (in terms of number of occupied sites), can be expressed as a function of the linear size L of the system. By using the definition of $P(p)$, we express $M_L = P(p)L^d$. In addition, percolation is produced in a finite system when $\xi(p)$ is of order L. In the vicinity of the percolation threshold we therefore have the relation:

$$(p - p_c) \sim L^{-1/\nu}. \tag{5.30}$$

By using this in the expression for the mass of the percolating cluster we obtain:

$$M_L \sim L^{d - \beta/\nu}. \tag{5.31}$$

Following its most direct definition, the fractal dimension is therefore:

$$d_F = d - \beta/\nu. \tag{5.32}$$

Starting from the values of exponents in Table 5.5, we obtain the values for the fractal dimension of the infinite cluster given in Table 5.6: These values have been verified by numerical simulations as shown in Fig. 5.10 of the data from [16].

Table 5.6 Fractal dimension of the infinite cluster

Dimension	2	3	Bethe: $d \geq 6$
Fractal dimension of the infinite cluster	91/48	2.52	4

Fig. 5.10 Size of the largest cluster (or percolating cluster) measured by numerical simulation on triangular lattices up to 10 million sites. The *straight line* through the data has a gradient of 91/48, which is the theoretical value of the fractal dimension of the infinite cluster in two dimensional space. After [16]

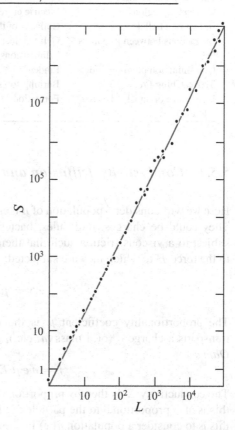

5.5 Dynamic Properties Near a Percolation Transition

Up to now, we have been interested in the static properties of percolation. In this final section, we will briefly address the essentials of what interests the user of percolation. For example, there must be a path for the fluid in a coffee percolator, but can we predict the flow of coffee as a function of the porosity p? In each of the physical situations mentioned at the beginning of this chapter, the same question concerning the transport properties can be posed, as summarised in Table 5.7. It can be shown that these dynamic quantities are often physically equivalent, in a first approximation. We recall briefly below how this is done for the conductance and the diffusion coefficient.

Table 5.7 Dynamic quantities in different physical situations of percolation

Small scale element	Large scale process	Dynamic quantity
Pores in ground coffee	Flow of liquid	Conductance
Conducting regions	Electric current	
Impact of a bullet	Collapse of the target	Elastic modulus
Interactions between polymers	Gel: "molecule" of macroscopic dimensions	
Contamination of an individual	Epidemic	Propagation speed
Tree catching fire	Burning forest	
Emergence of an island as sea level drops	Formation of a continent	Diffusion law of a random walker

5.5.1   Conductivity, Diffusion and Walking Ants

Here we will consider a population of n elementary mobile objects per unit volume. They could be charges, molecules, bacteria etc. Their constitutive movement is subject to a viscous friction such that their equilibrium velocity v is proportional to the force F to which they are subjected:

$$v = \mu F. \tag{5.33}$$

The proportionality coefficient μ is the mobility of the object. If each object transports a charge e (or a mass m, etc.), for a population ensemble this leads to *Ohm's law*

$$j = n e \mu E = \Sigma E. \tag{5.34}$$

The conductivity Σ is thereby proportional to the mobility μ. Einstein proved that this is also proportional to the particles' diffusion coefficient D. One way to show this is to consider a population $n(x)$ in equilibrium in a constant electric field E on the x axis producing a potential $V(x) = V_0 - Ex$. If we can neglect interactions between particles $n(x)$ is given by Boltzmann statistics:

$$n(x) = n_0 \exp\left(-\frac{eE}{kT}x\right). \tag{5.35}$$

This equilibrium also results from the balance, at every point, between the drift current j, given by (5.34), and the diffusion current j_D, resulting from the concentration gradient:

$$j_E + j_D = 0 \quad \text{and so} \quad \Sigma E = n e \mu E = D \nabla n = \frac{DeE}{kT}n. \tag{5.36}$$

This gives the Einstein relation $D = \mu kT$ and the proportionality between the diffusion coefficient D and the conductivity Σ in a homogeneous medium.

Physicists have therefore focused on describing the diffusion of the *walking ant* on percolating clusters since it also reflects the behaviour of the conductivity. We will discuss this situation in the following subsections. However, one should be prudent in generalising this relation for heterogeneous media, such as systems at the percolation threshold, because the effective values of D and Σ can be produced by different averages at large scales. At very long times, in fact, diffusion involves only the infinite cluster, whereas the conductivity should be averaged over clusters of all sizes, as we will show in the following subsection.

5.5.2   Diffusion and Conduction in Percolating Clusters

In the 1970s, there was a heated debate, regarding the critical behaviour of conductivity and diffusion, over a common scaling hypothesis:

$$D \sim \Sigma \sim (p - p_c)^\mu. \tag{5.37}$$

The debate was so lively that a new scaling relation for μ (often also called t) was proposed almost every year! The objective was to link μ to other exponents, particularly to the fractal dimension of the infinite cluster. Light was notably shed on this complex situation by the work of Alexander and Orbach [1]. It turned out that the dynamic properties are particularly complex because of the *multifractal* nature of the infinite cluster [2, 17]. This means that in theory an infinite number of dimensions is needed to rigorously describe transport in an infinite cluster at the percolation threshold. We refer the reader to [2] for more details. In the following, we will successively consider two types of initial conditions for our random walker: either the walker starts in the infinite cluster, or at any occupied site.

5.5.2.1   Diffusion in the Infinite Cluster

First of all we will consider the case of a random walker who starts from within the infinite cluster, assuming p is slightly larger than p_c. Suppose that after a time t, measured in number of steps from one site to another, the walker has travelled a mean square distance R^2 from its starting point. The walker may have visited each of the $M(t)$ sites within the sphere of radius R zero or more times. In an ordinary object, the normal diffusion laws lead to the relations:

$$M(t) \sim R(t)^d \sim t^{d/2}. \tag{5.38}$$

In a scale invariant object, we would be tempted to replace d in these two relations by the fractal dimension of the object. However, that would keep a normal diffusion law whereas experience shows that this is not valid at the percolation threshold. Here we use what is called the *spectral dimension* d_s (already introduced in Chap. 4) to characterise the temporal dependence of the space visited by the walker:

$$M(t) \sim R(t)^{d_F} \sim t^{d_s/2}. \tag{5.39}$$

As we have already discussed in Chap. 4, the name spectral dimension is justified by the low frequency ω phonon density in the percolating cluster [4], which is shown to be of the form ω^{d_s-1}. Equation (5.39) leads to an anomalous diffusion law:

$$R(t) \sim t^{\nu_D} \qquad \text{where} \qquad \nu_D = d_s/2d_F. \tag{5.40}$$

The characteristic length R can be considered as a coherence length of diffusion, where the inverse of time is equivalent to the distance to the critical point $t \to \infty$. Hence the choice of name is ν_D for the exponent characterising this anomalous diffusion. Remarkably differing significantly from the fractal dimension, the spectral dimension of the infinite cluster is practically independent of the spatial dimension (the conjecture by Alexander and Orbach [1] predicts that d_s is strictly independent of d). In two dimensions $d_s = 1.32$ and in three dimensions, $d_s = 1.33$, which are very close to the mean field value (for $d \geq 6$) of $d_s = 4/3$.

Diffusion in the infinite cluster remains anomalous up to spatial scales of the order of ξ, and becomes normal on longer times. The characteristic time τ of this change in regime is:

$$\tau(\xi) \sim \xi^{1/\nu_D} \sim (p - p_c)^{-\nu/\nu_D}. \tag{5.41}$$

The effective diffusion coefficient $D(p)$ in the infinite cluster at large scale ($R \gg \xi$), that is to say at very short times, is then expressed in the following way:

$$D(p) \sim \frac{\xi^2}{t} \sim (p - p_c)^{-\nu(2-1/\nu_D)} \sim (p - p_c)^{2\nu\left(\frac{d_F}{d_s}-1\right)}. \tag{5.42}$$

We could be tempted to deduce the conductivity by the simple proportionality of (5.42) due to the Einstein relation. In practice, it is necessary to take into account the contribution of the clusters of finite size.

5.5.2.2 Conductivity of a System Near the Transition: Starting From any Occupied Site

The calculation in the previous paragraph assumed that the starting point of the walker is situated in the infinite cluster. The conductivity of the system should be averaged over all the possible initial sites, however it is zero if the starting point is not in the infinite cluster. It is therefore necessary to weight the diffusion coefficient by the mass of the infinite cluster:

$$\Sigma \sim P(p)D(p) \sim (p - p_c)^{\nu\left(d+2\frac{d_F}{d_s}-d_F-2\right)}. \tag{5.43}$$

So the critical exponent of the conductivity obeys:

$$\mu = \nu \left(d + 2\frac{d_F}{d_s} - d_F - 2 \right). \tag{5.44}$$

Table 5.8 shows the critical exponents corresponding to the conductivity and diffusion compared to the fractal dimension of the infinite cluster in the case of two and three spatial dimensions ($d = 2$ and $d = 3$) as well as for the mean field on the Bethe lattice ($d \geq 6$).

5.5.2.3 Young's Modulus of a Gel

Gelification is an important application of percolation. The mechanical properties of a polymer solution undergo a sharp transition when the degree of crosslinking of molecules forming the gel passes the percolation threshold. The viscosity η diverges at the threshold (approached from below) while Young's modulus becomes finite above this level of crosslinking (see Fig. 5.11).

Crosslinking may be brought about by irradiation or chemical reaction. In general the rate of crosslinking increases linearly with time. In practice the transition shown in Fig. 5.11 is observed as a function of reaction time.

Table 5.8 Critical exponent μ corresponding to the conductivity as a function of the dimension

Dimension	$d = 2$	$d = 3$	Bethe: $d \geq 6$
d_F fractal dimension of the infinite cluster	91/48	2.52	4
d_S spectral dimension of the infinite cluster	1.32	1.33	4/3
$\nu_D = dS/2dF$ anomalous diffusion	0.348	0.263	1/6
μ conductivity	1.30	2.00	3

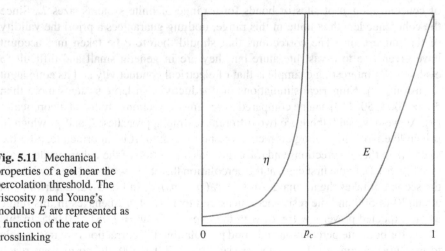

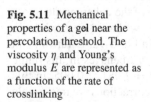

Fig. 5.11 Mechanical properties of a gel near the percolation threshold. The viscosity η and Young's modulus E are represented as a function of the rate of crosslinking

In the early 1980s, it was thought, by formal analogy, that this elastic transition was directly linked to the conductivity transition. This analogy, which would be valid in the case of continuous percolation is not valid in the case of bond percolation. In fact, a lattice of connected bonds above the percolation threshold can have a zero resistance to shearing. For example a cubic full lattice has no resistance to shearing because it can be completely folded without changing the length of any of the individual bonds. Hence, in practice, it is necessary to triangulate such structures to give them rigidity. Even in the case of a triangular lattice of bonds, a completely different rigidity percolation transition is observed from the normal percolation transition [10]. Not only is the percolation threshold different (0.6602 instead of $1/2$) but the critical exponents belong to a new universality class. In this case the coherence length exponent takes the value $\nu = 1.21$ (instead of $4/3$) and that of the order parameter is $\beta = 0.18$ (instead of 0.139). This comes from the fact that a triangle must be complete to be rigid, while only one bond is needed to carry the current to neighbouring bonds, The rigidity transition occurs when a path of connected complete triangles forms.

5.5.2.4 Scaling Corrections

The precise calculation of percolation exponents, as for all other critical exponents, requires taking into account correlations to the scaling, of diverse origins. Remember that universality, resulting in power law behaviours, is a direct consequence of the scale invariance of the system at the transition, which is itself a result of the divergence of a coherence length. We expect to observe this universality only for large systems, in the immediate vicinity of the transition, where the linearisation of the renormalisation transformation is valid. Certain correlations or scaling come from non linearities of the renormalisation transformation. However, determining an exponent numerically required tracing the expected power law in a finite region of concentration p of sites or bonds for a range of finite systems sizes L. Since the coherence length is finite in this range, nothing guarantees a priori the validity of the universality. The corrections that should therefore be taken into account have given rise to a vast literature but they are in general small and difficult to evaluate. An interesting example is that of electrical conductivity and its equivalent diffusion [17]. Numerical simulations of conductivity in large systems more than $500 \times 500 \times 500$ [13] can be compared to scaling corrections predicted theoretically [3]. As we have said, there are two different scaling approaches, L and p, which in principle enable the scaling corrections to be measured. It is, in principle, possible to compare these corrections and trace them as a function of the coherence length ξ (see Fig. 5.12). In the first case, at the percolation threshold we have $\xi = L$ whilst in the second, it takes the normal form $\xi = \xi_0 (p - p_c)^{-\nu}$. In Fig. 5.12, L varies from about 30 to 500 and the resistance, measured exactly at p_c, behaves in this range as the expected power law for L, with the exponent $t/\nu = 2.281 \pm 0.005$. In this range, for both site percolation and bond percolation, the corrections are impossible to measure. In contrast, for a cube of size $500 \times 500 \times 500$, the corrections reach

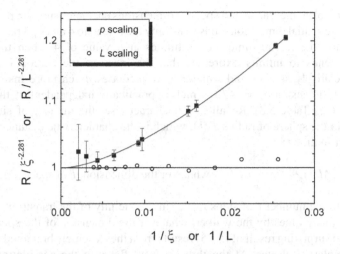

Fig. 5.12 Scaling corrections to the resistance near the percolation threshold in a simple cubic lattice. The resistance divided by the expected power law is represented as a function of the inverse of the side of the cube $1/L$ (open circles), in this case up to 500, for the case of site percolation (in the case of bond percolation the corrections are not measurable in this range of sizes) and as a function of the inverse of the coherence length $1/\xi(p)$ (filled squares), for the case of bond percolation for a cube of $500 \times 500 \times 500$ (after [13])

20% as the distance from the threshold $(p - p_c)$ becomes 10%, in other words ξ is of order 30. The form of the observed variation is in excellent agreement with the corrections predicted theoretically [3], but their magnitude is at least an order of magnitude larger than that observed by finite size scaling. This difference is no doubt related to the multifractal nature of the infinite cluster. As soon as we move away from the percolation threshold, the topological properties of the infinite cluster, such as its connectivity or the density of *red bonds*, are seriously affected.

5.6 Conclusion

Percolation is a unifying concept in describing nature [9], permeating all fields of science, We live in a three dimensional space where we are surrounded by many quasi one dimensional and two dimensional objects. Compared to the critical dimension $d = 6$, characterising percolation, these are rather small dimensions in which fluid does not flow easily in a heterogeneous material. In this particular space of ours, do we find that percolation is important? Far from the transition, there is no doubt it is. A single blockage makes a 1D pipe useless and gruyère cheese does not fall apart although it is more than 80% holes. Near the threshold, the universal character seems even wider.

Let us relook at the fractal and spectral dimensions characterising the percolating cluster: as the spatial dimension tends to infinity these tend to 4 and $4/3$ respectively. The fact that the fractal dimension saturates at a value of 4 when the spatial dimension tends to infinity expresses that the mass of the percolating cluster becomes relatively smaller and smaller as d increases. Even more instructive is the spectral dimension $d_s = 4/3$, which is practically independent of the spatial dimension (see Table 5.8). Recall that it characterises the number of sites $M(t)$ contained in the sphere of radius $R(t)$, which is the characteristic distance covered by a walker in time t:

$$M(t) \sim t^{d_s/2} \sim t^{2/3} \quad \text{whatever the dimension } d \text{ of space,} \qquad (5.45)$$

compared to the number t of steps made. In the vicinity of the transition, the sites are visited many times by the walker, whatever the dimension of the space of the system. The surprising result of (5.45) results from the extremely branched topology of the percolating cluster. At the threshold, all flows in the percolating cluster occur as if the *local dimension* was about 1 and independent of the dimension of space. Conductance and diffusion are dominated by obligatory passage points (bottlenecks) that are called *red bonds*. The percolating path or *backbone* of the infinite cluster, the only part useful for transport once the dead ends have been deleted, is formed of a string of such red bonds alternating with loops (*blobs*) of all sizes. Dynamic properties such as diffusion and conduction therefore show a "hyper universal" slowness near the transition, in the sense that they depends very little on the dimension of space. Once final point that the reader will not fail to notice is the simplicity of implementation of numerical simulations and the diversity of model systems imaginable. The reader may rightly wonder why we have mainly discussed numerical results and not presented experimental results, as we have done for phase transitions. In general, physical experiments are much more difficult to interpret, due to correlations in position, finite size effects, heterogeneity of elements and quality of contacts between them. In large systems, scaling approach predictions are accurately verified, but the precision obtained in the value of the exponents does not have the same quality as that obtained in the case of phase transitions.

References

1. S. Alexander, R. Orbach, Density of states on fractals-fractons. J. Physique Lett., **43**, L625 (1982)
2. L. De Arcangelis, S. Redner, A. Coniglio, Anomalous voltage distribution of random resistor networks and a new model for the backbone at the percolation threshold. Phys. Rev. B **31**, 4725 (1985)
3. H.G. Ballesteros et al., Scaling corrections: site percolation and Ising model in three dimensions. J. Phys. A **32**, 1 (1999)
4. J.P. Bouchaud, A. Georges, Anomalous diffusion in disordered media – Statistical mechanisms, models and physical applications. Phys. Rep. **195**, 127 (1990)

5. S.R. Broadbent, J.M. Hammersley, Percolation processes. Proc. Cambridge Philos. Soc. **53**, 629 (1957)

6. G. Deutscher, R. Zallen, J. Adler, *Percolation Structures and Processes*, about twenty papers in *Annals of the Isral Physical Society,* vol. 5 (Hilger, Bristol, 1983)

7. P.D. Eschbach, D. Stauffer, H.J. Hermann, Correlation-length exponent in two-dimensional percolation and Potts model. Phys. Rev. B **23**, 422 (1981)

8. P. Flory, Molecular size distribution in three-dimensional polymers: I, gelation.s J. Am. Chem. Soc. **63**, 3083 (1941)

9. P.G. De Gennes, La percolation, un concept unificateur. La Recherche **7**, 919 (1976)

10. D.J. Jacobs, M.F. Thorpe, Generic rigidity percolation: the peeble game. Phys. Rev. Lett. **75**, 4051 (1995)

11. P.W. Kasteleyn, C.M. Fortuin, Phase transitions in lattice systems with random local properties. J. Phys. Soc. JPN Suppl. **16**, 11 (1969)

12. H. Kesten *Percolation Theory for Mathematicians* (Birkhäuser, Boston, 1982)

13. B. Kozlov, M. Laguës, Universality of 3D percolation exponents and first-order corrections to scaling for conductivity exponents, Physica A, 389, 5339 (2010)

14. M. Laguës, Electrical conductivity of microemulsions: a case of stirred percolation. J. Physique Lett. **40**, L331 (1979)

15. J. Schmittbuhl, G. Olivier, S. Roux, Multifractality of the current distribution in directed percolation. J. Phys. A-Math. Gen. **25**, 2119 (1992)

16. D. Stauffer, A. Aharony, *Introduction to Percolation Theory*, 2nd edn. (Taylor & Francis, London, 1992)

17. O. Stenull, H.K. Janssen, Noisy random resistor networks: Renormalized field theory for the multifractal moments of the current distribution. Phys. Rev. E **63**, 036103 (2001)

Chapter 6
Spatial Conformation of Polymers

6.1 Introduction

6.1.1 Remarkable Scaling Properties

Although their synthesis takes place in chemistry, or even biology (for example DNA), polymers are widely studies by physicists, mainly because they form materials with remarkable properties; just think of the variety and usefulness of plastics in our day to day environment. However, it is not so much their interest and importance that preoccupies us here, but the fact that polymer physics is a domain in which scaling laws are omnipresent, as well as being conceptually fundamental and extremely useful in practice [5]. Polymers possess a natural scale variable: their degree of polymerisation N, that is to say the (very large) number of monomers making up each polymer. N is directly related to the mass of the polymer, which is the quantity measured experimentally. In addition, the large size of polymers legitimises their observation, description and simulation at supermolecular scales, where details of atomic interactions are only indirectly involved in terms of apparent geometric constraints and characteristic averages.

In this chapter, we will limit ourselves to presenting the scaling laws encountered when studying the spatial conformation of *one of the polymers in the system*. Due to the concurrence, or more often the competition, between the *connectivity* of the polymer (linear chain of monomers) and its *three dimensional structure* (interactions between monomers depending on their separation distance in real space), the N dependence of the observables characterising this conformation will be non trivial and very different from the usual spatial extension which would be observed in a cloud of free monomers.[1]

---

[1]The limit $N \to \infty$ is often called the "thermodynamic limit" but the nature and theoretical foundations of these two limits are different, as we will see in Sect. 6.3.3. However, these limits both have the effect of "coarsening" physical properties by revealing their dominant behaviours

A. Lesne and M. Laguës, *Scale Invariance*, DOI 10.1007/978-3-642-15123-1_6,
© Springer-Verlag Berlin Heidelberg 2012

The conformation of one of the polymers of the system will depend crucially on the relative weight of the interactions felt between one of its monomers and the other monomers, other polymers and solvent. It is therefore important to distinguish between three different concentration regimes. The first is that of very dilute solutions in which each polymer can be considered isolated and therefore oblivious to the presence of other polymers but very sensitive to the solvent. The second is that of very concentrated solutions in which interactions between different polymer chains play a dominant role. The final regime is that of solutions with intermediate concentrations, in which the chains interact with each other and the solvent. We will see that in these different situations, the N dependence of the observables takes the form of universal scaling laws, in the sense that their dependence on the specific atomic structure of the polymer (its "chemical formula") is only through a few effective parameters.

We will focus our attention mainly on the spatial conformation of an *isolated* polymer (which is observed in a very dilute solution) and on the scaling laws describing the *statistical* properties of this conformation, for example the average end-to-end distance and its probability distribution. By interpreting a polymer conformation as a trajectory of a random walk (where N plays the role of time), these scaling laws connect with the diffusion laws presented in Sect. 4.5. Their exponents are universal within classes determined by the physical context controlling the interactions between monomers and their influence of the conformation of the polymer chains. For example, specific behaviours are observed for *polyelectrolytes* (polymers in which the monomers are charged, Sect. 6.2.4). The solvent plays an equally essential role in the conformational properties. The affinity of solvent for the monomers can change when the temperature changes; if it decreases below that of the affinity of monomers for each other (we say that we pass from a good solvent to a bad solvent), a *conformational transition* is observed, during which the polymers collapse onto themselves adopting a compact globular conformation. The transition point, in this case a temperature, is called the Θ *point* of the polymer/solvent pair (see Sects. 6.2.3 and 6.3.2).

At the other extreme lie scaling laws describing properties of very concentrated solutions and *polymer melts*. The statistics of a chain within the ensemble is then that of an ideal, freely jointed chain in which the monomers in the chain do not exert any constraints on each other (Sect. 6.4.1). Finally, scaling laws appear in *semi-dilute* solutions, which are intermediate in concentration between very dilute solutions and polymer melts. The scaling variable is then given by the volume fraction of monomers (see Sect. 6.4.2).

We will see that it is possible, as in the case of percolation, to take into account the effect of all the physical ingredients by using essentially geometric models: the ideal chain (Sect. 6.2.1), self-avoiding chain (Sect. 6.2.2) and chain with excluded volume (Sect. 6.2.3). The universality of the scaling laws follows from this. The first

---

at the expense of dampening higher order corrections. For example, taking such a limit $N \to \infty$ transforms peaks into Dirac functions and smooth steps into discontinuities.

approaches used to determine conformational properties of a polymer were mean field type approaches (Flory theory, Sect. 6.3.1 or Flory–Huggins theory including the solvent, Sect. 6.3.2). We will present their successes but also their failures, which led to them being replaced by renormalisation methods (Sect. 6.3.5).

The large diversity of sizes, N, accessible both experimentally and numerically have enabled the validification and exploitation of the corresponding scaling laws. The important point is that they are valid in a whole range of temperatures and not just at a singular value T_c. In other words they govern the *typical* behaviour of polymers. They will therefore be robust, widely observed and used in all experimental and technological approaches. Polymer physics is a domain in which scale invariance is very strong and not limited to the existence of exponents: it is expressed through universal functions (*scaling functions*). In practice, experimental or numerical curves obtained for different values of N superimpose on a single curve after *rescaling* the variable and observables: for example, the end-to-end distance R is replaced by the scaling variable $RN^{-\nu}$.

Interest in studies of conformational properties of a polymer has recently been reinforced by remarkable experimental progress such that it is now possible to observe and even manipulate a single macromolecule in isolation (for example by fluorescent labelling and manipulation using micropipettes, atomic force microscope cantilevers or lasers used as "optical tweezers" after grafting beads to the ends of the molecule). In this way, theoretical models can be directly validated by experiment and conversely, their predictions form a valuable guide in interpreting the results of these experiments. They are used in a very fruitful way in biological contexts (for example for DNA molecules).

6.1.2 Persistence Length

All of polymer physics begins with a modelling step, consisting of "screwing up ones eyes" so as to see only the averaged out *complex* and *specific* atomic structure of each polymer in terms of only a *small number* of, far more universal, ingredients, which we believe are the dominating factors involved in the observed physical property. This trimming down is further reinforced when we are only looking for the possible scaling laws associated with the property. This is the approach adopted, in an exemplary way, when we are interested in the three dimensional shape (called the conformation) of linear polymers, that is to say polymers consisting of a linear chain of molecular units.

The first stage of modelling is to consider each monomer unit as an elementary entity (the smallest unit) and not to describe the more detailed structure: details of interatomic interactions or the presence of side chains will be taken into account in the effective size l_{eff} of the monomer unit and in the interaction, also effective, with neighbouring monomer units and more generally with the environment (non consecutive monomers, solvent, other polymers). Depending on the precision required, these effective parameters can be obtained from a molecular dynamics simulation or fitted a posteriori by comparing the parameter dependent predictions and the corresponding observations.

A typical example of an effective parameter is that of the persistence length of a polymer. We must first distinguish between two types of polymers; *flexible* and *semiflexible*.

The family of flexible polymers includes the simplest linear polymers, for example polyethylene, in which two successive monomers are linked by a single covalent bond. The dihedral angle between two monomers n and $n + 1$ is fixed by the nature of the bond, however, monomer $n + 1$ can take (with equal probability) b orientations relative to the chain of the first n monomers. These possible orientations correspond to b minima in the total energy with respect to the angular variable describing the orientation in space of the additional monomer. Figure 6.1 shows the case of $b = 2$, observed for example with the *cis* and *trans* configurations of organic polymer chains. The number of possible conformations therefore grows exponentially fast, in b^N, with the number N of monomers. As N increases the end of the chain will very quickly map out the surface of a sphere, which will be more dense, homogeneous and isotropic the larger N.

In other words, we observe a decorrelation of orientations. The chain length beyond which the orientation of the first and the last monomers can be considered independent is called the *persistence length l_p* of the polymer. To dominant order, it does not depend on temperature. It is convenient to redefine what we call a monomer and take this length l_p as a single unit. At this scale the polymer appears as a perfectly supple, flexible chain, called "freely jointed". This step, reducing the specificity of the initial model by incorporating it into a single parameter l_p, is particularly interesting in numerical models of polymers, on a lattice. In such a model a monomer will represented simply by a bond separating two adjacent nodes

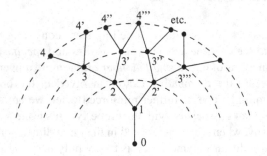

Fig. 6.1 *Entropic* persistence length l_p due to the existence of b possible relative orientations between two successive monomers. For convenience, the drawing was done with $b = 2$. It illustrates the fact that the points where one could find the end of the chain will fill the circle, in a more and more homogeneous and dense way as the length N of the chain increases (the radius grows as N whereas the number of points grows as 2^N). l_p is the length beyond which the orientation of a monomer has become independent from that of the first monomer in the chain. l_p is practically independent of temperature

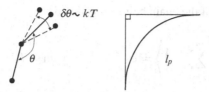

Fig. 6.2 *Left*: angular freedom $\delta\theta$, allowed by thermal fluctuations, between two successive chemical bonds along a semiflexible polymer. *Right*: graphical interpretation of the persistence length as the (average) characteristic length beyond which thermal fluctuations allow a curve in the chain corresponding to an angle of $\pi/2$ between the tangents at each end. This length typically has a $1/T$ dependence on temperature T

on the lattice (Fig. 6.4). The length of this bond is identified with l_p to return to real sizes.

The family of semiflexible polymers consists of polymers with more complex chemical structures, involving stronger bonds than a single covalent bond between monomer units, for example double bonds or additional hydrogen bonds or other physical bonds between successive monomers. A typical example is that of DNA. For DNA the monomers are the base pairs and they link together atoms in each strand of the double helix with covalent bonds ("phosphodiester bonds") forming the "backbone". To this, hydrogen bonds are added between the complementary base pairs (linking the two strands together) and *stacking* interactions between the base pair steps.[2] In such complex polymers, the orientation between successive monomers is well defined and the only flexibility allowed lies in the thermal fluctuations of the dihedral angle θ_j between monomers j and $j+1$ (see Fig. 6.2). Fluctuations in the angle θ_j and those of the angle θ_{j+1} are independent, so their variances, proportional to kT, are therefore additive. The angular freedom between monomer unit 1 and n will therefore behave as $\sqrt{nkT}$. When this freedom is quite large, for example when the typical angular deviation exceeds $\pi/2$, we can consider the orientations of the monomers statistically independent (Fig. 6.2). We therefore see the appearance of a minimal number $n_p \sim 1/kT$ (where $\sim$ means up to a constant factor) of monomers above which the chain of monomers has lost memory of its initial orientation (orientation of the first monomer). The corresponding length $l_p = n_p l_{\text{eff}}$ is called the persistence length (arc length) of the semiflexible polymer. It is written $l_p = A/kT$ where A depends on the atomic structure of the linear polymer under consideration. In practice, we can calculate l_p as the angular correlation length *via* the following relation ($l = n l_{\text{eff}}$ being the

---

[2]Incidentally it is the mismatch between the natural spacing of the bases along the strands, of the order of 7Å, and that of 3.4Å imposed by these stacking interactions that causes the molecule to adopt its double helix shape.

length of the segment of the chain being considered):

$$\left\langle \cos\left(\sum_{j=1}^{n} \theta_j\right)\right\rangle = e^{-l/l_p} \qquad \text{with} \qquad l_p = \frac{A}{kT}. \tag{6.1}$$

An elastic bending energy is therefore associated with each polymer configuration. If ρ is the local curvature of the chain, the elastic energy density per unit length[3] is written $A\rho^2/2$. The coefficient A, independent of temperature T to dominant order, is interpreted as the bending elastic constant. Such a model is called the *worm-like chain* model and it is used when the chain is not very flexible (l_p large compared to the molecular scale) and its rigidity is an essential conformational parameter.

6.2 Conformations of an Isolated Flexible Polymer

6.2.1 Polymers and Random Walks

It is fruitful to envisage the spatial conformation of a linear flexible polymer as the trajectory of a random walk. Then, mathematical results obtained for random walks (as in Sect. 4.3) can be interpreted directly in terms of statistical properties of a polymer. The number of monomers N corresponds to the number of steps of the random walk. Physical properties of the assembly of monomers (e.g. persistence length, excluded volume, attractive interactions) result in rules governing the displacement of the walker. From now on we will denote the root mean square length of a monomer as a (in practice chosen to be equal to the persistence length l_p to avoid angular constraints between successive steps). Saying that a random walk asymptotically follows the diffusion law $R(t) \sim t^{\gamma/2}$ means that the end-to-end distance $R(N) \equiv \langle|X_N - X_0|^2\rangle^{1/2}$ (see Fig. 6.3) follows the scaling law $R(N) \sim N^{\gamma/2}$ in the limit $N \to \infty$, up to a constant factor depending on the units chosen.[4] The simplest polymer model is that corresponding to an *ideal random walk* (identical and statistically independent, unbiased steps, Fig. 6.4). We have seen in Sect. 4.3.1 that in this model:

$$R(N) \sim a\sqrt{N}. \tag{6.2}$$

---

[3]By adopting a description of the chain as a continuous curve, of arc length coordinate s and local tangent $t(s)$, the bending energy of a length l of chain is written $(A/2)\int_0^l (dt/ds)^2(s)ds = (A/2)\int_0^l \rho(s)^2 ds$.

[4]Note that it is not at all obvious that the end-to-end distance of a real linear polymer chain follows a scaling law with respect to the number of monomers. At this stage it is only a working hypothesis and this present chapter describes the steps of its validation. The best evidence is the experimental observation of such a scale invariance. Some theoretical clues are given by the scaling laws mathematically demonstrated for normal and several anomalous walks (Chap. 4).

Fig. 6.3 End-to-end distance
R and radius of gyration R_g

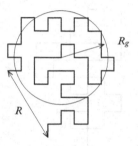

The distribution $P_N(r)$ is Gaussian,[5] given by

$$P_N(r) \sim e^{-dr^2/2Na^2}, \tag{6.3}$$

where d is the dimension of space. From this we deduce the entropy of an ideal chain:

$$S_N(r) = S_N(0) - k_B \frac{dr^2}{Na^2}. \tag{6.4}$$

Most models and results presented in Chap. 4 will have an equivalent in the context of polymer physics, in particular anomalous diffusion and its critical nature.

6.2.2 Self-Avoiding Random Walks

The ideal random walk model has the major deficiency of not taking into account the *excluded volume* constraints between monomers in the chain. In reality, monomers cannot interpenetrate each other, nor can they approach each other too closely. Consequently, polymer chains cannot cross themselves nor even touch themselves. We therefore talk of *self-avoiding chains*[6] shortened to SAW, (see Fig. 6.4) [18]. Quantitatively, this property is modelled by an infinitely repulsive interaction, of range the diameter of the exclusion zone, in other words by introducing an excluded volume v, also written $v = a^d w$ where w is a nondimensional parameter. This excluded volume is the analogue of the volume of the spheres in hard sphere models. In lattice models, it is sufficient to simply forbid double occupancy of sites. It should be emphasized that this model remains *purely geometric*: the physical property of short range repulsion between monomers is entirely taken into account in a constraint imposed on the trajectory of the chain, which results in reducing the space of allowed configurations.

---

[5]We talk about *Gaussian chains* when $P_N(r)$ is asymptotically Gaussian. This is the case whenever the range of correlations along the chain remains bounded.

[6]In mathematical terms, this corresponds to non Markovian random walks: the walker must keep an infinite memory of its history to avoid repassing a site it has already visited.

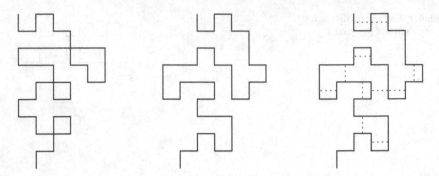

Fig. 6.4 Three polymer models on a (here *square*) lattice. *Left*: An ideal walk with independent successive steps (the length of a step is equal to the persistence length of the polymer, see Sect. 6.1.2); *Middle*: self-avoiding walk in which the track cannot cross; *Right*: self-avoiding walk with interactions, in which an attractive energy $-J$ is attributed to each of the contacts represented by *dashed lines*

In the model of a random walk that does not cross itself (self-avoiding walk), the chain has a complete memory of its preceding track. In this sense, temporal correlations have a range N which diverges with the length of the chain: *in the limit $N \rightarrow \infty$, such a polymer appears like a critical object.* Experimental[7] and theoretical results show that the end-to-end distance behaves as:

$$R(N) \sim N^{\nu(d)}. \tag{6.5}$$

We recover $\nu(d) = 1/2$ when the spatial dimension d is higher than a critical dimension $d_c = 4$, above which excluded volume constraints have a negligible probability of being felt and are not sufficient to change the ideal walk scaling law. If $d < 4$, we have $\nu(d) > 1/2$. This anomalous exponent is called the *Flory exponent*.[8] The values considered are $\nu(1) = 1$, $\nu(2) \approx 3/4$ and $\nu(3) \approx 3/5$. By definition, the fractal dimension of self-avoiding walks is equal to $d_f(d) = 1/\nu(d)$, and so less than the dimension 2 of ideal chains (Sect. 4.1.3). The radius of gyration[9] $R_g(N)$ follows a similar scaling law, with the same exponent ν, but with

---

[7]The experimental results mostly concern the case of three dimensions $d = 3$; two dimensions $d = 2$ can however be achieved by using thin films (for example Langmuir monolayers).

[8]We sometimes use this in a more restricted sense, to designate the value of this exponent when it is obtained in the Flory theory, Sect. 6.3.1.

[9]By writing the position of N monomers as $(r_i)_{i=1...N}$, the radius of gyration is defined by:

$$R_g^2(N) \equiv \frac{1}{N} \langle \sum_{i=1}^{N} |r_i - r_g|^2 \rangle = \frac{1}{2N^2} \langle \sum_{i,j}^{N} |r_i - r_j|^2 \rangle,$$

where r_g is the position of the centre of mass of the chain: $r_g = N^{-1} \sum_{i=1}^{N} r_i$. So, the radius of gyration is the radius of the sphere, centred at r_g, with the same moment of inertia as the chain would have if the total mass N of the polymer was redistributed on its surface (see Fig. 6.3).

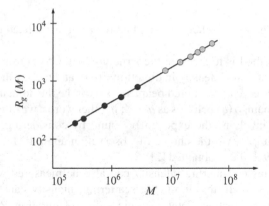

Fig. 6.5 Log-log graph of the radius of gyration R_g as a function of molecular weight M, directly proportional to the number of monomers N. The slope gives the value $\nu = 0.586 \pm 0.004$ for the Flory exponent. The *black* and *grey circles* correspond to two series of observations, by light scattering of dilute solutions of polystyrene in benzene (after [3])

a different prefactor. Experimental data (from light or neutron diffraction) show that the exponent $\nu(d)$ is invariant within classes of polymers, over a large range of temperatures. The value obtained (in three dimensions) is $\nu_{exp} = 0.586 \pm 0.004$ [3] (see Fig. 6.5). The most accurate values are obtained theoretically (Sect. 6.3.3). For example, we have $\nu(3) = 0.5880 \pm 0.0010$ [17]. Tracing $R(N)N^{-\nu}$ or $R_g(N)N^{-\nu}$ as a function of temperature T for different values of N leads to a universal curve for $T > T_\theta$ where the threshold T_θ is called the Θ point of the polymer (Sect. 6.2.3).

Measuring the exponent ν

A solution, dilute enough that the chains do not interact with each other and individually take the conformation they would if they were really isolated, can be observed by neutron diffraction.

Since the diffraction is elastic, the wave vector k_i of incident neutrons and the wave vector k_f of scattered neutrons have the same wavenumber k and their difference $q = k_f - k_i$ is related to the diffraction angle θ obeying $q = k \sin(\theta/2)$. We use "thermal" neutrons, that is neutrons whose kinetic energy is reduced by thermalisation to bring their wavelength (directly linked to the resolution) to values of the order of a nanometer, typical of polymers. In order to extract the signal due to a single polymer when we observe a solution, half of the chains are labelled with deuterium and a mixture of water (H_2O) and heavy water (D_2O) is taken as the solvent. By adjusting the proportions of the mixture ("contrast variation" method), certain terms in the scattering intensity can be cancelled, leaving only the term describing the contribution of individual chains, and therefore the behaviour of an isolated chain can be deduced.

The first method is to measure the structure factor $S(q)$ (directly proportional to the scattering intensity in direction $\theta(q)$), and to vary the direction θ in which we observe the scattered beam so as to probe the domain $a < q^{-1} < R_g$. In this domain, $S(q)$ behaves as $q^{-1/\nu}$; in other words, we directly observe a scaling law involving the exponent ν found by considering segments of variable length q^{-1}, *via* the choice of observation angle θ. In this way the value $\nu = 0.59 \pm 0.2$ is obtained [2].

The small angle scattering intensity can also be measured, which corresponds to $q \rightarrow 0$. In this limit, the scattering intensity can be predicted theoretically as a function of the average radius of gyration $R_g(N)$ of the chains. By fitting the theoretical curve to the experimental data the unknown quantity $R_g(N)$ can be determined. By successively considering different values of N, Flory's prediction can be tested and an estimation of the exponent ν can be given. This second method, apparently more direct, rests on the assumption of monodispersity of the chains, which in practice decreases its accuracy and reliability. However the systematic error coming from polydispersity can be corrected, leading to a value of $\nu_{exp} = 0.586 \pm 0.004$ [3].

6.2.3   The Role of Solvent: The Θ Point

A remarkable phenomenon, known by the name of the Θ *point*, is the transition, observed at a particular temperature $T = T_\theta$, between scaling laws with different exponents for the end-to-end distance or the radius of gyration of a polymer:

$$\begin{cases} R(N) \sim N^{\nu(d)} & \text{if } T > T_\theta \\ \\ R(N) \sim N^{\nu_\theta(d)} & \text{if } T = T_\theta \\ \\ R(N) \sim N^{1/d} & \text{if } T < T_\theta. \end{cases} \qquad (6.6)$$

with $1/d < \nu_\theta(d) < \nu(d)$ where d is the dimension of space. The fact that different scaling laws are observed either side of the temperature T_θ reveals that a conformational transition occurs here. The value of T_θ depends, not only on the polymer, but also on the solvent, showing that it plays an essential role in the phenomenon. In three dimensions $d = 3$, it is theoretically predicted that $\nu_\theta(3) = 1/2$ and that the asymptotic behaviour $R(N) \sim \sqrt{N}$ of an ideal random walk must be recovered. Note that, despite all that, the chain is not an ideal chain: the end-to-end distribution (or radius of gyration distribution), as well as the correlations between steps, differ from that on an ideal chain. In two dimensions $d = 2$, the theories agree on the value $v_\theta(2) = 4/7$ [8].

We can qualitatively understand the existence of this Θ point. It comes from the fact that the relative weight of interactions between, on the one hand two monomers, and on the other hand a monomer and the solvent, varies with temperature. At high temperature ($T > T_\theta$), the repulsion between monomers dominates and a monomer "prefers" to be surrounded by solvent: we call this regime "*good solvent*". In this case, the self-avoiding random walk model correctly describes the conformations of the chain (so it is called a *random coil*) and we have $R(N) \sim N^v$.

At low temperature ($T < T_\theta$), the repulsion between a monomer and the solvent molecules is so strong that the monomer "prefers" to be surrounded by other monomers. This causes the chain to collapse on itself, with $R(N) \sim N^{1/d}$: we call this regime "*bad solvent*" and the chain is in a *globule state*. The geometry of typical conformations is compact, with fractal dimension $d_f = d$. At $T = T_\theta$, the excluded volume repulsion between two monomers is exactly compensated by the apparent attraction between monomers coming from the solvent which repels them. The self-avoiding walk model only describes the chain in good solvent. To take into account the globular phase and the coil-globule conformational transition occurring at the Θ point, we need to add to the model a short range attractive interaction $-J$, which manifests itself whenever two non consecutive monomers become neighbours on the lattice. This model is called an *interacting self-avoiding walk*, shown in Fig. 6.4.[10]

The theoretical questions are to determine the nature of this transition and, if possible, to unify the three scaling laws observed in the three different phases into a compact form (the analogue of that describing tricritical points)[11]:

$$R(N, T) \sim N^{v_\theta} \, f(N^\phi(T - T_\theta)). \qquad (6.7)$$

The behaviour of the universal function f at infinity is fixed by the requirement to recover the scaling laws of the coil state and the globule state:

$$
\begin{cases}
f(z \to +\infty) \sim z^{\frac{v-v_\theta}{\phi}} \\[2ex]
f(z \to -\infty) \sim |z|^{\frac{1/d-v_\theta}{\phi}}.
\end{cases}
\qquad (6.8)
$$

---

[10]The self-avoiding walk is recovered in the limit $T \to \infty$. It describes the entropic contribution in the conformation distribution of the interacting self-avoiding walk (the effective parameter controlling the weight of the interaction is $K = \beta J$, where here $\beta = 1/kT$).

[11]In the phase transitions framework presented in Chap. 1, *tricritical points* are observed in systems that have a second control parameter ϵ (other than the temperature) governing, for example, the relative weight of two contributions in the Hamiltonian belonging to different universality classes. A line of critical points $T_c(\epsilon)$ is then observed and the tricritical points correspond to the singular points of this line and mark the transition from one universality class to another.

The exponent ϕ, called the *crossover exponent*, will in particular describe the way in which the transition temperature $T_\theta(N)$ depends on finite size N i.e. $T_\theta(N) - T_\theta \sim N^{-\phi}$. To experimentally observe the Θ point, we can for example use dilute polystyrene in cyclohexane, a solvent whose affinity for styrene monomers varies strongly with temperature. A "Θ solvent" is obtained for this particular polymer at a temperature of 34.5°C. Measurements give the value $\nu_\theta = 0.500 \pm 0.004$ after correction for the effect of polydispersity [3], in agreement with the theoretical prediction $\nu_\theta = 1/2$ in three dimensions [5].

6.2.4   Scaling Laws for a Polyelectrolyte

Let us continue this presentation of scaling laws shown by isolated polymers with the case of a naturally charged polymer, which we call a *polyelectrolyte*. One example is DNA, in which each phosphate group in its backbone carries a negative charge, so there are two negative charges per base pair,[12] which is extremely high (it gives a total of about 6×10^9 charges for the DNA contained in one of our cells!). A polyelectrolyte in aqueous solution with added salt (which is the situation for DNA in a cell) will strongly attract oppositely charged ions, which will partially neutralise it. We call these *counterions*.

It is possible to understand what will be the effect of charges on the conformation of a linear polyelectrolyte, when all the charges it carries are identical, without performing any calculations. Each charge repels its neighbours and the polymer will therefore maximise the distance between its monomers, optimally achieved when it is straight.[13] Let us consider the case of a semiflexible polymer, for example DNA. The molecule's natural rigidity is reinforced by a contribution coming from Coulomb repulsion between the charges. This increase of the chain rigidity can be taken into account by adding an "electrostatic" persistence length l_{el} to the structural persistence length l_p^0, such that $l_p = l_p^0 + l_{el}$ [16]. For example, the total persistence length of DNA in physiological conditions is 53 nm (for a radius of 1 nm), or around 150 base pairs, with comparable structural and electrostatic contributions. Polyelectrolytes are therefore very rigid molecules.

The global behaviour of a polyelectrolyte and the way in which it will interact with its more distant environment will clearly depend on how much it is neutralised by counterions, so on the number of counterions available in the solution and also the valency z of the ions. The capacity of a salt solution to neutralise charged objects placed in them is evaluated by their *ionic strength* $\psi = \sum_i c_i^2 z_i^2$ where the sum is over all the ionic species,[14] c_i being the concentration of species i of valency z_i.

---

[12]Base pairs are 3.4Å apart measured along the central axis of the double helix, giving a negative charge $-e$ per 1.7Å of DNA.

[13]Note that in solution Coulomb interactions are screened, so the straightening and stiffening of the polymer is limited and the total persistence length l_p is finite.

[14]The valency appears in squared form, the contributions from different species are all positive.

When the interaction of a polyelectrolyte with solvent and ions contained in it is studied in more detail, three characteristic length scales appear:

- The first is the *Bjerrum length* $l_B = e^2/4\pi\epsilon kT$, where $\epsilon = \epsilon_0\epsilon_r$ is the dielectric constant (or permittivity) of the solvent ($\epsilon_r = 78.5$ for water). From this definition, it is clear that this is the length at which the Coulomb energy between two charges is equal to the thermal energy kT. Therefore, l_B is the distance beyond which thermal motion takes over from Coulomb interactions.
- The second is the *Debye length* $l_D = \sqrt{kT\epsilon/e^2\psi} = 1/\sqrt{4\pi l_B\psi}$. It depends on the solvent (*via* ϵ) but also on the salt (*via* ψ). It is the characteristic screening length of the Coulomb interactions: in salt water the electrostatic potential takes the form $v(r) = \frac{1}{4\pi\epsilon r}\,e^{-r/l_D}$, with finite range l_D.
- The third, depending only on the polyelectrolyte, is the length λ per charge along the chain; the charge line density is therefore $\pm e/\lambda$, depending on the sign of the charges.[15] We distinguish weakly charged polyelectrolytes, for which $\lambda < l_B$, and strongly charged polyelectrolytes for which $\lambda > l_B$.

If the polyelectrolyte is weakly charged, its neutralisation by counterions will depend on the ionic strength. When salt is added to the solution the polyelectrolyte will soften because oppositely charged ions (released by disassociation when the salt dissolved) come and partially neutralise it and thereby reduce the electrostatic repulsion and the corresponding contribution l_{el} to the persistence length. This decrease in persistence length $l_p = l_p^0 + l_{el}$ is show experimentally when salt is added, until the value l_p^0 of the neutral chain is reached. The experimental results imply that the dependence of l_{el} on the concentration of dissolved salt obeys a scaling law but its exponent is still controversial.

If the polyelectrolyte is charged enough, a remarkable phenomenon occurs: *Manning condensation*. This corresponds to the formation of a layer of counterions practically in contact with the polyelectrolyte. The number of ions which condense onto the polyelectrolyte and neutralise it is determined by the temperature T, the charge per unit length of the polyelectrolyte (if it is high enough) and the valency z of the ions, but it is *independent of the concentration of ions* in the solution. We will go into more details below of this phenomenon and its similarity to a phase transition [19]. The lengths l_B and λ control the transition point, whereas l_D controls the thickness of the counterion layer.

Manning condensation

In the "mean field" treatment of the problem by Onsager [21], we start by assuming that there is only one counterion situated at a distance less than l_D and we calculate the partition function of N counterions. If $zl_B > \lambda$, it

---

[15]Here we are only considering the simple case of homopolymers, in which all the monomers carry the same charge.

diverges when N tends to infinity, showing that the starting hypothesis is false and that there are in fact a finite fraction of counterions which condense onto the filament, at a distance less than l_D. The condition $zl_B > \lambda$ means that the condensation is produced more easily as the temperature is lowered, that the polyelectrolyte is charged and that the valency of the counterions present in the solution is high. The threshold $zl_B = \lambda$ is interpreted as a phase transition point corresponding to the condensation of ions on the polyelectrolyte.[16]

A more rigorous approach, based on the Poisson-Boltzmann equation, confirms that this result is valid qualitatively: counterion condensation appears for $zl_B > \lambda$. The remarkable point is that the number of counterions is tuned in such a way that the resulting charge per unit length λ_{layer} of the covered polyelectrolyte, (i.e. the charge seen from outside, the effective charge felt be the environment) is lowered to the value e/zl_B, the threshold at which condensation begins to occur. The counterion layer grows until the value zl_B/λ_{layer} reaches the threshold, which stops further recruitment of counterions. The average distance between two effective charges on the covered polyelectrolyte is then equal to zl_B. This effect is common to all polyelectrolytes and in this sense it is universal. It will of course be involved in the ionic strength dependence of the electrostatic persistence length. An approximate calculation gives $l_{el} = l_D^2/4l_B$.

This condensation plays a crucial role in biology. Most biological macromolecules, for example DNA and RNA, are highly charged objects and the cellular environment where they are found is very rich in ions, including multivalent ions $(z > 1)$. The first effect of the condensation is to concentrate ions important for biological function in the immediate vicinity of the macromolecules. The second effect is to reduce the apparent charge density to a uniform value, so that all the polyelectrolytes present will seem to have the same charge density when "seen from afar". Special effects, when the polyelectrolyte can no longer be described as a uniformly charged filament, locally modify this general mechanism [10].

In conclusion, polyelectrolytes form a very special category of polymers, because in addition to the short range interactions (excluded volume, Van der Waals forces) polyelectrolytes have Coulomb interactions between the charges they carry. These interactions, even though screened by the solvent and the ions it contains,

---

[16]Since the solvent remains globally neutral (adding cooking salt in water introduces Na^+ ions and Cl^- in equal quantities), one could ask what has become of the other ions, those having the same charge as the polyelectrolyte? It can be shown that the play no role in the condensation phenomenon, which is localised in the immediate vicinity of the polyelectrolyte (and, in the end, involves not many ions compared to the total quantity present, but their specific localisation gives them important consequences) [19].

are far more long ranged. Scale invariant behaviours are still observed, but the exponents are radically different. So polyelectrolytes form a new universality class, alongside Brownian random walks (Sect. 6.2.1), self-avoiding walks (Sect. 6.2.2) and interacting self-avoiding walks (Sect. 6.2.3).

6.3 Theoretical Tools

6.3.1 Flory Theory

The first approach to calculate the exponent ν was developed by Flory. It is a mean field type approach declaring as order parameter the average monomer density $c = N/R^d$ inside a volume R^d occupied by the chain. It is traditionally presented as a procedure minimising the free energy $F(N, R)$ of the polymer with respect to R (at fixed N) leading to the expression of the radius of gyration $R_g(N)$. The Flory free energy is written (a being as before the length of a monomer):

$$\frac{F(N, R)}{k_B T} = wa^d \frac{N^2}{R^d} + \frac{dR^2}{a^2 N} + \text{constant.} \tag{6.9}$$

The first term is the total average repulsive energy, proportional to the average number of pairs of monomers close enough to each other to feel the repulsion. The average density of "contacts" is proportional to the average of the density squared, which we identify with the square of the average density c^2. The number of contacts involved in the average repulsive energy is therefore proportional to $R^d (N R^{-d})^2 = N^2 R^{-d}$. This typical mean field approach comes down to neglecting correlations between interacting pairs of monomers, in other words cooperative effects due to the fact that the monomers are assembled in a linear chain. Excluded volume effects, quantified by the excluded volume parameter are uniformly redistributed between the pairs of monomers and the configuration of the chain is only involved *via* the *average density*. The second term is the entropic contribution, also estimated in an approximate way by identifying it with its value for an ideal chain (Sect. 6.3.1). The constant added in F, independent of R, plays no role here. The radius of gyration is then obtained:

$$R_g(N) \sim N^{\nu_F} \quad \text{with} \quad \begin{cases} \nu_F(d) = \frac{3}{d+2} & (\text{if } d \leq 4) \\ \\ \nu_F(d) = \frac{1}{2} & (\text{if } d \geq 4). \end{cases} \tag{6.10}$$

The Flory formula is exact in dimension $d = 1$, very good in $d = 2$ (polymers in thin films), as well as in $d = 3$ ($\nu_F = 3/5$ compared to $\nu_{exp} = 0.586$ and $\nu_{th} = 0.5880$). The estimation obtained is remarkably correct, despite the approximations involved. We usually explain this success by the compensation of two errors involved in estimating the repulsive energy and the entropic contribution. In dimension $d = 4$ the value $\nu = 1/2$ of an ideal chain is recovered. In addition, the ratio of the excluded volume term to the entropic term behaves as $N^{2-d/2}$, which shows that excluded volume effects are a small perturbation to the ideal chain once $d \geq 4$, thus explaining the origin of the critical dimension $d_c = 4$ above which the statistics are that of an ideal chain ($\nu = 1/2$).

Flory's formula and long range correlations of self-avoiding walks

A better explanation is a new interpretation of the Flory theory proposed by Bouchaud and Georges [1]. By considering the polymer as a trajectory of a random walk of successive steps $(a_i)_i$, the end-to-end distance is calculated directly from its definition $R^2(N) = \langle (\sum_{i=1}^{N} a_i)^2 \rangle$. Since $\langle a_i \rangle = 0$, the cross terms $\langle a_i.a_j \rangle$ are simply the correlations $C(j - i)$, which gives the explicit formula for the end-to-end distance: $R^2(N) = \sum_{n=1-N}^{N-1}(N - |n|)C(n) = Na^2 + 2\sum_{n=1}^{N-1}(N - n)C(n)$.

The critical character of a self-avoiding walk results in the decrease of these correlations as a power law (remember that the power law decrease replaces the exponential decrease when the range of correlations becomes infinite): $C(n) \sim n^{-\alpha}$. These correlations result from the excluded volume constraint: the nth monomer is correlated to the previous monomers, since it must avoid them. If $\alpha > 1$, the sum of correlations is absolutely convergent and $R^2(N)$ behaves as N, with a proportionality factor which is no longer equal to $a^2 = C(0)$ but to the sum $\sum_{n=-\infty}^{\infty} C(n)$, as we have seen in Sect. 4.3.1, (4.25). Let us now suppose that $\alpha < 1$. The sum of correlations, behaving as $N^{1-\alpha}$ for a polymer of length N, will be proportional to the number of contacts (average number since we are considering the statistical quantity $R(N)$). The mean field approximation is involved in calculating this number, estimated as before to be N^2/R^d. Therefore $\alpha = \nu d - 1$ where ν is the exponent we were looking for. The second term appearing in the expression for $R^2(N)$ behaves as $N^{2-\alpha}$, in other words as $N^{3-\nu d}$. It is only involved if $2 - \nu d > 0$, that is to say if $\alpha < 1$, which is consistent with the hypothesis made for α. So in low dimensions it is the correlation term that dominates the behaviour of $R(N)$. Then the expression $R^2(N) \sim N^{2\nu}$ defining ν leads to the coherence relation $N^{2\nu} \sim N^{3-\nu d}$, leading to Flory's formula $\nu_F = 3/(d + 2)$ and specifying the value of the critical dimension $d_c = 4$.

In large enough dimensions, we have $\alpha > 1$ and the average number of contacts remains bounded, as does the sum of the correlations. The excluded volume constraints are always present but they are not felt, the relative weight of potential crossings tends to zero as N tends to infinity. Correlations will therefore play a negligible role in large enough dimensions and it is then the first term which dominates and $R(N) \sim \sqrt{N}$.

This approach, rooted in the statistical description of a polymer as a random walk and not involving the free energy, better explains the remarkable result of Flory theory (but it does not validate the method Flory used to obtain it).

6.3.2 Good and Bad Solvents, Θ Point and Flory–Huggins Theory

To take into account the Θ point of the polymer solution, the Flory approach must be generalised to explicitly take into account the solvent. The theory developed in this way is known by the name Flory–Huggins theory. We introduce the volume fraction $\Phi = ca^d$ of monomers (dimensionless); the volume fraction of solvent is then $1 - \Phi$. The relevant free energy is the free energy of mixing $F_{mix}(\Phi) = F(\Phi) - \Phi F(1) - (1 - \Phi)F(0)$. Described at a site it is written:

$$F_{mix}(N, \Phi) = kT \left[\frac{\Phi}{N} \log \left(\frac{\Phi}{N} \right) + \frac{1}{2} (1 - 2\chi)\Phi^2 + \frac{t\Phi^3}{6} + \dots \right]. \quad (6.11)$$

The term $\frac{1}{2}(1 - 2\chi)\Phi^2$ describes the pairwise interactions. $(1 - 2\chi)$ is called the second virial coefficient. The parameter χ is the sum of three contributions:

$$\chi = \frac{\chi_{MM}}{2} + \frac{\chi_{SS}}{2} - \chi_{MS} \quad (6.12)$$

coming from the interaction between monomers, between two solvent molecules and between a monomer and the solvent respectively. The monomer-solvent repulsion ends up resulting in an apparent attraction between monomers. In general, χ decreases when the temperature T increases and $\chi \geq 0$ (for interactions independent of T, we have $\chi \sim 1/T$). At high temperature, $v = a^3(1 - 2\chi(T))$ is positive and we call this a "good solvent". If the temperature is low, v becomes negative and we call it a "bad solvent". This theory justifies describing the effect of the solvent together with that of interactions between monomers, through a single excluded volume parameter:

$$v = a^3(1 - 2\chi(T)). \quad (6.13)$$

The point $\chi = 1/2$ where the excluded volume parameter cancels corresponds to the Θ point in this theory. We sometimes talk about a "Θ solvent" but we have already emphasised that the Θ point is a characteristic of a solvent-polymer pair. At the Θ point, since $v = 0$, the next term $t\Phi^3/6$ in the expansion of $F_{mix}(N, \Phi)$ can no longer be neglected. This term is called the "three body term" because it involves the cube of the volume fraction. It describes the average effect of correlations between interacting pairs.[17]

6.3.2.1   Analogy with Ferromagnetism: n-Vector Model with $n \to 0$

A discerning advanced technique for the calculation of scaling laws satisfied by isolated linear polymers and self-avoiding walks, was carried out in 1972 by De Gennes [4]. The idea is a formal analogy between statistical properties of polymers and critical properties of particular spin systems, which had the notable advantage of having been already exhaustively studied and classified in universality classes with known critical exponents (see Chap. 3). Before going into the details of this analogy and its exploitation, we can propose a preamble by showing the parallel scaling laws:

$$\xi \sim |T - T_c|^{-\nu} \qquad \text{and} \qquad R \sim N^\nu. \qquad (6.14)$$

A formal correspondence between spin systems to the left and self-avoiding walks to the right is given by:

$$\xi \longleftrightarrow R \qquad (6.15)$$

$$t \equiv \frac{T - T_c}{T_c} \longleftrightarrow 1/N. \qquad (6.16)$$

We have seen in Chap. 3 that a system of spins $(S_i)_i$ of constant modulus S, placed at nodes on a regular lattice and in ferromagnetic interaction (restricted to nearest neighbours) belongs to a universality class entirely determined by the dimension d of the lattice and the number n of components of the spins. The spins being of constant modulus, we *choose* to normalise them $S^2 = n$. In the presence of an external magnetic field h, the Hamiltonian of the system, if it consists of Q spins, is written:

$$\mathscr{H} = -J \sum_{<i,j>} S_i . S_j - \sum_{i=1}^{Q} h . S_i, \qquad (6.17)$$

where the sum $\sum_{<i,j>}$ is over nearest neighbour pairs of spins. By denoting $\beta = 1/kT$ (not to be confused with one of the critical exponents defined in Chap. 1) and $d\Omega_i$ the angular integration over all possible orientations of spin i (the phase space is the set of angular coordinates of the spins since their modulus is constant), the partition function is written:

---

[17]The probability of "real" three body interactions (that is to say between three monomers) is very weak and these interactions do not have observable consequences.

$$\mathscr{Z}(\beta) = \int e^{-\beta \mathscr{H}} \prod_{i=1}^{Q} d\Omega_i \qquad (\beta = 1/kT). \tag{6.18}$$

The series expansion of the integrand and term by term integration[18] involves the moments of spins with respect to the natural weight $\int \prod_{i=1}^{Q} d\Omega_i$ in the phase space of angles of Q spins. The integrals over different spins, that is to say over different angular degrees of freedom, will decouple and factorise: $\int \int S_1.S_2 d\Omega_1 d\Omega_2 = \sum_{\sigma=1}^{n} \int S_{1,\sigma} d\Omega_1 \int S_{2,\sigma} d\Omega_2$. Since different spins play an identical role in the problem (translational invariance), we can consider the moments of one spin S, with components denoted by the index σ_j:

$$\langle S_{\sigma_1}...S_{\sigma_q} \rangle_0 = \int S_{\sigma_1}...S_{\sigma_q} \frac{d\Omega}{\Omega_{tot}}. \tag{6.19}$$

The notation $\langle \rangle_0$ indicates a simple average, [19] with respect to the natural volume of phase space of spin angles, with total volume Ω_{tot}. The odd moments are zero by symmetry. It can be shown that if we let n tend to zero,[20] once the calculation is complete the only moments that remain are the second order moments:

$$\langle S_{\sigma_1} S_{\sigma_2} \rangle_0 = \delta_{\sigma_1 \sigma_2}. \tag{6.20}$$

Higher order moments are zero in the limit $n \to 0$. Consequently, the series expansion of the integrand gives only a finite number of non zero terms, which validates calculating the partition function $\mathscr{Z}$ by series expansion and term by term integration (moment expansion).

Let us make the calculation of $\mathscr{Z}$ more explicit to understand how random self-avoiding walks appear in calculating statistical properties of a lattice of spins with $n \to 0$ components. By introducing coupling constants J_{ij} such that $J_{ij} = J$ if sites i and j are nearest neighbours and $J_{ij} = 0$ if not, the partition function $\mathscr{Z}(\beta)$

---

[18]The validaty of this term by term integration is a priori problematic and often remains so a posteriori, apart from in the case considered here where we take $n = 0$, in which the series only contains a finite number of non zero terms, which ensures its convergence and the validity of the calculation.

[19]Be careful not to confuse $\langle \rangle_0$ with the average $\langle \rangle = \langle \rangle_\beta$ taken with respect to the Boltzmann distribution of Q spins. The average $\langle \rangle_0$ corresponds to $\beta = 0$ and describes the purely entropic contribution.

[20]The number n, initially a positive integer, appears as a parameter in the result of the moment calculation, so at this stage we can assign any real number to it, including here the rather difficult to conceive value $n = 0$. It is the same trick used in assigning the value $d = 4 - \epsilon$ to the dimension of space (see Sect. 3.3.5).

in zero field is written:

$$\frac{\mathscr{Z}}{\Omega_{tot,Q}} \equiv \left\langle \prod_{i>j} e^{\beta J_{ij} S_i . S_j} \right\rangle_0$$

$$= \left\langle \prod_{i>j} \left(1 + \beta J_{ij} \sum_{\alpha=1}^n S_{i,\alpha} S_{j,\alpha} + \beta^2 J_{ij}^2 \sum_{\alpha=1}^n \sum_{\sigma=1}^n S_{i,\alpha} S_{j,\alpha} S_{i,\sigma} S_{j,\sigma} \right) \right\rangle_0 .$$
(6.21)

The higher terms in the expansion contribute zero. Each term can be represented by a graph by associating a bond (i,j) to each non zero J_{ij} appearing in this term. Note that the averages over different spins factorise. $S_{i,\alpha}$ must appear 0 or 2 times with the same component α, in a term for its average $\langle\rangle_0$ to give a non zero contribution in the limit $n \to 0$. In other words, a given site must belong to exactly two sites, or not be involved. The corresponding graphs of the terms giving a non zero contribution are therefore closed loops, do not overlap and involve the same component α in each of their sites.

By carrying out the calculation, it can be shown that $\mathscr{Z}/\Omega_{tot}^Q = 1$ in the limit $n \to 0$. Calculating the correlation function of spins at finite temperature is addressed in the same way, showing that:

$$\langle S_{i,\alpha} S_{j,\sigma} \rangle_\beta = \sum_{N=0}^\infty (\beta J)^N \, \aleph_N(ij) \qquad\qquad (n \to 0), \qquad (6.22)$$

where $\aleph_N(ij)$ is the number of self-avoiding paths of N steps connecting the sites i and j. This expression is the discrete form of a Laplace transformation with respect to the variable N, evaluated at the value $\log(1/\beta J)$. Remember that $kT_c = J$, so that the reduced variable t is expressed $\log(1/\beta J) = t$ and the formula becomes:

$$\langle S_{i,\alpha} S_{j,\sigma} \rangle_{\beta,n=0} = \sum_{N=0}^\infty e^{-Nt} \, \aleph_N(ij). \qquad (6.23)$$

This explicitly gives the link between the number of paths $\aleph_N(ij)$ and the correlation function of spins $\langle S_{i,\alpha} S_{j,\sigma} \rangle_\beta$ calculated in the limit $n \to 0$. The number $\aleph_N(ij)$ of self-avoiding paths connecting sites i and j leads to the Boltzmann distribution of a chain of length N at infinite temperature, which is nothing other than the entropic contribution to the distribution at finite temperature, normally called the *density of states*. Since the site i is fixed,

$$\frac{\aleph_N(ij)}{\aleph_N} = P_N(r = r_{ij}) \qquad \text{with} \qquad \aleph_N = \sum_j \aleph_N(ij). \qquad (6.24)$$

The relation (6.23) is the basis of the calculation of exponents associated with random self-avoiding walks knowing those of the spin system. The point of this roundabout approach is that the scaling properties of spin systems will be directly transferable to self-avoiding paths due to the explicit link between their statistical properties.[21] For example, the existence of a unique characteristic length $\xi \sim t^{-\nu}$ will give the typical linear path size, expressed as a function of the variable $1/N$ as: $R \sim N^{\nu}$, and hence the value of the Flory exponent $\nu = \nu(d, n = 0)$. This approach gives the value 0.5880 ± 0.0010, more in agreement with experimental results than Flory's value 0.6 and considered today as the best estimate of ν [17]. From this the scaling $P_N(r) \sim N^{-\nu d} \phi(rN^{-\nu})$ is also drawn. The asymptotic behaviour of the distribution ϕ is obtained, starting from knowing the asymptotic behaviour of the spin correlation function, by inverse Laplace transform, giving $\phi(x) \sim e^{-x^{1/(1-\nu)}}$.

In the same way the scaling law describing the probability that a self-avoiding walk returns to its origin (closed path) can be determined. The number of closed chains of $N + 1$ steps behaves asymptotically as $N^{-2+\alpha}$, where α is the specific heat exponent (for $n = 0$ and d corresponding to the lattice being considered). We also obtain that the critical dimension of isolated linear polymers is the same as that of spin systems, i.e. $d_c = 4$. Above this dimension, the mean field like behaviour (behaviour of an ideal chain in the case of a polymer) applies.

Within the framework of this analogy the Θ point, described in Sects. 6.2.3 and 6.3.2 in the limit $N \to \infty$ appears as the end of a line of critical points (traversed on varying the excluded volume, i.e. the temperature), which we call a *tricritical point*. In three dimensions $d = 3$, it lies next to the region in which the mean field is valid, justifying the law $R(N) \sim \sqrt{N}$ observed at this point.

6.3.2.2 Polymers as Unusual Critical Objects

In the framework of lattice models, we can predict the critical properties of an isolated linear polymer (in good solvent) by introducing the generating function:

$$\mathscr{G}(\omega) \equiv \sum_{N=0}^{\infty} \omega^N \aleph_N(d), \qquad (6.25)$$

where $\aleph_N(d)$ is the number of possible conformations of a chain of N steps and in d dimensions. An exact enumeration follows from a mathematical extrapolation procedure (Padé approximant) determining the expression of $\aleph_N(d)$, confirmed by numerical simulations (by the Monte Carlo sampling method):

$$\aleph_N(d) = \lambda \, \mu_d^N N^{\gamma-1}. \qquad (6.26)$$

---

[21] A completely analogous approach, the *Fortuin-Kasteleyn representation*, is used in the context of spin glasses, but in the opposite direction, to access statistical properties of spin systems from those of the associated geometric system, in this case the model of directed percolation [15].

In comparison, we have $\aleph_N(d) = (2d)^N$ for an ideal chain. The quantity μ_d is an effective connectivity constant ($\mu_d < 2d - 1$). Substituting this expression for $\aleph_N(d)$ into the definition of $\mathscr{G}(\omega)$, gives:

$$\mathscr{G}(\omega) \sim \left(\frac{1}{\omega - \omega_c}\right)^\gamma \quad \text{with} \quad \omega_c = \frac{1}{\mu_d} \tag{6.27}$$

showing a critical point $\omega = \omega_c$. In the framework of the previous analogy with a spin system, the generating function $\mathscr{G}(\omega)$ is interpreted as a magnetic susceptibility χ_m:

$$kT \; \chi_m = \sum_j \langle S_{i_0,\sigma} S_{j_0,\sigma}\rangle_{\beta,n=0} = \mathscr{G}(\beta J). \tag{6.28}$$

Expanding χ_m in t gives $\chi \sim t^{-\gamma}$, meaning the exponent γ in $\aleph_N$ is a susceptibility exponent. It depends on the dimension: $\gamma(d = 2) = 43/32$ and $\gamma(d = 3) \approx 1.1608$.

This generating function concept can be generalised to interacting self-avoiding chains eventually leading to the Θ point and properties of the coil–globule transition:

$$\mathscr{Z}(K,\omega) = \sum_{N=0}^{\infty} \sum_m e^{NKm} \, \omega^N \, \aleph_N(m), \tag{6.29}$$

where m is the number of contacts with a monomer and $\aleph_N(m)$ the number of chains of N monomer steps with m contacts. This expression is nothing but the grand canonical partition function, where ω is the *fugacity* controlling the length distribution of chains. At infinite temperature or zero interaction ($K \equiv \beta J = 0$), the simple self-avoiding walk and function $\mathscr{G}(\omega)$ are recovered. By analogy with the behaviour of $\mathscr{G}(\omega)$, the expression of $\mathscr{Z}(K,\omega)$ suggests there exists a line of critical points $\omega = \omega_c(K)$. These preliminary results strongly encourage us to turn to renormalisation methods to complete the phase diagram in parameter space (K,ω).

One may doubt the validity of models of polymers on lattices since the positions of monomers in a real polymer do not only take discrete values corresponding to nodes on a lattice. We expect that the discrepancies in the statistical properties with respect to continuous polymer models to disappear as the length of the chain increases. The effect can on the other hand be important for short chains and it should be evaluated and taken into account in the analysis of numerical results (lattice simulations). Comparisons between lattice simulations and simulations in which the positions are allowed to vary (almost) continuously indicate that the effect is only serious for very short chains ($N \leq 20$). Renormalisation method will also enable us to unify discrete and continuous models and show that their asymptotic properties are identical.

6.3.2.3   Criticality and Extensivity of an Isolated Polymer

Let us emphasise that the previous analogy between spin systems and isolated polymers is indirect and should not be pushed too far. There actually exist crucial differences between the two types of system. The size N of a polymer is, at the same time, both the reference extensive variable, directly proportional to the molecular weight of the polymer, and a parameter controlling the state of the polymer since $1/N$ plays the same role as the temperature difference $T - T_c$ in a spin system.[22] In the case of an isolated polymer, letting N tend to infinity is a very unusual procedure, because at the same time as the system size tends to infinity, the control parameter $\epsilon = 1/N$ tends to its critical value $\epsilon_c = 0$. Therefore the "thermodynamic limit" and the critical point are reached *simultaneously*, as if the number of spins and the temperature were varied together. This confirms the statement already made above that: *in the limit in which its size tends to infinity, a polymer is a critical object at all temperatures and it will show scaling laws at all temperatures.*

The role of real space is also different in a spin system and a system of an isolated macromolecule, which stops us being able to simply transfer the concepts of extensibility, thermodynamic limit and all the thermodynamic formalism which follows [22]. We therefore do not have the same conceptual framework at our disposal for conformational transitions of isolated macromolecules as for phase transitions. In particular, the scaling law satisfied by a polymer chain as a function of its length N presents only a formal analogy with finite size scaling laws encountered for phase transitions and percolation. It does not have the same physical meaning nor origin. The difficulty can be identified as the fact that although the volume V and density N/V of the spin system are fixed once and for all at the moment of "fabrication" of the system (even if a mental fabrication), the volume R^d and density NR^{-d} of an isolated polymer are the *observables*, arising spontaneously from the statistics of the polymer conformations and therefore varying with the control parameters. Depending on the temperature, we have a radius of gyration $R_g \sim N^\nu$ if $T > \theta$, $R_g \sim N^{\nu_\theta}$ if $T = \theta$, and $R_g \sim N^{1/d}$ if $T < \theta$. In one sense, everything happens as if the polymer "lived" in a space of dimension $d_{\text{eff}} = 1/\nu_{\text{eff}}$ (with $\nu_{\text{eff}} = \nu$, ν_θ or $1/d$), which would give it a volume $V_{\text{eff}} = R^{1/\nu_{\text{eff}}} \sim N$ and therefore an apparent constant density (remember that the thermodynamic limit of a spin system is defined and calculated at constant density). From this fact, the typical fractal geometries of polymers cannot be ignored when studying the statistical properties and in particular in defining the thermodynamic limit. For example, the scale invariance of the end-to-end distance distribution $P_N(R)$ will be radically different:

---

[22]This point comes from the fact that N measures the importance of the non Markovian nature of the random walk; the memory becomes infinitely long as N becomes infinite.

- In the coiled phase, $d_{\mathrm{eff}} = 1/\nu$, the polymer shows a non trivial fractal scale invariance $P_N(R) \sim p_c(RN^{-\nu})$.
- In the θ phase, $d_{\mathrm{eff}} = 1/\nu_\theta$ and the polymer shows another non trivial fractal scale invariance $P_N(R) \sim p_\theta(RN^{-\nu_\theta})$.
- In the globular phase, $d_{\mathrm{eff}} = d$, the polymer locally fills all space and the dependence of its properties on N is that of an extensive system in d dimensions: $P_N(R) \sim p(R/N^{1/d})$.

Note that the auxiliary functions p (different in the different phases) appearing here are not equal to $P_{N=1}$. *These functions p are the renormalised distributions including in an effective way the correlations at small scale*[23] [13]. These scaling arguments can be pushed further to obtain an expression unifying the distribution of the end-to-end distance. The idea is to construct an expression rewriting the scaling above in different limiting situations where we expect that a single phase contributes to the statistics. This gives [14]:

$$P_N(R) \sim \exp\left[-A_1 \left(\frac{N^\nu}{R}\right)^{\frac{d}{\nu d - 1}} - A_2 \left(\frac{R}{N^\nu}\right)^{\frac{1}{1-\nu}} \right] \tag{6.30}$$

(A_1 and A_2 are numerical constants).

6.3.3   Renormalisation Approaches

We have just seen that from the point of view of conformational properties, a polymer behaves like a critical object. It is therefore natural to attempt to implement general principles of renormalisation to take into account as best as possible the effects of excluded volume and the "anomalous" correlations they introduce in the random walk representing the polymer conformation. Several approaches have been developed:

- In real space
- In the grand canonical ensemble
- Using the analogy with the n-vector model ($n \to 0$)
- Using a perturbation method about the Gaussian chain

---

[23]That is to say before the scaling behaviour emerges. It is precisely the fact that the scaling behaviour is not yet confirmed for small values of N which requires the redefinition of the effective distributions p. A temporal analogue would be a random walk for which the asymptotic regime emerges transiently failing to show the same scale invariance. For example, a transitory trap, with characteristic time $\tau_{\mathrm{eff}} > 1$ does not destroy a normal diffusion but simply leads to a "renormalisation" of the diffusion coefficient D to the weaker $D_{\mathrm{eff}} = D/\tau_{\mathrm{eff}}$, or equivalently in terms of an effective number of steps $N_{\mathrm{eff}} = N/\tau_{\mathrm{eff}}$ ($ND_{\mathrm{eff}} = N_{\mathrm{eff}}D$). Implementing this explicit calculation of the distributions p in the case of self-avoiding walks needs a numerical sampling [13].

6.3.3.1 Approaches in Real Space

The most direct approach consists of redefining the elementary unit as a "macromer" of k consecutive monomers. The renormalisation transformation leading to a size a_1 and dimensionless excluded volume w_1 ($v_1 = a_1^d w_1$) of these $N_1 = N/k$ macromers can be written:

$$\mathscr{R}_k : \quad \begin{cases} N_1 = N/k \\ a_1 = a\sqrt{k}\,[1 + A_k(w)] \\ w_1 = wk^{2-(d/2)}\,[1 - W_k(w)] \end{cases} \tag{6.31}$$

showing the "reference" transformation $a_1^0 = a\sqrt{k}$ and $w_1^0 = w_0 k^{2-d/2}$ that we would obtain with an ideal chain, by estimating the number of contacts inside a macromer of length k to be $k^{2-d/2}$. A_k and W_k have to be determined numerically. For $d \geq 4$, it can be shown that $w = 0$ is the only fixed point and that it is a stable fixed point. This *demonstrates* that the asymptotic behaviour of an ordinary chain is actually that of an ideal chain: above the critical dimension $d_c = 4$, excluded volume effects play a negligible role, simply because the probability of a chain crossing itself is too small for it to be felt. For $d < 4$, a fixed point w^* exists. At step j (therefore there are N/k^j macromers of length a_j and excluded volume w_j in the renormalised chain), the end-to-end distance is written:

$$R_j = a_j \; f\left(\frac{N}{k^j}, w_j\right), \tag{6.32}$$

where f is some function independent of j. For large enough j, we can replace w_j by its limit w^* in f and in the renormalisation relation for a giving: $a_{j+1}/a_j \approx \sqrt{k}(1 + A_k(w^*))$. In general, a renormalisation transformation takes into account the change of a *model* when the scale at which the real system is described, is changed.[24] Renormalisation does not affect the system itself and must consequently conserve all its observable properties. So in the present case, the observable R_j must not depend on j (if the renormalisation has been done correctly!). Therefore the j dependence of a_j must cancel with that coming from the argument N/k^j of $f(., w^*)$. The form of the scaling $f(x, w^*) \sim x^\nu$ adequately ensures the cancellation and provides the scaling law $R(N) \sim N^\nu$ and the value of the exponent ν in the given context involving the functions A_k and W_k.

The reasoning we have just presented relies on a very strong scale invariance hypothesis: the functions A_k and W_k must be the same at all levels of iteration (i.e. be independent of a) and be exact enough that the radius of gyration of the renormalised chains remains effectively equal to the real radius of gyration. In practice, we have simply moved the difficulty to lie now in the (numerical) determination of

---

[24]It is by relating two "subjective" views of the system that allows us to access the "objective" properties of the critical exponents.

the functions A_k and W_k and the method is all in all not very different from a direct numerical determination of ν.

One of the advantages of the previous approach is its ability to be extended to take into account the Θ point [20]. In the version above, interactions between non consecutive monomers are written only implicitly by the single parameter w (dimensionless excluded volume). Yet we have seen in Sect. 6.3.2 that the Θ point was characterised by the cancellation of the coefficient w and that therefore the next term, $t\Phi^3/6$, in the expansion of F_{mix}/kT as a function of the volume fraction Φ became dominant. So we will take the same approach[25] but include the transformation of t in $\mathcal{R}$. The renormalisation flow, that is to say the set of trajectories under the action of $\mathcal{R}$, is represented in Fig. 6.9. Iterating $\mathcal{R}$, in other words moving along the trajectories, comes down to considering longer and longer initial lengths N, tending to infinity as we approach the points A and B, which therefore give the asymptotic behaviour $N \to \infty$.

The fixed point A $(w^*, t = 0)$, identical to the fixed point of the reduced transformation, describing the universality class of self-avoiding walks. The extended approach therefore brings nothing new concerning the exponent ν, but the fact that

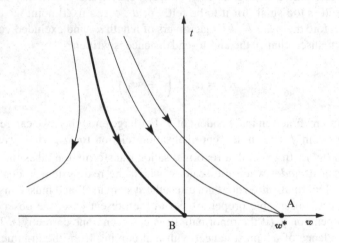

Fig. 6.6 Renormalisation flow in (w, t) space (in $d = 3$ dimensions). Iterating the renormalisation comes down to letting the length N of the chain tend to infinity. The fixed point A is associated with the universality class of random self-avoiding walks whereas the fixed point B corresponds to the Θ point, i.e. to the transition to compact configurations situated to the left of the separating line in bold (after [4])

---

[25]It is a general procedure: the more parameters we consider (the more terms in the Hamiltonian or evolution law), the more accurately we describe the change of scale in the space of models, so the more chance we have of discriminating correctly between different universality classes. In this way these are coherent arguments that determine the validity of a renormalisation method. This shows that the parameters that could be included to enlarge the space in which the renormalisation acts play no role in the large scale properties ("irrelevant" parameters).

$t = 0$ in the class shows that the "three body term" $t\Phi^3/6$ is effectively negligible asymptotically. The renormalisation approach therefore allows us to *demonstrate* the validity of the approximation of truncating the expansion of $F_{mix}(\Phi)$ after the second term when we are only interested in the coiled phase (in which the typical conformations are self-avoiding walks). We say that the term $t\Phi^3/6$ is *irrelevant* (see Chap. 3). This type of result, demonstrating that certain terms play no role in the asymptotic properties, is another success of renormalisation, just as remarkable as the explicit calculation of critical exponents. We can then rigorously justify the use of minimal models to describe large scale behaviour.

The trajectories to the left of the separation line reaching B correspond to the globular phase. The three body term here is also asymptotically negligible[26] and the pairwise interactions become more and more attractive, driving the collapse of the chain on itself as we observe in bad solvent.

The fact that $w = 0$ at fixed point B means it is identical to the Θ point. In the limit $N \rightarrow \infty$, this point is situated at the end of a dividing line which qualifies it as a *tricritical point* by analogy with diagrams obtained for spin systems. Analysis of the action of $\mathscr{R}$ in the vicinity of B leads to a scaling law $R^2(N) \sim a^2 N$ (in three dimensions).[27]

6.3.3.2 Grand Canonical Approach

The grand canonical approach is a method built on geometrical bases similar to those used in the context of percolation (so still in real space). It was devised for a polymer model on a square or cubic lattice with excluded volume (by forbidding double occupancy of a site) and attractive interactions between monomers that are neighbours on the lattice without being consecutive in the chain (interacting self-avoiding walk model, Sect. 6.2.3). The difficulty emerging when we carry out a direct geometric renormalisation of a configuration is to *preserve the connectivity of the chain and its self-avoiding property*. It seems, to do so we need to allow a certain flexibility in the number of monomers. The method must therefore be developed in grand canonical space. The model then has two control parameters[28]:

- The coupling constant J of the attractive interaction, or equivalently the dimensionless coefficient $K = \beta J$.
- The fugacity ω controlling the chain size distribution, the number of monomers N here being no longer fixed but is instead one of the *observables* characterising

---

[26]Even though the three body term is asymptotically negligible in the coiled and globular phases, it should be emphasised that it plays a key role near the Θ point, entering in the expression of the separatrix leading to point B.

[27]The next term in this scaling law can also be evaluated and shown to behave as $N/\log N$ and we therefore talk about *logarithmic corrections*.

[28]This approach was first of all developed just for self-avoiding walks, bringing into play only the fugacity [23].

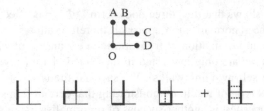

Fig. 6.7 Calculation of the probability ω_1 of the existence of a renormalised bond, by exact enumeration on a 2×2 motif. The final configuration takes into account the possibility of encountering very compact conformations at low temperature. We obtain $\omega_1 = \omega^2 + 2\omega^3 + e^K \omega^4 + e^{2K} \omega^4$

the state of the chain. The probability of observing a chain of N steps is proportional to ω^N.

The renormalisation is then performed in this generalised conformational space. The idea is still the same. First we "decimate" the conformation, in this case redefining the elementary units (motifs), which boils down to changing the resolution with which we look at the conformation. Then we determine the parameters $(K_1, \omega_1) = \mathscr{R}(K_0, \omega_0)$ involved in the statistical description of the decimated configurations. The geometrical operation implied in defining the renormalisation transformation $\mathscr{R}$ here is a bit more complicated than the decimations in spin systems and percolation networks. It is known by the name *corner rule* due to the shape of the motifs on which we define $\mathscr{R}$ by exact enumeration (see Figs. 6.7 and 6.8). Let us first of all determine ω_1, defined as the probability of a bond existing in the renormalised chain. It will be equal to the probability with which the chain entering the motif by the corner O passes through it, leaving by A or B; exits by C or D should not be counted to avoid ending up with a branched or crossed renormalised configuration. The different contributions and resultant transformation are given in Fig. 6.7. We see that the transformation of the number of monomers is not one-to-one, hence the necessity of working in the grand canonical space. It is ω not N that is the parameter transformed by $\mathscr{R}$. The same approach of exact enumeration, implying this time two motifs, is used to determine the transformation K_1. We actually determine the transformation $\omega_1^2 e^{K_1}$ of the probability of a contact interaction (Fig. 6.8).

We then exploit renormalisation in the classical way. Determining the fixed points of $\mathscr{R}$ shows three non trivial fixed points[29] identified with the fixed point associated with self-avoiding walks (point A, $K = 0, \omega_c = 1/\mu_d$) already seen in Sect. 6.3.2, the Θ point (point B) and the globular phase (point C at infinity) respectively. For $K > 0$, there exists another critical value $\omega_c(K)$ of the fugacity:

---

[29]These fixed points should not be confused with the points A and B on Fig. 6.9. The parameter space is actually different, as is the theoretical context – interacting self-avoiding walks (Fig. 6.9) and Flory–Huggins theory (Fig. 6.9). The analogy in results obtained by these two different approaches strengthens the validity of their conclusions.

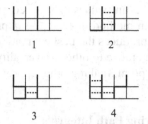

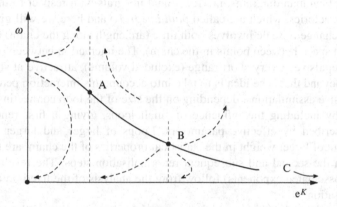

Fig. 6.8 Calculation of the probability $\omega_1^2 e^{K_1}$ of a contact between two renormalised bonds. We have represented a few starting chain configurations similar up to one contact at the level of the renormalised chain. We must exclude counting configurations that will lead to branches or crossings in the renormalised configuration. We obtain: $\omega_1^2 e^{K_1} = \omega^4 + 2\omega^5 + 3\omega^6 + e^K\omega^5(1 + \omega)^2 + e^{2K}\omega^4(\omega + 2\omega^2 + 4\omega^3 + \omega^4) + 2e^{3K}\omega^7 + e^{4K}\omega^6(1 + \omega + 2\omega^2) + e^{6K}\omega^8$

Fig. 6.9 Renormalisation flow in (e^K, ω) parameter space. Point A is associated with the universality class of self-avoiding walks (scaling law $R(N) \sim N^\nu$), point B corresponds to the Θ point (scaling law $R(N) \sim N^{\nu_\theta}$) and point C, pushed back to infinity with the choice of coordinates (e^K, ω), corresponds to the globular phase (scaling law $R(N) \sim N^{1/d}$)

therefore a line of critical points appears, terminating in the Θ point. The globular phase is well described by the renormalisation procedure we have just presented. We find the exact value $1/d$ of the exponent ν. The results can be improved by considering larger motifs, for example 3×3 (so a scale factor $k = 3$). The complexity of the enumeration quickly limits the size of motifs.

6.3.3.3 Approach "By Analogy"

Another renormalisation approach is that developed for the system of spins with $n \to 0$ components, to calculate the exponent $\nu(d, n \to 0)$ in the framework of the analogy developed in Sect. 6.3.2. The calculation of the critical exponents has been conducted for all relevant values of n, in conjugate space and within the framework of the perturbation method in $\epsilon = 4 - d$ presented in Chap. 3. It is a very technical method but it produces the best estimation of $\nu = 0.5880 \pm 0.0010$ [17]. However it does not generalise to other universality classes and does not give access to properties of the Θ point, making other approaches necessary.

6.3.3.4 Approaches Involving Path Integrals

The most powerful methods are the very mathematical methods developed by putting the polymer and its conformations in the context of (continuous time) stochastic processes. It involves perturbative methods taking as zeroth order the Wiener process associated with ideal chains. They are technically difficult (involving Feynman diagrams, i.e. functional integrals with respect to the Wiener process trajectories, which are called *path integrals*) and here we will give just an idea. The change in scale involves both time (arclength along the chain) and space (Euclidean space between points in the chain). The interaction between two sites i and j is repulsive at very short range (excluded volume), attractive at short range and zero beyond that. The idea is to take into account this interaction peturbatively by successive assimilations, depending on the size of the loop connecting i and j. We start by including the influence of small loops, giving a first renormalised chain, described by effective parameters. Loops of larger and larger size, less frequent but of larger weight in the statistical properties of the chain, are taken into account in the second and subsequent renormalisation steps. The results (scaling laws and associated exponents) follow from the analysis of the fixed points of this renormalisation.

Note that it is a situation in which one must introduce an "ultra-violet cutoff" discarding very small scales. Here the singularity, comes from using the "ideal" object limit, that is the Weiner process: the associated velocity diverges at small scales. We must therefore truncate the wavevector space, in practice as soon as we reach the molecular scale, where the physics becomes regular again while the Wiener process becomes singular. The theory of self-avoiding walks is renormalisable in the sense that the macroscopic results do not depend on the value of the cutoff.

6.4 Polymer Solutions

To complete our tour, we will very briefly mention the case of less dilute solutions in which different chains influence each other. In this vast domain, we will focus on a particular point, the concept of a *blob*, because it illustrates the concept of correlation length, essential throughout this book, in the context of polymers.

6.4.1 Polymer Melts

Let us briefly consider the case in which the system made up of only polymers, called a *polymer melt*. Due to the interpenetration of different chains, the probability that a monomer has a neighbouring (non consecutive) monomer from the same chain is negligible compared to the probability it is surrounded by monomers belonging to other chains. These other chains will, in a way, screen the intra-chain excluded volume interactions, in the sense that these interactions have almost no opportunity to be felt. The consequence is that the statistics of a chain in a melt is that of an ideal chain: $R(N) \sim \sqrt{N}$. This is verified experimentally by fluorescently marking the chain, or labelling it with deuterium if we observe it by neutron diffraction (Sect. 6.2.2).

Let us point out that remarkable phenomena, still not well understood, appear when the temperature of such a liquid polymer is lowered. A glass transition is observed leading to a metastable amorphous gel phase. Slow relaxation phenomena occur associated with properties of nonlinear response (e.g. aging and violation of the fluctuation-dissipation theorem) [24].

6.4.2 Semidilute Solutions

In a solution, there are two concentrations (number of monomers per unit volume) involved: the average concentration c_0 of the solution, adjustable by the observer, and the concentration $c = N/R_g^d$ of the interior of the volume occupied by the chain, which is an observable indirectly adjustable by changing the quality of the solvent and the length of the chains. We also use the volume fraction $\Phi = ca^3$ (and $\Phi_0 = c_0 a^3$), which is a dimensionless quantity.

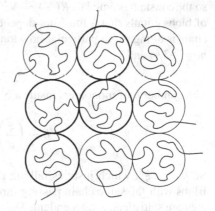

Fig. 6.10 The concept of "blobs" in a semidilute solution: the statistics of the segment of chain inside a blob is that of a self-avoiding walk; the chain of blobs (*highlighted blobs*) however follows the statistics of an ideal walk

As long as $\Phi_0 \ll \Phi$ (a dilute solution), the conformation of a chain is not affected by the presence of other chains and the arguments made for an isolated polymer describe this real situation well. If we increase Φ_0, the situation changes when the chains start to interpenetrate. The threshold, Φ^*, is reached when $\Phi_0 = \Phi$ and we therefore have $\Phi^* \sim N^{1-\nu d}$ in good solvent. We call a solution in which $\Phi^* \ll \Phi_0 \ll 1$ a *semidilute solution*. Here a chain will feel its own influence, but also those of the other chains ($\Phi_0 \gg \Phi^*$) and that of the solvent ($\Phi_0 \ll 1$).

We can establish (and observe) various scaling laws with respect to Φ_0. For example, the osmotic pressure $\Pi = -\partial F/\partial V$ (variation in the free energy when the volume of the solvent is varied while the number of monomers is kept fixed) in a monodisperse solution behaves as:

$$\frac{\Pi a^3}{kT} = \text{constant } \Phi_0^{9/4}. \tag{6.33}$$

The exponent $9/4$ reflects the connectivity of chains. We would have an exponent of 1 in a dilute solution of disconnected monomers (i.e., a perfect gas of monomers) of the same concentration [5]. More generally, the *observables will be universal functions of Φ/Φ^**. In a semidilute solution, a key concept is that of a *blob*, which is a small volume containing a segment of polymer of mass b monomers and linear size ξ (e.g. its radius), such that the segment "ignores" everything happening outside of this volume (see Fig. 6.10). In other words, the configuration of the segment is that of an isolated polymer and two blobs are statistically independent. ξ is naturally interpreted as the correlation length of the system. The complete system appears as a homogeneous assembly of blobs, in the interior of which we observe the scaling laws of an isolated polymer.

Let us limit ourselves to the case of three dimensions $d = 3$. The length ξ, which is an observable of the particular system, can also be put in the form $\xi = \text{constant } (\Phi/\Phi^*)^m$ where m is an exponent to determine. By definition, there is only one blob per chain for $\Phi = \Phi^*$ (they are only just beginning to interpenetrate) so the constant is equal to $R(N) \sim N^{3/5}$ (taking $\nu \approx \nu_F = 3/5$). The same concept of blobs entails that ξ must not depend on N. A blob ignores what the rest of the chain is doing, so in particular how long it is in total (N). Since $\Phi^* \sim N^{-4/5}$, then $m = -3/4$ and

$$\xi \sim a \, \Phi^{-3/4}. \tag{6.34}$$

The number b of monomers inside a blob is:

$$b \sim \left(\frac{\xi}{a}\right)^{5/3} \sim \Phi^{-5/4}. \tag{6.35}$$

So $b = c \, \xi^3$, which justifies talking about "homogeneously close packed blobs". Blobs with the same chain passing through them will perform an ideal walk since they are statistically independent. Therefore:

$$R^2(N, \Phi) \sim \xi^2 \left(\frac{N}{b}\right) \sim a^2 N \Phi^{-1/4}. \tag{6.36}$$

We can also directly find R in the form $R = \text{constant}\ (\Phi/\Phi^*)^m$ and determine the constant and the exponent m, which leads to the same result. This blob picture is in agreement with the behaviour of the autocorrelation function $g_{auto}(r)$ of a chain: at short range ($r < \xi$), we recover the correlation function $g_{saw}(r)$ of a self-avoiding walk whereas at larger distances ($r > \xi$), the correlation function $g_{auto}(r)$ will reflect the ideal behaviour of the chain of blobs. In three dimensions, it becomes:

$$g_{auto}(r) \sim \begin{cases} g_{saw}(r) \sim r^{-4/3} & \text{if } r < \xi \\ c\xi/r & \text{if } r > \xi. \end{cases} \tag{6.37}$$

The pair correlation function $g(r)$ of the solution (the monomers of the pair no longer necessarily belong to the same chain) also involves the characteristic size ξ of the blobs:

$$g(r) \sim \frac{c\xi}{r} e^{-r/\xi} \qquad \text{(Ornstein–Zernike)}. \tag{6.38}$$

6.5 Conclusion

We conclude that the universality of conformational properties of a polymer occurs due to the following reason, already invoked in the case of percolation: we can reduce all the physics into geometric parameters, here the persistence length and the excluded volume. Polymer physics is therefore one of the areas where scaling laws are the best verified and the most used to concretely understand the *typical* behaviour of the systems under consideration. Here we have given only a very brief overview of the range of scaling theories in polymer physics, by addressing only the case of statistical properties related to *solutions of linear homopolymers*. Many other polymeric systems have been studied following the same scaling methods, for example copolymers (i.e. polymers made up of segments of different chemical composition), polymers grafted or adsorbed on a surface, branched polymers etc. We refer the reader to the reference book on the subject [5], an introductory book [12] and more technical books [6, 11] without forgetting the historic work [9]. In addition we have only considered conformational properties, that is to say the static properties. Dynamical properties (e.g. modes of deformation, reputation, *depinning*) and stress response properties (e.g. force-extension curves, shear, viscoelasticity) also show remarkable scaling properties. These dynamical aspects are treated in for example [7] and [11].

References

1. J.P. Bouchaud, A. Georges, Anomalous diffusion in disordered media: statistical mechanics, models and physical applications. Phys. Rep. **195**, 127 (1990)
2. J.P. Cotton, D. Decker, B. Farnoux, G. Janninck, R. Ober, C. Picot, Experimental determinations of the excluded volume exponent in different environments. Phys. Rev. Lett. **32**, 1170 (1974)
3. J.P. Cotton, Polymer excluded volume exponent ν: an experimental verification of the n-vector model for $n = 0$. J. Physique. Lettres **41**, L231 (1980)
4. P.G. De Gennes, Exponents for the excluded-volume problem as derived by the Wilson method. Phys. Lett. **38A**, 339 (1972)
5. P.G. De Gennes, *Scaling Concepts in Polymer Physics*, 2nd edn. (Cornell University Press, Ithaca, 1984)
6. J. Des Cloizeaux, G. Janninck, *Polymers in Solutions* (Oxford University Press, Oxford, 1990)
7. M. Doi, S.F. Edwards, *The Theory of Polymer Dynamics* (Oxford University Press, Oxford, 1986)
8. B. Duplantier, H. Saleur, Exact tricritical exponents for polymers at the Θ-point in two dimensions. Phys. Rev. Lett. **59**, 539 (1987)
9. P.J. Flory, *Principles of Polymer Chemistry* (Cornell University Press, Ithaca, 1953)
10. W.M. Gelbart, R.F. Bruinsma, P.A. Pincus, V.A. Parsegian DNA-inspired electrostatics. Phys. Today **53**, 38 (2000)
11. A.Y. Grosberg, A.R. Khokhlov, *Statistical Physics of Macromolecules* (AIP, New York, 1994)
12. A.Y. Grosberg, A.R. Khokhlov, *Giant Molecules, Here, There, Everywhere* (Academic, Princeton, 1997)
13. J.B. Imbert, A. Lesne, J.M. Victor, On the distribution of the order parameter of the coil-globule transition. *Phys. Rev. E* **56**, 5630 (1997)
14. D. Lhuillier, A simple model for polymeric fractals in a good solvent and an improved version of the Flory approximation. J. Phys. France **49**, 705 (1988)
15. P. Kasteleyn, C.M. Fortuin, Phase transitions in lattice systems with random local properties. J. Phys. Soc. Jpn. (Suppl.) **26**, 11 (1969)
16. M. Lebret, B. Zimm, Distribution of counterions around a cylindrical polyelectrolyte and Manning's condensation theory. Biopolymers **23** 287 (1984)
17. J.C. Le Guillou, J. Zinn Justin, Critical exponents for the n-vector model in three dimensions from field theory. *Phys. Rev. Lett.* **39**, 95 (1977)
18. N. Madras, G. Slade, *The Self-Avoiding Walk* (Birkhauser, Boston, 1993)
19. G.S. Manning, Limiting laws and counterion condensation in polyelectrolyte solutions. I. Colligative properties. J. Chem. Phys. **51**, 924 (1969)
20. A. Maritan, F. Seno, A.L. Stella, Real-space renormalization group approach to the theta point of a linear polymer in 2 and 3 dimensions. Physica A **156**, 679 (1989)
21. L. Onsager, The effects of shape on the interaction of colloidal particles. Ann. N. Y. Acad. Sci. **51**, 627 (1949)
22. D. Ruelle, *Thermodynamic Formalism* (Addison-Wesley, London, 1978)
23. H.E. Stanley, P.J. Reynolds, S. Redner, F. Family, in *Real-Space Renormalization*, ed. by T.W. Burkhardt, J.M.J. van Leeuwen. Position-space renormalization group for models of linear polymers, branched polymers and gels, chapter 7 (Springer, Berlin, 1982), p. 169
24. G. Strobl, *The Physics of Polymers* (Springer, Berlin, 1997)

Chapter 7
High Temperature Superconductors

7.1 Superconductivity and Superconductors

This chapter is dedicated to the study of the superconductor–insulator transition in high temperature superconductors. We saw, in Sect. 1.3.3, that metal superconductors were not so interesting from the point of view of critical phenomena. This is because their coherence length at $T = 0$ is of the order of a micron, so the critical region is around 10^{-14} K, which is obviously impossible to observe. However this is not true for high temperature superconductors, which exhibit a critical temperature of the order of 100 K. The most important high temperature superconductor family is that of the superconducting cuprates discovered in 1986 by Bednorz and Müller. Since then other families have been discovered for example C60, MgB2 and AsFe. However the most important family is still the one that was discovered first, the cuprate family, which is the only one with superconducting properties above 100 K. This is also the most documented family both experimentally and theoretically. For these reasons, in this chapter we will focus only on superconducting cuprates. Their coherence length at $T = 0$ being of the order of 15Å, the critical region can reach a few tens of kelvins! So these materials represent, in principle, an ideal case in which to study critical phenomena, even more so given that the transition can be induced by a magnetic field (as for all superconductors) or by doping (a unique property of high temperature superconductors). In principle, these two parameters, can induce *quantum* transitions at $T = 0$. This physical system certainly seems interesting to test new descriptions of critical phenomena.

However, there are two significant difficulties physicists face. One is that the physical system in question is one of exceptional microscopic complexity. Despite 100,000 or so publications on the subject the underlying mechanisms have still not been reliably identified. In addition the structure and composition of high temperature superconductors are complex and subjected to uncontrollable variations. As a consequence, the other difficulty arises which is that whilst there is a large amount of experimental data, it is not very reproducible. The data are rarely able to distinguish between different theoretical descriptions. Whatever it is,

A. Lesne and M. Laguës, *Scale Invariance*, DOI 10.1007/978-3-642-15123-1_7,
© Springer-Verlag Berlin Heidelberg 2012

this domain is a strong driving force for research. We could say it gave birth to a new condensed matter physics since the proposed mechanisms differ so much from that of the solid state physics founded in the 1930s by Felix Bloch.

Here we will concentrate on properties of superconductor–insulator transitions as a function of temperature T and doping δ (charge injected into the copper oxide CuO_2 planes), summarised in a (T, δ) plot called a phase diagram. Physicists try to explain it by two approaches; microscopic or phenomenological. *Microscopic* approaches rely on the existence of pairs of charges, Cooper pairs, which are actually observed in high temperature superconductors. The methods consist of describing the density, properties and coherence of these pairs as a function of position in the phase diagram (see Fig. 7.3). One of the crucial questions, still not yet resolved when this book was in press, is the probable, but not demonstrated, existence of non condensed (non coherent) pairs in the region of the phase diagram called the *pseudogap*. The theme of this chapter picks up on the second type of approach, *phenomenological* descriptions. The question is simple: transitions belong to which universality class in each region of the phase diagram? First of all however, we give a brief introduction to superconductivity and superconducting materials.

7.1.1   Mechanisms and Properties

7.1.1.1   A Challenge for Physicists' Imaginations

In the superconducting state, discovered in 1911, matter shows strange electric and magnetic properties. Superconductivity remained an enigma for 50 years. The name "superconducting state" comes from the sudden complete disappearance of electrical resistance in a material when its temperature is lowered below a certain value T_c called the critical temperature. The first superconductor to be discovered (by Holst and Kammerling Onnes) was mercury, which is a superconductor when its temperature is below 4.2 K i.e. $- 269°C$. At the beginning of 1986, the superconductor with the highest critical temperature, 23 K, was niobium-germanium Nb_3Ge. It was therefore a real surprise in 1986 when Müller and Bednorz discovered cuprates, in which the current record for the highest critical temperature is $T_c = 138 K$ for mercury barium calcium copper oxides composed of HgBaCaCuO. A priori, superconductivity seems to be a violation of the principle of the impossibility of perpetual motion. It allows an extremely stable electric current in a closed superconducting ring. Such a current could persist longer than the age of the universe! This kind of perpetual motion therefore actually does exist (it is used today in thousands of hospitals in MRI scans), however, like the movement of electrons in Niels Bohr's model of an atom, it is quantum in nature. Since the 1920s, physicists have known that superconductors are not perfect conductors (normal metals in which the resistance could become extremely low for some unknown physical reason). If so, the resistance would increase with the smallest defects, material perturbation, surfaces, electrical contacts etc. But, the exact opposite is seen. Superconductivity is a robust state, sometimes even reinforced by disorder (see Fig. 1.12 in Chap. 1)! In the 1930s, physicists understood that superconductivity is also forbidden in the quantum description of a normal metal due to the fact that the quasi-continuous spectrum of excitations excludes any state with zero resistance.

In 1933, the physicist Alexander Meissner discovered that superconducting material does not let magnetic flux penetrate it. This property, the Meissner–Ochsenfeld effect, was a key to the description of the mechanisms of superconductivity. The London description can predict the (very small) depth λ magnetic flux can penetrate into a superconductor. It also contains the principle of a property that would not be observed and explained until 30 years later: magnetic flux can penetrate a superconductor only in the form of quanta, of a universal value of $\Phi_o = h/2e$.

7.1.1.2 A Powerful Mean Field Description

The thermodynamic description of a superconducting system by Ginzburg and Landau [7] led to a calculation of another characteristic length, the coherence length ξ. This second length characterises the thickness over which superconductivity gradually disappears near an interface with a non superconducting material. From this we understand that the superconducting state leads to a large energy gain. This gain in energy explains why the exclusion of magnetic flux, the Meissner effect, is possible despite its high energy cost. If we increase the magnetic flux density B, beyond a value B_c (corresponding to an external magnetic field H_c) such that the gain and loss in energies are equal, the material suddenly becomes normal and lets the magnetic flux penetrate. The Ginzburg–Landau theory also shows that two types of superconductors exist, depending on whether ξ is larger than $\lambda\sqrt{2}$ (type I superconductors) or smaller than $\lambda\sqrt{2}$ (type II superconductors). In this second case, by far the most common in practice in materials of interest, the onset of superconductivity is favoured at interfaces with regions in the normal state.

Between two values H_{c1} and H_{c2}, type II superconducting materials are in a mixed state, where the superconducting material is pierced by narrow regions in the normal state, called vortices, containing exactly one quantum of magnetic flux. If the magnetic field is less than H_{c1} then the material contains no vortices. Whereas above H_{c2} the material is in a normal state without any magnetic properties.

The superconductors that interest us particularly here, high temperature superconductors, are very much type II, ξ being about 100 times smaller than λ. The Landau description predicts that the ratio H_{c2}/H_{c1} is of the order of the square of the ratio λ/ξ. In high temperature superconductors, H_{c2} is of the order of hundreds of teslas, whilst H_{c1} is of the order of 0.01 tesla. In the great majority of practical situations, these superconductors are in the mixed state where they contain vortices.

Paradoxically, the thermodynamic description of Ginzburg–Landau theory, constructed *blindly* without knowing the microscopic mechanisms, remains particularly useful in describing high temperature superconductors, exactly because it *assumes nothing* about the *microscopic mechanisms*.

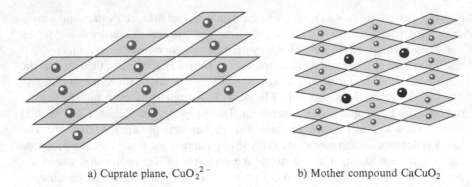

a) Cuprate plane, $CuO_2{}^{2-}$ b) Mother compound $CaCuO_2$

Fig. 7.1 Basis structure of cuprate compounds. (**a**) Cuprate CuO_2 planes have a chessboard structure where the corners of the squares represent oxygen atoms and one square in two has a copper atom in the centre. (**b**) The neutral, insulating, mother compound $CaCuO_2$, consists of alternating cuprate planes and divalent ions Ca^{2+}

Ginzburg–Landau theory predicts the thermal dependence of all quantities. In this mean field approach, the characteristic lengths both diverge with the inverse of the square root of the distance from the critical temperature:

$$\lambda(T) \sim \xi(T) \sim (T_c - T)^{-1/2}. \qquad (7.1)$$

This description is completely valid in the case of metal superconductors. We saw in Sect. 1.3.3 that the Ginzburg criterion predicts a particularly small critical region (10^{-14} K) because of their long coherence length. This is not true for high temperature superconductors (see Fig. 7.1).

7.1.1.3 BCS Theory: Finally a Microscopic Description

In 1957, the physicists Bardeen, Cooper and Schrieffer proposed a mechanism that agreed excellently with the superconducting properties of many metals. Their key idea was that electrons are associated in pairs, *Cooper pairs*. In this state they form *bosons* (particles of integer spin) which obey different rules from those obeyed by single electrons which are *fermions* (of half integer spin). The paradox of an electronic perpetual motion contradicting physics disappears in this description, the legitimacy of which was quickly confirmed by direct observation of electron pairs and their effects.

The first ingredient involved in this theory is the long distance interaction between the electrons and vibrations (or phonons) of ions in the crystal lattice. John Bardeen observed in 1955 that this can indirectly lead to an attraction between electrons. The exchange of phonons between electrons can be illustrated by a very schematic classical picture. An electron passing through distorts the crystal lattice by repelling negative charges and attracting positive charges. A second electron passing through in its wake does so more easily due to this displacement of ions and therefore behaves as if it was attracted by the first electron.

A second essential step was made by the young theoretical physicist Leon Cooper, employed by John Bardeen, in 1954, to establish a theoretical framework for this system. Surprisingly Cooper showed that the attraction between electrons, regardless of its smallness, must completely mess up the states accessible to the electrons close to the Fermi surface. Powerful perturbation calculations using Feynman diagrams were unable to predict this effect and yet Leon Cooper's demonstration of this instability of electronic states under the effect of an attraction, even if very weak could not be ignored.

A third pillar was necessary to construct the theory: *condensation* of electron pairs assembling in the same quantum state at low temperature. After having actively searched for a wavefunction description, the young student John Robert Schrieffer finally tried an expression presented during a seminar on a totally different domain by the nuclear physicist Tomonaga. The result was stunning. In a few weeks the BCS trio recalculated all the properties of this novel state by replacing the *Bloch functions* normally used for electrons in a metal by Schrieffer's wavefunction. They obtained many results in perfect agreement with superconductor properties, including the following:

- The critical temperature is given by the expression:

$$T_c = \Theta_D \exp\left(-\frac{1}{n(E_F)V}\right), \qquad (7.2)$$

where Θ_D is the Debye temperature of the material, characterising the phonons, $n(E_F)$ is the density of states at the Fermi level in the normal state and V is the attractive potential between electrons.
- The density of states in the superconducting state has a *band gap* of 2Δ, proportional to the number of condensed pairs. The Cooper pairs are stable as long as they are not given an energy greater than or equal to 2Δ. Below this value, the pairs stay in the ground state (possibly moving and so carrying a current in this state), without causing even the tiniest energy dissipation. This characteristic is what explains the robustness of the superconducting state.
- The energy 2Δ characterises the typical interaction energy between two electrons in a pair and is therefore of the same order of magnitude as the thermal energy kT_c that destroys superconductivity by breaking up the pairs. For superconductors with weakly bound Cooper pairs, the relation $2\Delta = 3.5\,kT_c$ predicted by BCS theory, is well established for metals for which $n(E_F)V < 1$. For other metals, a generalisation of BCS theory by Eliashberg accounts for $2\Delta/kT_c$ well.

By noticing that the Debye temperature and the electron–phonon coupling (measured by V) are related, we find the not so encouraging result [13] that within the BCS framework we cannot hope for values of T_c higher than 30 K.

BCS theory bears a hallmark of the greatest theories: a single new quantity introduced by Cooper, the attractive potential V between electrons, leads to quantitative predictions of dozens of seemingly independent properties.

7.1.1.4   Superconductors with Electrical Resistance

One important consequence of the Ginzburg–Landau description concerns the maximum electrical current density, the *critical current*. In the "mixed" state, in which a type II superconductor is penetrated by vortices, the resistance is not strictly zero if the electrical current density is larger than a critical value J_c. Unlike type I superconductors, *the resistance of type II superconductors can be non zero*, due to the motion of vortices under the influence of the electric current. This motion causes an induction (due to the Lorentz force), which in turn results in an electric field and finally the Joule effect. Everything happens as if the motion of vortices was submitted to a viscous drag. Note that even if most of the volume of the material is in the superconducting state with a local zero resistivity, the overall resistance is non zero. Another paradox is that some material defects can considerably increase the critical current density by blocking vortex motion. For most applications, the critical current density needs to be larger than a value of order 10,000 to 100,000 A/cm^2. In general such values are largely achieved or exceeded in good quality thin films. However they are harder to obtain in cables. In 2011, commercial superconducting wires using Y(Dy)BaCuO thin films reached about 200 kA/cm^2.

7.1.1.5   Josephson Junctions

The quantum nature of the superconducting current results in diverse characteristic properties. The macroscopic coherence of the quantum wavefunction can produce interferences like that produced by light through a thin slit. The equivalent here is superconducting current crossing a thin insulating region interrupting the super-conductor circuit. We call this superconductor–insulator–superconductor junction a Josephson junction. It can be made in a number of different ways. The properties of such a junction are very unusual. For example, Josephson junctions can store binary information and can therefore be used in logic components for computers.

Another property is the variation of the critical current of the junction with applied magnetic field, which is used to measure extremely weak magnetic fields, for example, up to femtoteslas. A SQUID (*Superconducting Quantum Interference Device*) is a combination of two Josephson junctions on a superconducting ring and is equivalent to Young's two slits for light. The voltage measured between the two superconducting regions varies periodically with the magnetic flux that penetrates the ring, each oscillation corresponding to a quantum of magnetic flux entering or leaving. Counting the quanta, or even smaller flux variations, penetrating the ring enables very refined measurements of magnetic fields. Actually magnetic flux is not quantised in the vacuum, but only inside materials in the superconducting state, and only those of type II.

7.1.2 Families of Superconducting Compounds

There are many superconducting materials belonging to diverse families. If we consider just the elements, 38 of them are superconducting (out of the 101 elements classified in the periodic table). A dozen or so other elements are superconducting under pressure or in thin films (see Figs. 1.10 and 1.11). Some, such as tungsten, are superconducting in amorphous form, that is when the atoms have disordered positions. Elements that are not superconductors are often magnetic. Superconducting and the magnetic states are in competition because they both lower the energy of the material. However, the magnetic order is particularly sensitive to the quality of the crystal whereas superconductivity is not very sensitive as the coherence length is much longer than the interatomic distances. So, at low temperature, magnetism can be favoured in a crystal, whilst superconductivity replaces it in the same material when it is quenched in a crystallograpically disordered state (Fig. 1.12).

As this book goes to press, the most studied high temperature compound is $YBa_2Cu_3O_7$ ($T_c = 92$ K), and the one with the highest critical temperature 138 K is $Hg_{0.8}Tl_{0.2}Ba_2Ca_2Cu_3O_{8+\delta}$ [3]. These compounds are layered perovskite structures of copper oxide with a quasi two dimensional character. This is the important characteristic for the scaling properties we will present in this chapter. They are made up of regular layers of CuO_2, in which the superconducting currents reside, alternating with layers of non superconducting atoms. The CuO_2 planes interact with each other by *Josephson coupling* across the non superconducting atomic layers. Copper plays an essential role which is not well understood, and all attempts to replace it with another element have not led to comparable superconducting properties (for instance, the Fe-As family discovered in 2008 only reaches critical temperatures in the 50 K range). In their normal state these superconducting cuprates are poor metals or insulators with mechanical properties of ordinary ceramics. A wide variety of theoretical models have been proposed to capture the specific superconducting properties of high temperature superconductors, such as their high critical temperature and their very short coherence length. So far, experiments have not been able to distinguish between the proposed descriptions.

In 1990, a new family of superconductors appeared, the fullerenes, also called buckyballs because of the shape of the molecule C_{60} which is the essential component of their architectures. C_{60} actually has the exact form of a football of 1 nm diameter! The critical temperature of the compound Cs_2RbC_{60} is 33 K. There are also many other families of superconductors, from organic compounds to uranides (elements with atomic numbers larger than 91), with sometimes very exotic properties.

7.1.3 Applications: From Telecommunications to Magnetoencephalography

Many applications have been imagined using the variety of unusual properties of superconductors. Few have actually been commercialised, because of technical difficulties in implementing them, particularly the necessity of cooling the material. However, superconducting electromagnets (for example for MRI medical equipment), made of niobium alloys, and measurement of very weak magnetic fields with SQUIDs are well established applications. Superconducting cuprates cooled by liquid nitrogen now seem to be usable for microwave frequency components of mobile communication. Superconducting materials should also increase the speed of computers and decrease the energy dissipated by components. It is now known how to make good quality Josephson junctions and how to make them work in a computer, in *RSFQ logic elements* invented by the physicist Likharev.

"Strong current" uses (magnets, alternators, motors etc) are still not established because of the mechanical properties and current densities required. But they represent a huge market starting to open in the 2010s. Cuprate ceramics are not as ductile and malleable as metal alloys. However cables several kilometers long, maintained at liquid nitrogen temperature, 77 K, have already been commercialised. They started to supply energy to big cities in 2010 (Manhattan).

7.2 The Phase Diagram

Despite their diversity, cuprate superconductors contain only one molecular superconducting component, the cuprate CuO_2 plane. Its structure is represented in Fig. 7.1a.

In the valence state 2 (the most common for copper), the cuprate plane is ionised, carrying two negative electric charges per copper atom. A simple structure, called the "mother", or infinite phase, is made of a neutral compound of alternating cuprate planes and divalent ions (see Fig. 7.1b). Due to their ionic radius, it is mainly calcium ions Ca^{2+} which are used to make superconducting compounds. However $CaCuO_2$ is an insulating antiferromagnet with Néel temperature 530 K [11]. Single electrons from copper (Cu) atoms in the $3d^9$ orbital are effectively localised by a strong Coulomb interaction. Displacing an electron from one copper atom to its neighbouring atom costs and energy of the order of 5eV. Although not our main topic, we should at least point out that this repulsion energy leads to strongly correlated electron properties. A priori Fermi liquid type descriptions leading to single electron bands are not valid. The microscopic description of high temperature superconductors is still an open problem (in 2011). The only well established fact is the extreme complexity of the mechanisms involved, which seem to locally combine charge and magnetic effects as well as superconducting effects all at similar order.

Since the compound $CaCuO_2$ is insulating, charges must be injected in the cuprate plane to *dope* them to obtain superconducting properties.

7.2.1 Doping

Doping in cuprates is the analogue of doping in semiconductors, which modulates their conductivity within a very large range. However there are significant differences between the two systems:

- Doping semiconductors is intended to inject a low concentration of carriers (10^{12} to 10^{18}charges/cm^3, or 10^{-10} to 10^{-4} charges per atom) in an empty conduction band. This shifts the Fermi level into the bandgap, but it does not penetrate the conduction band, which remains sparsely populated. The charges therefore remain very dilute, so a *rigid band* model can be used, identical to that of non doped material.
- Doping cuprates must reach concentrations of tenths of charge per atom to display superconducting properties. This causes a significant *chemical* transformation of the compounds. The valency of copper is changed in a range from 1.7 (doping with electrons) to 2.3 (doping with holes). The resulting doped compounds are very different from the non doped compounds. Any rigid band model is only a very simplistic approximation of the new electron density.

The particularly high level of concentration of charges necessary to achieve superconductivity leads to the following property, as a direct consequence of Poisson's equation:

> Injected charges are confined in one or two atomic layers around the charge reservoir.

The principle of this effect is that the non doped material is insulating, but doping locally establishes metallic properties. The screening length, or Debye length, is very small in metals. Consequently, all sources of doping are only active in one or two atomic layers around it due to the large change in electron density required.

Doping consists of inserting into the structure electroactive molecular blocks, called *charge reservoir blocks*, which will inject charges to their immediate surroundings. The point made above indicates that:

> To be effective, reservoir blocks must be of the size of a few atoms, at least in one direction.

Superconducting properties of different cuprate compounds essentially depend on the level of doping of the CuO_2 planes, as well as the coupling between different cuprate planes in the structure. Here we are just interested in the first point, namely,

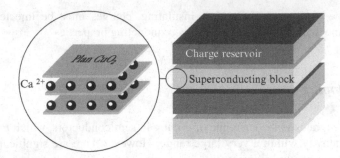

Fig. 7.2   Cartoon of the layered structure of superconducting cuprates

in a phenomenological approach to the dependence of properties as a function of doping. We first specify how cuprates are doped in practice.

7.2.1.1   Doping Chemically

The structure of standard cuprate compound superconductors can be visualised schematically as alternating layers of two types of block, superconducting and reservoir (Fig. 7.2). The superconducting block is composed of n alternating CuO_2 planes separated by $n - 1$ atomic planes (in general Ca^{2+} and Y^{3+} or a trivalent rare earth element for the $RBa_2Cu_3O_7$ family). The charge reservoir block is made of a succession of layers $[AO]\,[BO]_m\,[AO]$, where m is 1 or 2. Due to their ionic radii, the atom A is Sr or Ba, whereas the atom B can be Cu (in another oxidation state), Bi, Tl, Hg, etc. There is a large diversity of charge reservoir blocks as many electroactive blocks satisfy the constraints of structural compatibility with the superconducting block.

When the nature and/or degree of oxidation of the reservoir block is changed while the superconducting block remains unchanged, the average doping of cuprate planes can be varied over a large range. The observed properties can be represented in the (*doping, temperature*) plane in a phase diagram (Fig. 7.3). This diagram has four regions and is qualitatively symmetric about zero doping:

- At small doping, cuprates are insulating, showing long range antiferromagnetic order. The Néel temperature, above which the system is disordered, rapidly decreases with the absolute value of doping to zero for a doping of a few hundredths of charges per copper atom.
- For higher doping, at low temperature we cross an insulating region with complex properties that are not well understood. In this region, long range order has disappeared, but large antiferromagnetic fluctuations persist. A *pseudogap* is observed which seems to be reminiscent at high temperature of the superconducting gap observed below the critical temperature. Cooper pairs could be formed in this region without showing an ordered phase. This region is bounded by a maximum

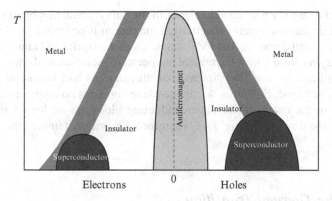

Fig. 7.3 Schematic phase diagram of a cuprate compound as a function of the doping of the CuO_2 planes

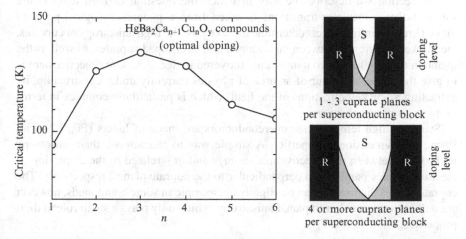

Fig. 7.4 As we change the number n of cuprate planes in each superconducting block (see Fig. 7.2), the critical temperature is maximum for $n = 3$. The diagram indicates that above this value, there is no longer enough doping (*grey zone*) of the cuprate planes by the reservoir blocks

temperature T^* (broad grey line in Fig. 7.3) which corresponds to a regime change rather than a sharp transition.

- At higher doping, a region is observed in which the compound is superconducting, within a particular range of doping. For positively charged doping (holes), the range is about 0.05 to 0.3 with a maximum critical temperature for a doping of about 0.18. For negative doping, a symmetric but narrower superconducting region is observed.
- For doping about an absolute value of 0.3, a metal region is observed.

If we vary the number n of cuprate planes in the superconducting blocks, whilst at the optimal doping for superconductivity, a maximum is observed for $n = 3$. This is particularly well demonstrated for compounds of the HgBaCuO family (Fig. 7.4).

For n varying from 1 to 3, the critical temperature increases as we would expect. The system becomes less two dimensional, fluctuations less important and order becomes more robust. From $n = 4$, cuprate plane doping is no longer homogeneous and planes in the centre of the superconducting block are no longer sufficiently doped (see the diagram in Fig. 7.4). Therefore the critical temperature decreases with n.

7.2.2   The Thermal Transition

In this section we describe the way in which the question of high temperature superconductor–insulator transitions is asked from a phenomenological point of view. Three essential aspects determine the analysis: the high anisotropy of cuprates, the marked difference between underdoped and overdoped cuprates, as well as the quantum transitions at zero temperature (covered in Sec. 7.3). Our objective here is to give the reader a flavour of an area of physics currently under construction, by extracting a few key directions of the field, which is particularly complex in terms of microscopic mechanisms.

Standard high temperature superconductors are made of layers (Fig. 7.2) with strong two dimensional properties. A simple way to characterise their anisotropy is the ratio between the effective masses $m_{//}$ and $m_\perp$ related to the dispersion of electronic states parallel and perpendicular to the cuprate planes, respectively. The cuprate planes themselves can be slightly anisotropic in some compounds, however we will omit this small intraplanar anisotropy, which only plays a small role in their properties.

7.2.2.1   Anisotropy

The anisotropy of the conductivity σ, gives a good idea of the ratio between the effective masses γ:

$$\gamma = \sqrt{\frac{m_\perp}{m_{//}}} \sim \sqrt{\frac{\sigma_{//}}{\sigma_\perp}}. \tag{7.3}$$

The anisotropy γ varies in a wide range depending on the high temperature superconductor structure. Table 7.1 shows the currently accepted values for typical cuprate compounds at optimal doping. The essential factor determining this anisotropy is the thickness and conductivity of the reservoir block. The anisotropy factor γ fixes the ratio between the corresponding characteristic lengths in each direction:

Table 7.1 Anisotropies of principal families of high temperature superconductors at optimal doping[1]

Compound	T_C (K)	γ	$\lambda_{//0}$ (Å)	$\lambda_{\perp 0}$ (μm)	$\xi_{//0}$ (Å)	$\xi_{\perp 0}$ (Å)
YBaCuO	92	9	1000	0.87	15	1.6
LaSrCuO	35	18	1700	3.2	30	1.7
$HgBa_2CuO_4$	94	29	800	2.2	26	0.9
$HgBa_2Ca_2Cu_3O_8$	133	50				
$Tl_2Ba_2CuO_6$	88	120				
$Bi_2Sr_2CaCu_2O_8$	85	≥ 250	2500	63	27	0.1

$$\frac{\xi_{//}}{\xi_{\perp}} = \frac{\lambda_{\perp}}{\lambda_{//}} = \gamma. \qquad (7.4)$$

Table 7.1 shows the values of these lengths at zero temperature for a few cuprate compounds.[1] Parallel to the cuprate planes, these lengths range from 15 to 30 Å, for $\xi_{//0}$, and from 1,000 to 2,500Å, for $\lambda_{//0}$. The spread of values is much larger for lengths perpendicular to the cuprate planes, ranging from 0.1 to 1.6 Å, for $\xi_{\perp 0}$, and 0.9 to 63¯m for $\lambda_{\perp 0}$.

7.2.2.2 2D and 3D Regimes for an Optimally Doped Cuprate

The dimensionality of superconducting behaviour is determined by the way in which $\xi_{\perp}(T)$ compares to the distance c_b between two superconducting blocks:

- If $\xi_{\perp}(T) \ll c_b$ then the superconductivity has a 2D character.
- If $\xi_{\perp}(T)$ is greater than or equal to c_b then the superconductivity has a 3D character.

Depending on the compound, c_b varies between 12 and 18 Å, values much larger than $\xi_{\perp 0}$. Consequently:

- At zero temperature, or clearly less than T_c, cuprate compounds all show a 2D behaviour.
- Near the critical temperature, the coherence length $\xi_{\perp 0}$ in the direction perpendicular to the cuprate planes diverges as:

$$\xi_{\perp} = \xi_{\perp 0}\, t^{-\nu} \quad \text{with} \quad t = \frac{T_c - T}{T}. \qquad (7.5)$$

---

[1]These lengths are only orders of magnitude. They have been measured by different teams, using diverse techniques and on samples of varying quality. Therefore the published values are quite varied.

Table 7.2 Values of the temperature T_{3D} of the change of regime such that for $T_c - T_{3D} <$ $T_c - T < T_c$ the superconducting behaviour is three dimensional

Compound	T_C (K)	γ	$\xi_{\perp 0}$ (Å)	c_b (Å)	t_{3D}	$T_C - T_{3D}$ (K)
$YBa_2Cu_3O_7$	92	9	1.6	12	0.05	4.5
$Bi_2Sr_2CaCu_2O_8$	85	≥ 250	0.1	15	0.0005	0.04

• There exists a region of temperature t_{3D} such that if $t < t_{3D}$ the superconducting behaviour is three dimensional:

$$\xi_{\perp}(t_{3D}) = c_b \qquad \text{or} \qquad t_{3D} = \left(\frac{\xi_{\perp 0}}{c_b}\right)^{1/\nu}. \tag{7.6}$$

Table 7.2 illustrates this regime change in two extreme cases, $YBa_2Cu_3O_7$ and $Bi_2Sr_2CaCu_2O_8$. In practice, $YBa_2Cu_3O_7$ is the only one for which we observe a real three dimensional superconducting behaviour in a range of a few kelvins. This behaviour is still only clearly demonstrated if we include the finite size effects of dynamics and "scaling corrections" [17]. For other cuprate compounds, the 3D regime only occurs in a region around T_c that is too narrow to see.

7.2.2.3 2D and 3D Regimes as a Function of Doping

So far, we have only considered optimally doped compounds. One of the significant advances in the study of high temperature superconductors was, around 1993, the observation of the radical difference between underdoped and overdoped compounds. It is one of the keys to elucidating the physical mechanisms of superconductivity in these structures (still in the future in 2011). Many changes in their properties as a function of doping are observed, including the symmetry of the order parameter (which we will not address), and the ratio $2\Delta/kT_c$, which reaches very high values (more than 20 compared to the BCS value of 3.5) at low doping, to around 5 for overdoped compounds. We are especially interested in the variation of the anisotropy and the characteristic lengths. Figure. 7.5 shows the characteristic evolution of anisotropy as a function of doping for a cuprate compound.

To date, the only detailed and reproducible measurements have been made on LaSrCuO and YBaCuO compounds. For YBaCuO, it turns out that the anisotropy γ increases by a factor of about three when we pass from optimal doping ($T_c = 92$ K) to highly underdoped ($T_c = 70$ K). In this range, the coherence length $\xi_{\perp 0}$ perpendicular to the cuprate planes remains roughly constant whilst the coherence length $\xi_{//0}$ parallel to the cuprate planes increases by a factor of three [9].

Fig. 7.5 Schematic evolution
of the anisotropy of cuprates
as a function of their doping

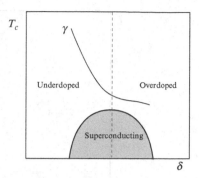

7.2.2.4 Two Schematic Pictures for the Phase Diagram

To close this brief presentation of the phase diagram of high temperature super-conductors, we offer the reader two schematic pictures that summarise the essential properties. For simplicity we assume that the doping is by holes. This is also the case that has been most widely studied. A priori, we expect a qualitatively symmetric behaviour for doping by electrons.

The first scheme summarises the *universality classes* that we expect from experimental observations, as a function of doping and temperature (Fig. 7.6). The following points summarise this figure:

- *Overdoped* compounds can be described as classical superconductors using mean field theory. Their transition can be described by a version of BCS adapted to the particular symmetry of the order parameter (d symmetry) for which we expect a ratio $2\Delta/kT_c$ of about 4.5. This has been confirmed by experiments [4].
- *Underdoped* compounds have an anisotropy that is larger the smaller the doping (Fig 7.5) and they are therefore essentially of the 2D–XY class.
- In the immediate vicinity of the thermal transition, there exists a 3D-XY region due to the divergence of the coherence length $\xi_\perp(T)$. We have seen that in practice this region is only observable for YBaCuO. For other cuprates, its extension is less than 1 K (Table 7.2).
- At zero temperature, there exists two quantum critical points; one for doping δ_u (u standing for *underdoped*) corresponding to the *superconductor–insulator* transition, and one for doping δ_o (o standing for *overdoped*) corresponding to the *superconductor–metal* transition.

Many microscopic or phenomenological descriptions have been proposed to explain the features of this diagram. We present below the description proposed in 1993 by Emery and Kivelson [5] which is a good pedagogical example (Fig. 7.7).

Order in a superconducting state is represented by a quantum wavefunction:

$$\Psi = |\Psi|e^{i\theta}. \tag{7.7}$$

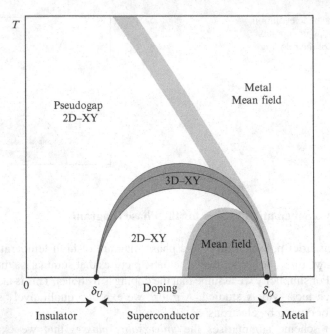

Fig. 7.6 Schematic picture of the expected universality classes in different regions of the phase diagram of cuprate compounds (after Schneider and Singer 2000). *Dark grey* areas correspond to the superconducting state and *light grey* to the critical region. The *wide grey lines* correspond to regions of regime change rather than sharp transitions. At zero temperature, the doping level δ_u is a quantum critical point for the superconductor–insulator transition, whilst doping δ_o corresponds to the superconductor–metal transition

The modulus and phase of this function are related to the density of Cooper pairs and the superconducting current respectively. The volume density n_s of paired electrons is equal to the squared modulus of this function:

$$n_s = |\Psi|^2. \tag{7.8}$$

The superconducting current density j_s can be expressed from the value of the momentum operator:

$$\widehat{p} = -i\hbar\nabla + \frac{e}{c}A, \tag{7.9}$$

where A is the vector potential. In a region where the density n_s of Cooper pairs is constant, the current has a part proportional to the gradient of θ and a part related to the vector potential:

$$j_s = \frac{n_s e}{m}\left\{\hbar\nabla\theta + \frac{e}{c}A\right\}. \tag{7.10}$$

In Fig. 7.7, the phase ordering dominates the superconductor–insulator transition for underdoped materials, whilst the formation of pairs dominates the transition for overdoped materials. The grey line represents the order–disorder transition for the

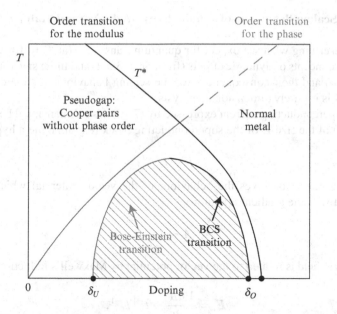

Fig. 7.7 Schematic diagram for high temperature superconductors, proposed by Emery and Kivelson [5]. The *grey line* represents the order–disorder transition of the phase "teta" of Cooper pairs, whilst the *black line* represents the transition of the wavefunction modulus, i.e. the transition of the appearance of Cooper pairs. For overdoped materials, as soon as the pairs appear they are simultaneously condensed in a coherent state, so there is only one transition, of BCS type. However for underdoped materials, there exists a pseudogap region where uncoherent pairs exist. These disordered pairs can then condense at lower temperature into a coherent state. The transition then behaves as a Bose–Einstein condensation

phase "teta" of Cooper pairs, whilst the black line represents the transition for the modulus of the wavefunction, that is to say the transition of formation of Cooper pairs:

- For *overdoped* materials, as soon as pairs appear, they are simultaneously condensed into a coherent state, i.e. there is only one transition, of BCS type.
- For *underdoped* materials below a temperature of T^*, a *pseudogap* state exists where pairs exist but without order in their phase. These disordered pairs can then condense at a lower temperature into a coherent state. So there is a Bose–Einstein condensation transition.

This schematic diagram, by Emery and Kivelson, is the basis for many microscopic descriptions of high temperature superconductors proposed since. The essential idea is that of a change in nature of the transition between the underdoped (Bose Einstein) and overdoped (BCS).

7.2.2.5   Scaling Behaviour of Conductivity (in the Normal State)

Before describing what we expect for quantum transitions (at $T = 0$), we recall a few basic concepts of dynamic effects (from Sect. 3.5.2) and finite size effects (from Sect. 3.5.4) and their consequences for the scaling behaviour of superconductors. Initially this is purely dimensional analysis.

The superconducting current expressed by (7.10) has a dimensional behaviour as a function of the size L of the superconducting block (e.g. imagine a hypercube of side L):

$$j \sim L^{1-d}. \tag{7.11}$$

The same expression gives the behaviour of the vector potential which must be consistent with the gradient operator:

$$A \sim L^{-1}. \tag{7.12}$$

The electric field is related to the vector potential by Maxwell's induction equation:

$$E = \frac{1}{c} \frac{\partial A}{\partial t} \sim t^{-1} L^{-1}. \tag{7.13}$$

The dimensional behaviour of the conductivity σ related to time and the system size:

$$\sigma = \frac{J}{E} \sim t L^{2-d}. \tag{7.14}$$

The dynamic exponent z, which we will define later when talking about growth mechanisms (Sect. 8.1.2), relates the characteristic time to the coherence length:

$$t \sim L^z \sim \xi^z. \tag{7.15}$$

Finally, the conductivity can be written in the following form:

$$\sigma(T, H) = \xi^{2-d+z} F \left(\frac{\phi}{\phi_0} \right) = \xi^{2-d+z} F \left(\frac{H\xi^2}{\phi_0} \right), \tag{7.16}$$

where H is the magnetic field, $\phi_0 = h/2e$ the quantum of magnetic flux and ϕ the flux crossing a "coherence area" ξ^2. $F(x)$ is a universal function, constant for $T > T_c$ and finite at T_c. From this we deduce the dependence of σ on magnetic field as H tends to zero, by replacing ξ^2 by $1/H$ in such a way that the argument of F remains constant:

$$\sigma(T_c, H) \sim H^{-\frac{2+z-d}{2}}. \tag{7.17}$$

In the same way the frequency dependence of the conductivity can be deduced from (7.14):

$$\sigma(T, \omega) = \xi^{\frac{2-d+z}{2}} \Sigma(\omega \xi^2), \tag{7.18}$$

Table 7.3 Exponents obtained from measurements of the conductivity of two YBaCuO compounds

Material	Mechanism of fluctuations	Characteristic T	$\nu(z-1)$	ν	z
Ceramics	σ continuous thermal and dynamic fluctuations	$T_c = 92.44$ K	2/3	$\nu_{3D} = \dfrac{2}{3}$	$z_{3D} = 2$
Monocrystal	$\sigma(\omega)$ Melting vortex glass	$T_{Glass} = 91$ K	8	$\nu_G = 1.6$	$z_G = 6$

where Σ is a universal function. σ and ω are both complex numbers with real an imaginary parts related to the resonant and damped components in the classical way. Equation (7.18) gives the critical behaviour of the modulus and phase as the frequency tends to infinity:

$$|\sigma(T_c,\omega)| \sim |\omega|^{-x} = |\omega|^{-\frac{2-d+z}{z}} \tag{7.19}$$

whilst the phase tends to:

$$\varphi(\omega \to \infty) = \frac{\pi}{2}\left(\frac{2-d+z}{z}\right) \tag{7.20}$$

Remarkably, in two dimensions, the phase tends to $\pi/2$ whatever the value of z.

In practice, experimental results are mostly for YBaCuO compounds. We present results obtained for two types of material, ceramics [16] and monocrystals [Koetzler 1999] of YBa$_2$Cu$_3$O$_7$. In both cases, we observe the material near to T_c where the behaviour is known to be 3D-XY, for example the variation of the magnetic susceptibility. In this case the exponent of (7.16) is simply $z-1$ and the conductivity of DC current of a 3D material behaves as:

$$\sigma \sim \xi^{z-1} \sim \left(\frac{T-T_c}{T_c}\right)^{-\nu(z-1)}. \tag{7.21}$$

Table 7.3 shows results obtained, in the first case for DC current and in the second case as a function of frequency. The data clearly show the existence of dynamic effects in the superconductor–insulator transition of YBaCuO.

7.3   Quantum Transitions in Superconductors

A "quantum transition" is a change of state at $T = 0$ driven by the variation of a physical quantity other than temperature. In the case of superconductivity, this can be the magnetic field H, film thickness d_F or doping δ. In principle, we expect to see an analogy between thermal and quantum transitions by replacing $T - T_c$ by $H - H_c$, $d_F - d_{FC}$ or $\delta - \delta_c$. In the same way a *divergence in the coherence length*, ξ, leads to universality classes characterised by universal critical exponents. Effects of the magnetic field on high temperature superconductors are complex and the subject of a large literature. We will not cover these in this chapter, but instead draw the attention of the reader to some published results as a function of thickness and doping. Even though the quality and extent of measurements is not in general sufficient to establish the existence of quantum transitions, they are not in contradiction with the corresponding model predictions.

The basic concepts of *thermal transitions* are valid for *quantum transitions*, however now the fluctuations are *quantum fluctuations*. In principle these are only relevant for a *strictly zero temperature*, otherwise thermal fluctuations should dominate. However in practice quantum fluctuations can extend to finite temperatures close to $T = 0$ (Sects. 7.3.1 and 7.3.2). This is due to the dynamics of quantum fluctuations whose effect is quickly averaged out at all finite temperatures. At zero temperature, only quantum fluctuations exist and their static value cannot be separated from their dynamics. Table 7.4 summarises these points.

Because of the importance of dynamics for quantum transitions, we define two coherence lengths: a normal coherence length ξ and a temporal coherence length ξ_τ. If we choose to study the critical quantum behaviour as a function of a quantity δ, which could be the magnetic field, doping or film thickness, the scaling behaviour of the lengths will be characterised by exponents ν and ν_τ respectively, as well as the dynamic exponent $z = \nu_\tau/\nu$. Deviation from the critical quantity δ_c will be called $\underline{\delta}$:

$$\underline{\delta} = \frac{\delta - \delta_c}{\delta_c} \qquad \xi \sim |\underline{\delta}|^{-\nu} \qquad \xi_\tau \sim \xi^z \sim |\underline{\delta}|^{-\nu_\tau} \qquad z = \frac{\nu_\tau}{\nu}. \qquad (7.22)$$

Scaling relations for different physical quantities are obtained from the standard relations for thermal transitions, by replacing the temperature deviation $t = (T - T_c)/T_c$ by $\underline{\delta}$ and $1/kT$ by ξ_τ. As an example, the free energy is characterised by a critical exponent $2 - \alpha$ related to the exponents given above by the (new) *hyperscaling* relation:

$$2 - \alpha = \nu(d + z). \qquad (7.23)$$

When the dimension of space d is less than or equal to two, it is necessary to consider spatial variations of the order parameter in the *longitudinal* (parallel to the order parameter) and *transverse* directions. Longitudinal variations are associated

Table 7.4 Thermal and quantum transitions

	Finite temperature	Zero temperature
Thermal fluctuations	Dominant	Absent
Quantum fluctuations	Not relevant because averaged out by dynamics	The only fluctuations present, divergence of relaxation time
Transition	**Thermal**	**Quantum**

with the usual coherence length and we introduce a transverse coherence length for the transverse variations. The details of this approach go beyond the scope of this chapter but the reader may find them in detailed references [17]. A direct consequence of the scaling behaviour of the transverse order parameter, and notably its possible *torsion*, is the relationship between the critical temperature and δ near the quantum critical point:

$$T_c \sim \underline{\delta}^{zv}. \tag{7.24}$$

In two dimensions a simple relation also exists between T_c and the London penetration length λ , when $\underline{\delta} \to 0$

$$T_c(\underline{\delta}) \sim \frac{1}{\lambda_x(T = 0, \ \underline{\delta})\lambda_y(T = 0, \ \underline{\delta})}. \tag{7.25}$$

Since high temperature superconductors have a practically square lattice (perfectly square in some compounds), the relationship becomes simply $T_c \propto 1/\lambda^2$. Here we present the scaling relations related to T_c as an example but it is possible to predict the behaviour of many physical quantities near the quantum critical point, such as conductivity, by replacing the temperature deviation $t = (T - T_c)/T_c$ by $\underline{\delta}$ and $1/kT$ by ξ_τ. This gives the conductivity in two dimensions tending to:

$$\sigma(\underline{\delta} \to 0) = A\sigma_Q = A\frac{4e^2}{h}, \tag{7.26}$$

where $4e^2/h = 6.45\mathrm{k}\Omega$ and A is a numerical factor that depends on the microscopic model.

7.3.1 Superconductor–insulator Transition with Respect to Film Thickness

Remarkable measurements on bismuth films have clearly shown a transition as a function of film thickness [12]. The film is deposited on a crystal plane at the atomic scale and covered with a 1nm thick amorphous layer of germanium. The bismuth film thickness is carefully controlled between 0.9 nm, when the film behaves as an insulator, and 1.5 nm when it is a superconductor. The transition occurs at a critical thickness of 1.22 nm for which the resistance is $R_c = 7{,}8\mathrm{k}\Omega$ (Fig. 7.8). Equation (7.24) leads to a value of the exponent zv of about 1.3.

Measurements of this type are also made on cuprate compound superconductors by two different techniques. The first consists of ultra thin films, for example of LaSrCuO [15], in which a superconductor–insulator transition is observed in agreement with an exponent zv of around 1. This technique is however very dangerous in the case of cuprates, because the critical thickness (1 nm for LaSrCuO) is of the order of the thickness of the structural unit ($c = 1.33$ nm for LaSrCuO).

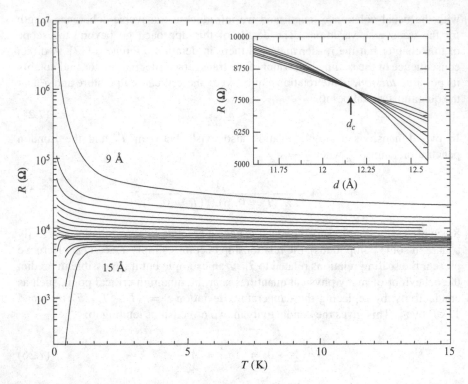

Fig. 7.8   Resistance "per square" of films of bismuth of different thicknesses as a function of temperature. Plotting resistance as a function of thickness (insert) shows all the curves cross at the critical thickness of 1.22 nm

The risk is we might observe a percolation transition (Chap. 5) rather than a quantum transition. In addition, cuprates are highly oxidised in order to by optimally doped but we observe a significant variation in their degree of oxidation near an interface.

The other technique used is to make superlattices with superconducting layers of variable thickness alternating with non superconducting layers. This procedure was used by several authors in the case of YBaCuO alternating with layers of PrBaCuO of identical structure where yttrium is replaced by praseodymium [8]. Unlike YBaCuO, PrBaCuO remains insulating whatever its degree of oxidation.

The results shown in Fig. 7.9 lead to a value around 1 for the exponent $z\nu$. In the case of cuprates, the quantities z, ν and A (7.26) have been evaluated with different microscopic models. As an example, we present below, in Table 7.5 predictions proposed by Cha et al. [1,2], Sorensen et al. [18] and those by Schneider and Singer [18]. The experimental results, although quite uncertain, appear to be in agreement with values of $z\nu \sim 1$ as predicted by some microscopic descriptions of underdoped cuprates summarised in Table 7.5.

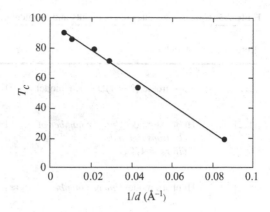

Fig. 7.9 Variation in critical temperature of a YbaCuO/PrBaCuO superlattice as a function of the thickness of alternate layers [8]

7.3.2 Doping Transitions

Here we are interested in the description of the high temperature superconductor phase diagram for which we presented experimental determinations in Sect. 7.2 (Fig. 7.3). In general, depending on the compound, this diagram has been only very partially explored experimentally due to chemical and structural constraints. We will discuss the case for which the diagram has been described in a reasonably complete way, that of $La_{1-x}Sr_xCuO_4$ compounds, in which substituting lanthanum by strontium provides a doping by holes in the whole superconducting regime [6, 14, 19, 20]. We are interested in two transitions observed at $T = 0$ when we increase the doping (see for example Fig. 7.6):

- *superconductor–insulator* transition, occurring in underdoped compounds for a doping δ_U.
- *superconductor–normal metal* transition, occurring in overdoped compounds for a doping δ_O.

7.3.2.1 Superconductor–Insulator and Superconductor–Normal Metal Transitions

In underdoped compounds, the quantum superconductor–insulator transition belongs, in principle, to the 2D-XY universality class (Fig. 7.6). The expected behaviour for $T_c(\delta)$ is predicted by (7.24):

$$T_c \sim (\delta - \delta_U)^{z\nu}. \tag{7.27}$$

The exponent z and ν can be evaluated from microscopic descriptions (Table 7.5). The models in which the predictions are closest to the experimental observations

Table 7.5 Calculated values of exponents and A calculated for different models

Reference	Model	z	v	zv	$A = \sigma/\sigma_Q$
[1,2]	Anisotropic $(2+1)D - XY$ model	≈ 1	$\approx 2/3$	$\approx 2/3$	0.285
[1,2]	Hubbard model (*integer number of electrons in band, filling* $= 1/2$)	≈ 1	$\approx 2/3$	$\approx 2/3$	0.52
[1,2]	Hubbard model (*integer number of electrons in band, filling* $= 1/3$)	≈ 1	$\approx 2/3$	$\approx 2/3$	0.83
[1,2]	Hubbard model + random interaction (*integer number of electrons in band, filling* $= 0$)	≈ 1.07	≈ 1	≈ 1.07	0.27
[1,2]	Hubbard model + random interaction (*integer number of electrons in band, filling* $= 1/2$)	≈ 1.14	≈ 1	≈ 1.14	0.49
[18]	Hubbard model + disorder + short range repulsion (*non integer number of electrons in band*)	≈ 2	≈ 1	≈ 2	0.14
[18]	Hubbard model + disorder + long range Coulomb interaction (*non integer number of electrons in band*)	≈ 1	≈ 1	≈ 1	0.55
[17]	Attractive Hubbard model	2	1/2	1	

predict an exponent zv of around 1 for the quantum superconductor–insulator transition.

In overdoped compounds, the superconductor–normal metal transition is mean field like and we can describe it as a classic BCS transition in a normal metal (Fermi

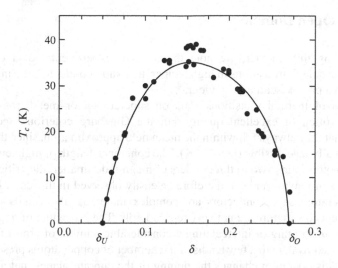

Fig. 7.10   Phase diagram of LaSrCaCuO. Experimental points provided by references [6, 14, 19, 20]. *Continuous line* corresponds to (7.29)

liquid) in which the superconductor–insulator transition occurs under the effect of decreasing doping. The mean field approach in this case leads to an exponent $z\nu \sim 1/2$ for the quantum superconductor–normal metal transition[2]:

$$T_c \sim (\delta_O - \delta)^{1/2}. \tag{7.28}$$

Equations (7.27) and (7.28) are only valid near $T_c = 0$. Expressions can be proposed that describe the entire superconducting region assuming that it never strays too far from the quantum regime, i.e. that the temperature remains low enough. For example, one of these is:

$$T_c = \left[a\,(\delta - \delta_U)^{-1} + b\,(\delta_O - \delta)^{-1/2} \right]^{-1}. \tag{7.29}$$

This expression accounts for existing experimental observations for LaSrCaCuO doped by the substitution Ca → Sr (Fig. 7.10). However, the number and diversity of values measured does not at all allow us to claim that the exponent $z\nu$ is actually near the values indicated theoretically.

---

[2]For the calculation see for example the reference [17].

7.4   An Open Domain

In concluding this chapter, we note that current experimental data on super-conductor–insulator transitions in high temperature superconductors are insufficient to establish a precise scaling behaviour.

In terms of thermal transitions, data on superconductor–metal transitions in metals are often of excellent quality, but the Ginzburg criterion (Sect. 1.3.3) indicates that they always fall within the mean field approximation, since the critical region is totally inaccessible ($\approx 10^{-14}$ K). The coherence length of high temperature superconductors being two to three orders of magnitude smaller, the critical region reaches tens of kelvins and can therefore be easily observed in this case. However, high temperature superconductors are complex materials, the details of whose composition and structure are not very reproducible. The procedure of doping itself consists of substituting or intercalating a considerable number of atoms into these compounds, equivalent to a few tenths of the number of copper atoms present on the structure. This operation changes the doping of the cuprate planes, but also many electronic and structural properties of the compound. It can be done in different ways, by oxidation or substitution of various cations, and we never obtain exactly the same result. If it is well established that the transitions observed in cuprates are significantly affected by fluctuations, thermal or quantum, their poor reproducibility prevents, for the moment, them being used to infer detailed scaling behaviours.

References

1. M.C. Cha, M.P.A. Fisher, S.M. Girvin, M. Wallin, A.P. Young, Universal conductivity of 2-dimensional films at the superconductor-insulator transition. Phys. Rev. B **44**, 6883 (1991)
2. M.C. Cha, S.M. Girvin, Universal conductivity in the boson Hubbard-model in a magnetic field. Phys. Rev. B **49**, 9794 (1994)
3. P. Dai, B.C. Chakoumakos, G.F. Sun, K.W. Wong, Y. Xin, D.F. Lu, Synthesis and neutron powder diffraction study of the superconductor $HgBa_2Ca_2Cu_3O_{8+\delta}$ Tl substitution. Physica C **243**, 201 (1995)
4. G. Deutscher, Coherence and single-particle excitations in th high-temperature superconductors. Nature **397**, 410 (1999)
5. V.J. Emery, S.A. Kivelson, Importance of phase fluctuations in superconductors with small superfluid density. Nature **374**, 434 (1995)
6. Y. Fukuzumi, K. Mizuhashi, K. Takenaka, Universal superconductor-insulator transition and Tc depression in Zn-substituted high-Tc cuprates in the underdoped regime. Phys. Rev. Lett. **76**, 684 (1996)
7. V.L. Ginzburg, L.D. Landau, *Zh. Eksperim. i Teor Fiz.* **20**, 1064 (1950)
8. R.G. Goodrich, P.W. Adams, D.H. Lowndes, D.P. Norton, Origin of the variation of the Tc with superconducting layer thickness and separation in YBa2Cu3O7-x/PrBa2Cu3O7 superlattices. Phys. Rev. B **56**, 14299 (1997)
9. M.A. Hubbard, M.B. Salamon, B.W. Veal, Fluctuation diamagnetism and mass anisotropy of YBa2Cu3O6+x. Physica C **259**, 309 (1996)
10. J. Koetzler, M. Kaufmann, Two-dimensional fluctuations close to the zero-field transition of Bi2Sr2CaCu2O8, Phys. Rev. B **56**, 13734 (1997)

11. M. Mali, J. Roos, D. Brinkmann, A. Lombardi, Temperature dependence of the sublattice magnetization of the antiferromagnet Ca0.85Sr0.15CuO2. Phys. Rev. B **54**, 93 (1996)

12. N. Markovic, C. Christiansen, A.M. Goldman, Thickness magnetic field phase diagram at the superconductor-insulator transition in 2D. Phys. Rev. Lett. **81**, 5217 (1998)

13. W.L. McMillan, Transition temperature of strong-coupled superconductors. Phys. Rev. **167**, 331 (1968)

14. T. Nagano, Y. Tomioka, Y. Nakayama, K. Kishio, K. Kitazawa, Bulk superconductivity in both tetragonal and orthorhombic solid-solutions of (La1-xSrx)2CuO4-Delta. Phys. Rev. B **48**, 9689 (1993)

15. H. Sato, H. Yamamoto, M. Naito, Growth of (001) LaSrCuO ultrathin films without buffer or cap layers. Physica C **274**, 227 (1997)

16. T. Schneider, H. Keller, Extreme type-II superconductors. Universal properties and trends. Physica C **207**, 366 (1993)

17. T. Schneider, J.M. Singer, *Phase Transition Approach to High Temperature Superconductivity* (Imperial College Press, London, 2000)

18. E.S. Sorensen, M. Wallin, S.M. Girvin, A.D. Young, Universal conductivity of dirty bosons at the superconductor-insulator transition. Phys. Rev. Lett. **69**, 828 (1992)

19. H. Takagi, T. Ido, S. Ishibashi, M. Uota, S. Uchida, Superconductor-to-nonsuperconductor transition in (La1-xSrx)2CuO4 as investigated by transport and magnetic measurements. Phys. Rev. B **40**, 2254 (1989)

20. J.B. Torrance et al., Properties that change as superconductivity disappears at high-doping concentrations in La2-xSrxCuO4. Phys. Rev. B **40**, 8872 (1989)

Chapter 8
Growth and Roughness of Interfaces

8.1 Introduction

Our daily lives are full of interfaces of various kinds. Be it through our skin or across the membranes of our cells, we are continually exchanging matter and energy with the environments we pass through. The shape of these interfaces is crucial for exchange processes, as is for example the fine structure of our lungs for exchanging oxygen, or the surface of glass for its transparence to light. In general their morphology depends on the scale at which we look, e.g. the smooth surface of Mars seen from the Earth is very different from the roughness of its soil seen by a hypothetical Martian.

Here we apply scaling approaches to the shape of interfaces, their formation and their evolution in nonequilibrium. We have already discussed in Chap. 4 a nonequilibrium process, diffusion, where we followed the position r of a random diffuser over time. Here we are interested in the position $h(r, t)$ of an interface between two media (Fig. 8.1), for example, during a *growth process* by aggregation of material from one of the media to the other. The generic term "growth" is in practice used in the general the sense of *evolution* for physical situations such as propagation of a front, soil erosion, spread of a forest fire, flow of liquid in a porous medium or growth of a bacteria colony.

As in the study of phase transitions, we are not interested here in the very detailed shape of the interface, or its evolution, but in its asymptotic scale invariant properties. Here the property of statistical[1] self-similarity is extended to *self-affinity*. An interface is self-affine if it has the same statistical properties under changes of scale with different factors, k for the height of the interface h and k' for the position along the interface r, such that $h(r)$ and $kh(k'r)$ have the same statistical properties.

---

[1] Here we will only encounter *statistical* scale invariance. When we use the expressions "scale invariance", "self-similarity" or "self-affinity" the subtext is that these terms should be understood in the *statistical* sense.

A. Lesne and M. Laguës, *Scale Invariance*, DOI 10.1007/978-3-642-15123-1_8,

Fig. 8.1 Evolution of the
interface between two media.
Material coming from
medium B can aggregate to
medium A, which can in turn
be eroded by medium B. For
example, A could be a fire
devastating B, or a liquid
soaking into a porous
medium B

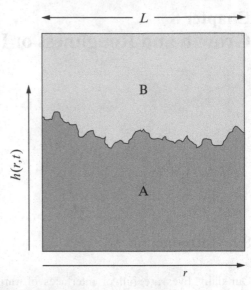

Fig. 8.2 Growth along the edge of, initially straight, atomic steps on a monocrystal of silicon
shown by Scanning Tunnelling Microscopy (IEMN, Laboratoire de Physique, Lille)

In simple systems r has only one component, as was the case for example of
the simulations of a burning forest that we talked about in Sect. 5.1, where the fire
was lit at time zero along one side of the forest. Here we assume that the forest
can be irregular, but that its density is always higher than the percolation threshold.
Another system in the same family is that of growth, atom by atom, along the edge
of an "atomic step" that is initially straight (Figs. 8.2 and 8.6).

A random walk in one dimension gives an analogy of an interface such as the
one in Fig 8.1. The trajectory of the position h of the walker over time plays
the role of r. An important application of this approach is to DNA sequences.
Genetic information code is made up of bases from two families, purines (A, G)

and pyrimidines (T, C). We can let one correspond to a step up and the other to a step down (see Sect. 11.3). In this way we obtain a graphical representation of the sequence (Fig. 8.3) which gives an idea of correlations in its underlying structure [6, 11].

In this chapter we present three models of growth, of increasing complexity, followed by the method of dynamic renormalisation which can explain their universality. We end with a presentation of different types of evolution equations and their corresponding universality classes.

8.1.1 Discrete Models, Continuous Equations

Many physical systems mentioned above are well described by discrete models, e.g. an atom joins the aggregate, a tree catches fire, a bacterium is born etc. Such

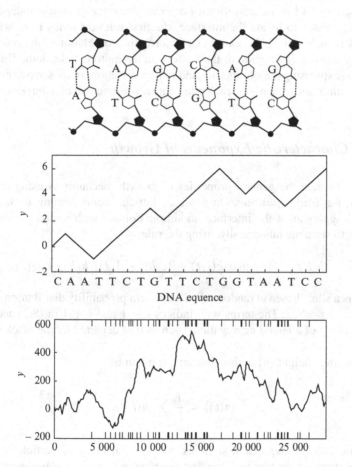

Fig. 8.3 A "DNA walk" exposing surprising correlations in the "non coding" regions (after [6, 11])

models are well adapted to numerical simulations which can start from a very simplified description at the elementary scale. As we have leant earlier in this book, the asymptotic behaviours are more sensitive to the way properties change with scale than to microscopic details of a model. At large scales, these details can be neglected and stochastic evolution equations can be used, of the following type:

$$\frac{\partial h}{\partial t} = G + \eta, \tag{8.1}$$

where G represents the deterministic part of the growth and η the noise to which it is subjected.

There are basically two approaches to establish such equations. We can either treat a particular growth problem by characterising its mechanisms by appropriate expressions for G and η, or we express the *symmetries* of the problem by these expressions. In this case we make use of the *Landau* approach used for the mean field description of phase transitions. For example, since growth is independent of the origin chosen to locate the interface, the first relevant terms in G will be the spatial derivatives of h. We know that renormalisation classifies the *relevant* and *irrelevant* terms as a function of their effect on asymptotic behaviour. This scaling approach (expressing the symmetries and renormalisation) in most cases establishes *minimal* continuous equations, each one being representative of a *universality class*.

8.1.2 Characteristic Exponents of Growth

We will illustrate the general properties of growth mechanisms using the simple model of *ballistic deposition*. On a square lattice, atoms coming from region B (Fig. 8.1) aggregate at the interface on simple contact with region A. This model is simple to simulate numerically, using the rule:

$$h(i, t + 1) = \max[h(i, t) + 1, \ h(i - 1, t), \ h(i + 1, t)], \tag{8.2}$$

where i is a site, chosen at random from a uniform probability distribution, at which an atom is adsorbed. The terms with indices $i - 1$ and $i + 1$ in (8.2) account for the possibility of a lateral aggregation, such as that depicted by the black square in Fig. 8.4.

The average height $\langle h \rangle$ at time instant t is given by:

$$\langle h(t) \rangle = \frac{1}{L} \sum_{i=1}^{L} h(i, t). \tag{8.3}$$

The name *ballistic deposition* signifies that the trajectory of particles in the gas suddenly stops on *contact* with the first interface site, and all subsequent diffusive

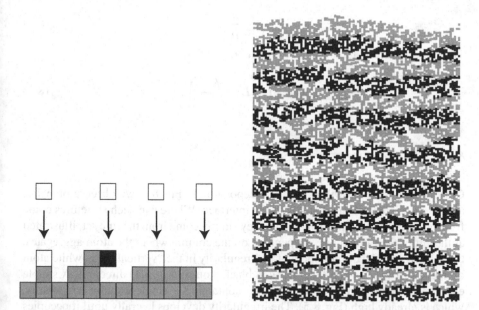

Fig. 8.4 Ballistic deposition. *Left*: mechanism for aggregation of atoms; *Right*: results of a simulation on an interface of $L = 100$ sites, on which 12 800 atoms are deposited. The colour was changed after each 800 atoms had been deposited

motion is forbidden. If the flux of atoms is constant, $\langle h \rangle$ is proportional to t. The *roughness* of the interface is described by a characteristic thickness $w(t)$, which measures the average distance from the average height:

$$w(L,t) = \left[\frac{1}{L} \sum_{i=1}^{L} [h(i,t) - \langle h(t) \rangle]^2 \right]^{1/2}. \tag{8.4}$$

At the beginning of the deposition process, the roughness is zero. If L and t are large, the following behaviour is observed experimentally:

- Initially, $w(L,t)$ grows with time as a power law t^β.
- For very long times, $w(L,t)$ saturates at a value depending on the size as L^α.
- The crossover between these regimes occurs at a time t_x which varies as L^z.

Growth is therefore characterised by three exponents α, β, and z. In the case of ballistic deposition in one dimension, experiments give values of $\alpha = 0.47 \pm 0.02$, $\beta = 0.33 \pm 0.006$ and $z = 1.42 \pm 0.09$. It is important to note that these three exponents are not independent, as indicated by determining the crossover time t_x in each of the three regimes:

$$w(L,t_x) \sim t_x^\beta \sim L^\alpha \tag{8.5}$$

which leads to a scaling relation:

$$z = \frac{\alpha}{\beta}. \tag{8.6}$$

Therefore, a universal relation between w, L and t is:

$$w(L,t) \sim L^\alpha f\left(\frac{t}{L^z}\right). \tag{8.7}$$

8.1.3 Correlations

Closely examining growth by ballistic deposition in Fig. 8.4, we observe branches and pores oriented roughly at 45° to the interface. Where can such structures come from in a totally random mechanism? They simply come from the rule set, illustrated at the left of the same figure. Depending on the column where the atom aggregates, the rule triggers the development of an irregularity in the "vertical" (e.g. white atom on the left of Fig. 8.4) or "lateral" (e.g. black atom in Fig. 8.4) direction. A simple experiment is to start with an interface containing a single column irregularity, which is already high (Fig. 8.5). The irregularity develops laterally until it occupies the whole width of the interface. This illustrates the existence of *correlations* in the growth process. The characteristic width of the "tree" in Fig. 8.5 is a *correlation length* $\xi_{//}$ in the direction parallel to the interface. We can also define a correlation length $\xi_\perp$ *perpendicular* to the interface which has the same scaling behaviour as the characteristic thickness w. Figure 8.5 illustrates well the saturation of w occurring once the correlation length $\xi_{//}$ is of the order of L. From (8.7) we can predict the scaling behaviour of $\xi_{//}$ before saturation:

$$w(L,t) \sim L^\alpha f\left(\frac{t}{L^z}\right) \sim L^\alpha g\left(\frac{L}{\xi(t)}\right) \tag{8.8}$$

which leads to:

$$\xi(t) \sim t^{1/z} \quad \text{for} \quad t \ll t_x. \tag{8.9}$$

Fig. 8.5 Ballistic growth from an interface initially containing a point irregularity. The initial interface consists of 100 sites and a column of size 35. The colour was changed after each 800 atoms had been deposited

8.1.4 Model of Random Growth

In this section, we present the solution of the model of random deposition. The mechanism, illustrated in Fig. 8.6, is the simplest one could imagine. The rule is simply to randomly (from a uniform distribution) choose one column i and increment its height.

Given that there are no correlations between columns, each one grows with a probability p, at each instant in time. The probability $P(h, N)$ that a column has height h when N atoms have been deposited is given by the binomial distribution. The value of w^2 can be extracted directly from this (independent of L since there are no correlations between columns):

$$w^2(t) = Np(1 - p). \tag{8.10}$$

For constant flux, we obtain that w is proportional to $t^{1/2}$ and $\beta = 1/2$. As the correlation length remains zero in this model, the roughness never saturates. Neither does the interface show self-affine properties since there is no characteristic scale parallel to the interface.

The continuous equation that describes the asymptotic behaviour of this model is:

$$\frac{\partial h(x, t)}{\partial t} = p + \eta(x, t). \tag{8.11}$$

The term η, which is zero on average, expresses the random nature of deposition. Its second moment depicts the absence of spatial and temporal correlations:

$$\langle \eta(x, t)\eta(x', t') \rangle = 2Da\delta(x - x')\delta(t - t'), \tag{8.12}$$

where a is the size of an atomic unit.

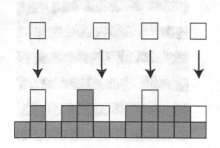 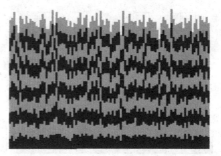

Fig. 8.6 Random deposition. *Left*: mechanism for aggregation of atoms (unlike the ballistic mechanism in Fig. 8.4 there is no lateral growth). *Right*: results of a simulation of an interface consisting of 100 sites on which 50,000 atoms are deposited. The colour was changed after every 5,000 atoms had been deposited

Integrating the evolution equation leads to $\langle h \rangle = p\,t$ and:

$$\langle h^2(x,t) \rangle = \left\langle \left[p\,t + \int_0^t dt'\eta(x,t') \right]^2 \right\rangle = p^2 t^2 + 2Dt. \tag{8.13}$$

We therefore obtain w^2:

$$w^2(t) = \langle h^2 \rangle - \langle h \rangle^2 = 2Dt. \tag{8.14}$$

On average the interface behaves as a biased one dimensional random walk, with a mean displacement proportional to time.

8.2 Linear Approach Including Relaxation

Here we consider a model of random deposition that includes a *relaxation* in the position of the deposited atom *towards the nearest potential well*, that is to say the site j nearest to i that is a local minimum in h (Fig. 8.7).

This relaxation process smooths the surface and all porosity disappears (Fig. 8.7). In one dimension, experiments lead to values of $\alpha = 0.48$ and $\beta = 0.24$. Our first

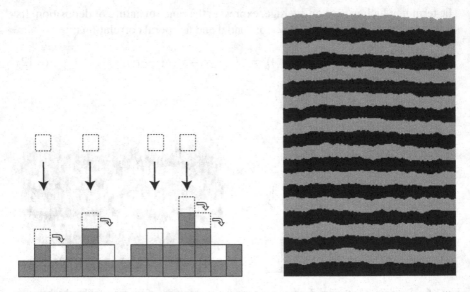

Fig. 8.7 Random deposition with relaxation. *Left*: mechanism for aggregation of atoms (unlike the random deposition illustrated in Fig. 8.6 the atoms can relax to a neighbouring site with higher coordination number). *Right*: results of a simulation on an interface of 100 sites on which 30,000 atoms are deposited. The colour was changed after each 1,500 atoms had been deposited

objective is to establish a continuous equation reflecting this process. The first step will be to study the symmetries of the growth model, symmetries that the equation must obey.

8.2.1 Symmetry Study

Unlike the ballistic or random deposition mechanisms, relaxation implies a *local equilibrium*. This means for example that the interface properties would be the same if we exchanged the media A and B (Fig. 8.1). This is not obvious a priori, but becomes clearer if we reformulate the process seen from each medium:

- Seen *from medium A* as we have just done:
- Randomly choose a column i
- Search for the nearest local minimum j
- Deposit an atom on this site

- Seen *from medium B*:
- Randomly choose a column i
- Search for the nearest local *maximum* j
- Remove an atom from this site

This is a process of *smoothing* the interface, which leads to maxima and minima that are geometrically statistically symmetric. As well as growth processes, this interface could also represent, for example, the border between two magnetic domains. In the following subsections we will in turn discuss the invariance of the mechanism, and therefore of its equation, with respect to spatio-temporal translation, rotation about the growth axis and by change of sign of h (exchanging the two media).

8.2.1.1 Translation in Time

We take the general form of (8.1) describing the evolution of the interface including the dependences we expect a priori:

$$\frac{\partial h(r,t)}{\partial t} = G(h,r,t) + \eta(r,t), \tag{8.15}$$

where G is the determinant part and η the stochastic part of the evolution mechanism. Invariance of the process with respect to changing the origin of time $t \rightarrow t + t_0$ forbids any explicit dependence on time. However, derivatives with respect to time such as $\frac{\partial h(r,t)}{\partial t}$ are in agreement with this symmetry.

8.2.1.2   Translation Parallel to the Interface

In the same way, growth is independent of the origin chosen locating r along directions parallel to the interface. The equation must therefore be invariant with respect to the transformation $r \to r + r_0$. Explicit dependencies on r are therefore not allowed and G can contain only combinations of differentials such as:

$$\frac{\partial}{\partial r_x}, \quad \frac{\partial^2}{\partial r_x^2}, \quad \frac{\partial^3}{\partial r_x^3}, \quad \cdots \quad \frac{\partial^n}{\partial r_x^n}$$

8.2.1.3   Translation in the Growth Direction

The equation must also take into account the fact that growth is independent of the origin chosen locating the position of the interface along the h direction. So it must be invariant with respect to transformations $h \to h + h_0$. Therefore, explicit dependencies on h are not allowed and G can only contain combinations of differentials such as:

$$\nabla h, \quad \nabla^2 h, \quad \cdots \quad \nabla^n h$$

8.2.1.4   Inversion of the Growth Direction

In the particular case of growth with relaxation, each incident atom is placed at an equilibrium position. We have seen that this leads to a form of the interface in which the media A and B can be exchanged. The equation must be invariant to inverting the interface $h \to -h$. Given that the first term $\frac{\partial h(r,t)}{\partial t}$ in the evolution equation is odd in h, even terms are not allowed and G can only contain combinations of differentials such as:

$$\nabla h, \quad \nabla^2 h, \quad (\nabla h)^3, \quad (\nabla^2 h)^3 \quad \cdots \quad (\nabla^n h)^{2p+1}$$

8.2.1.5   Rotation about the Growth Direction

Rotational symmetry excludes all odd order differential forms, so G can only contain terms such as:

$$\nabla^2 h, \quad \cdots \quad \nabla^{2n} h$$

8.2.2   The Edwards–Wilkinson Equation

From the rules in the previous subsection, we obtain the general form of the equations respecting the symmetries of random growth with relaxation:

$$\frac{\partial h(r,t)}{\partial t} = a_1 \nabla^2 h + a_2 \nabla^4 h + \cdots + a_n \nabla^{2n} h$$

$$+ \left[b_1 \nabla^2 h + b_2 \nabla^4 h + \cdots + b_p \nabla^{2p} h \right] \left[c_1 (\nabla h)^2 + c_2 (\nabla h)^4 + \cdots + c_q (\nabla h)^{2q} \right]$$

$$+ \cdots + \eta(r,t) \tag{8.16}$$

The asymptotic properties ($t \to \infty$ and $L \to \infty$) are most sensitive to lowest order terms. A renormalisation calculation can rigorously show that only the lowest order term is *relevant*. This means that it is the only term influencing the value of the exponents characterising the growth. In this way the Edwards–Wilkinson equation is obtained [3], which is the simplest equation describing the mechanism of random relaxation of a surface:

$$\frac{\partial h(r,t)}{\partial t} = \nu \nabla^2 h(r,t) + \eta(r,t) \qquad \text{[Edwards–Wilkinson]}. \tag{8.17}$$

The main characteristics of this equation are the following:

- It conserves the average height $\langle h \rangle$. To describe the growth mechanism of Fig. 8.7 an extra term $F(r,t)$ should be added which measures the flux of atoms adsorbed at point r and time t.
- When the flux F is uniform and constant, it does not affect the asymptotic properties of roughness. The first order equation (8.17) contains all the physics of this growth model. It is sufficient to represent the corresponding universality class.
- It has the same form as the diffusion equation (see Chap. 4) if the noise term is omitted. Like the diffusion equation, it gradually erases irregularities in interfaces.
- The coefficient ν plays the role of surface tension. The larger ν, the faster the smoothing of the system.
- It correctly describes random growth with relaxation, as long as the gradient ∇h of the interface remains small. This means that:

$$|\delta h| \ll |\delta r|. \tag{8.18}$$

- We also know that $|\delta h| \sim |\delta r|^\alpha$ from the definition of α. For long distances this condition therefore constrains $\alpha < 1$. This is actually the case, as we will show in the following subsection.

8.2.2.1 Solution by Scaling Arguments

It is possible to calculate the exponents of growth by scaling arguments, but we will also present an exact solution in the next subsection. Assume that the interface is self-affine, we subject it to a change in scale with parameter b:

$$r \to r' = br \qquad h \to h' = b^\alpha h \quad \text{and} \quad t \to t' = b^z t. \tag{8.19}$$

The interface obtained by such a transformation should be statistically identical to
the original. This means that the equation must be invariant under this transformation. We should also evaluate the scaling behaviour of the noise η. Here we assume
that the noise has no correlations:

$$\langle \eta(r,t)\eta(r',t')\rangle = 2Da^d\delta^d(r-r')\delta(t-t'), \tag{8.20}$$

where a is the size of an atomic unit. From this relation, we can obtain the scaling
behaviour of the noise:

$$\eta(br,b^zt) = b^{-\frac{d+z}{2}}\eta(r,t). \tag{8.21}$$

By substituting (8.19) and (8.21) in the Edwards–Wilkinson equation (8.17) we
obtain:

$$b^{\alpha-z}\frac{\partial h(r,t)}{\partial t} = vb^{\alpha-2}\nabla^2 h + b^{-\frac{z+d}{2}}\eta(r,t). \tag{8.22}$$

By expressing the fact that this equation is independent of the value of b, we obtain
the values of the exponents:

$$\alpha = 1-\frac{d}{2} \qquad \beta = \frac{1}{2}-\frac{d}{4} \qquad z = 2. \tag{8.23}$$

Note that $\alpha = 0$ and $\beta = 0$ for the critical dimension $d = d_c = 2$.

8.2.2.2 Exact Solution

Since the equation is linear, it is possible to solve it exactly. To do so, we use the
Fourier transform $(r,t) \rightarrow (q,\omega)$:

$$h(q,\omega) = \frac{\eta(q,\omega)}{vq^2 - i\omega}. \tag{8.24}$$

We express the absence of noise correlations in this space as:

$$\langle \eta(q,\omega)\eta(q',\omega')\rangle = \frac{2D}{\tau^2 a^d}\delta(q+q')\delta(\omega+\omega'), \tag{8.25}$$

where a is the size of an atomic unit and τ is an elementary time step. Combining
these two equations (8.24) and (8.25) and returning to real space, we obtain[2]:

---

[2]a is the elementary size parallel to r and τ the elementary time, included here for the homogeneity
of the expression.

$$\langle h(r,t)h(r',t')\rangle = \frac{D}{2v}\,|\,r-r'\,|^{2-d}\,f\left(\frac{v\,|\,t-t'\,|^{1-\frac{d}{2}}}{|\,r-r'\,|^{2-d}}\right),\qquad(8.26)$$

where the function $f(x)$ is proportional to x at short times, and saturating to 1 at long times. This expression recovers the exponents obtained in the previous subsection by simple scaling arguments. In one dimension the values of the exponents, $\alpha = 1/2$ and $\beta = 1/4$, are very close to the values $\alpha = 0.48$ and $\beta = 0.24$ observed experimentally for random deposition with relaxation. Here we obtain confirmation that the Edwards–Wilkinson equation (8.17), constructed based on symmetry arguments, does belong to the same universality class as the discrete model that we started out with.

We note that for $d = 2$ we obtain $\alpha = 0$ and $\beta = 0$, which correspond to logarithmic variations in w as a function of t and L. These exponents are negative when the interface is in three or more dimensions, that is to say when the growth takes place in four or more dimensions. This means that all irregularities introduced by noise are smoothed out very quickly by the surface tension and the interface is flat.

The essential feature of the Edwards–Wilkinson equation (8.17) is that it is linear and therefore invariant under inversion of the direction of growth. It describes mechanisms close to equilibrium well, but it is incapable of describing mechanisms far from equilibrium, such as ballistic deposition. For that, nonlinear terms need to be included. Kardar et al. [8] were the first to introduce a nonlinear equation, which now carries their name (KPZ equation).

8.3 The Kardar–Parisi–Zhang Equation

We now want to construct an equation that obeys the same symmetries as the Edwards–Wilkinson equation except invariance to inversion of the growth direction. We simply include in the Edwards–Wilkinson equation (8.17) the lowest order term that was forbidden by this symmetry, i.e. the term $(\nabla h)^2$. This give the Kardar–Parisi–Zhang (KPZ) equation:

$$\frac{\partial h(r,t)}{\partial t} = v\nabla^2 h + \frac{\lambda}{2}\,(\nabla h)^2 + \eta(r,t).\qquad(8.27)$$

8.3.1 Construction of the KPZ Equation by Physical Arguments

We want to explain, among other mechanisms, ballistic growth in a continuous description that is more general than the model on a square lattice presented above (Sect. 8.1.2). In ballistic growth atoms are frozen as soon as they make contact

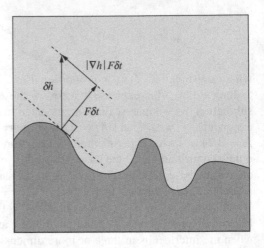

Fig. 8.8 Transposition of the ballistic growth mechanisms into a system of continuous growth. The growth increment has a components normal and parallel to the local normal of the interface

with the surface. We assume that the line passing through two atoms in contact at the interface can have a component parallel to the interface (black atom on Fig. 8.4) unlike the random deposition mechanism (Fig. 8.6). This microscopic model therefore results in a *lateral* growth, the effects of which we attempt to characterise in this new equation. This mechanism of growth by ballistic deposition can be transformed to a continuous description in which a growth increment δh *normal* to the average direction of the interface is made up of two components; one, $F\delta t$, *parallel to the local normal* of the interface and the other lateral, $|\nabla h| F\delta t$, *perpendicular to the local normal* (Fig. 8.8).

An expression for δh can be extracted from this:

$$\delta h = F\delta t \left[1 + (\nabla h)^2\right]^{1/2} \approx F\delta t \left[1 + \frac{(\nabla h)^2}{2}\right] \qquad (8.28)$$

illustrating that the lateral growth can be taken into account by a term of the form $(\nabla h)^2$. If $\lambda > 0$ in the KPZ equation, this term has the effect of reinforcing the local gradient of the interface. This effect is opposite to that of $\nabla^2 h$ which smooths out irregularities.

8.3.2   KPZ Exponents from Scaling Arguments

As for the Edwards–Wilkinson equation we can use the fact that the KPZ equation (8.27) must be invariant under changes of scale:

$$\frac{\partial h(r,t)}{\partial t} = \nu b^{z-2}\nabla^2 h + \frac{\lambda}{2}b^{\alpha+z-2}(\nabla h)^2 + b^{-\alpha+\frac{z-d}{2}}\eta(r,t) \qquad (8.29)$$

where the two variables have been divided by $b^{\alpha-z}$ (remember that $z = \alpha/\beta$). We now have three relations to determine two exponents. Since the nonlinear term dominates over the linear term, one response is to ignore it, leading to $\alpha = \frac{2-d}{3}$ and $\beta = \frac{2-d}{4+d}$. Unfortunately this result is wrong because we cannot ignore the linear term. The reason for this is that here we have to account for the coupled variations of the coefficients ν, λ and D when we change scale. We will see that a renormalisation approach can overcome this difficulty. In particular it shows that the nonlinear term as it appears in (8.29), is actually invariant under change of scale and that therefore:

$$\alpha + z = 2. \tag{8.30}$$

This relation is valid whatever the dimension. An application of the fluctuation-dissipation theorem shows that $\alpha = 1/2$ in one dimension, meaning that the interface evolves over time like a non correlated random walk. In one dimension, we therefore obtain:

$$\alpha = 1/2 \quad \beta = 1/3 \quad \text{and} \quad z = 3/2 \qquad \text{[KPZ 1D]}$$

This result, which uses (8.30) established by renormalisation, is in excellent agreement with experimental values measured for ballistic deposition in one dimension. To study the behaviour in other dimensions, a more general renormalisation approach is needed.

8.4 Dynamic Renormalisation

After their application to phase transitions at equilibrium, renormalisation methods were quickly adapted to address nonequilibrium phenomena [4, 7]. The formalism using diagrammatic methods is hard work, but the principle is the same as that we presented in Chap. 3. We illustrate the principle of this approach for the case of the KPZ equation (8.29) expresses the transformation of the KPZ equation by a change in scale by a factor b. We said that it is not possible to ignore the variation of the parameters ν, λ and D in this change in scale. Below we propose a dynamic renormalisation approach which takes into account the coupling between these three parameters. Equation (8.29) gives the following equations:

$$\nu \to b^{z-2}\nu$$

$$D \to b^{z-d-2\alpha}D \tag{8.31}$$

$$\lambda \to b^{z+\alpha-2}\lambda.$$

We know the exact solution of the Edwards–Wilkinson equation obtained for $\lambda = 0$. Here the objective is to study the KPZ equation by an expansion in powers of λ about the solution to the Edwards–Wilkinson equation. Below we give an outline of this approach.

8.4.1 Renormalisation Flow Equations

Let us take (8.24), which expresses the Fourier transform of the Edwards–Wilkinson equation, and add the KPZ nonlinear term:

$$h(q,\omega)=\frac{1}{vq^2-i\omega}\left[\eta(q,\omega)-\frac{\lambda}{2}\int\int\frac{d^dk\,d\Omega}{(2\pi)^{d+1}}k(q-k)h(k,\Omega)h(q-k,\omega-\Omega)\right].$$

(8.32)

8.4.1.1 Expansion About the Edwards–Wilkinson Equation

This expression can be expanded to express h in terms of powers of λ. The Feynman diagram method significantly lightens the calculation. We define a *propagator* $P_\lambda(q,\omega)$ relating h and the noise η:

$$h(q,\omega) = P_\lambda(q,\omega)\eta(q,\omega).$$

(8.33)

For $\lambda = 0$, the propagator is that of the Edwards–Wilkinson equation (see (8.24)):

$$P_0(q,\omega) = \frac{1}{vq^2-i\omega}.$$

(8.34)

The perturbation calculation to third order in λ, leads to:

$$P_\lambda(q,\omega) = P_0(q,\omega)-$$

$$2\lambda^2 D P_0^2(q,\omega)\int\int\frac{d^dk\,d\Omega}{(2\pi)^{d+1}}[kq]\,[k(q-k)]$$

$$P_0(k,\Omega)P_0(q-k,\omega-\Omega)P_0(-k,-\Omega)+O(\lambda^4).$$

(8.35)

Calculating the integral allows us to evaluate effective parameters, v_{eff} and D_{eff}, such that:

$$P_\lambda(q,\omega,v,D) = P_0(q,\omega,v_{\text{eff}},D_{\text{eff}}) + O(\lambda^4)$$

(8.36)

giving:

$$v_{\text{eff}} = v\left(1-\lambda^2\frac{D}{v^3}K_d\frac{d-2}{d}\int dq\,q^{d-3}\right)+O(\lambda^4)$$

(8.37)

$$D_{\text{eff}} = D\left(1 + \lambda^2 \frac{D}{v^3} K_d \int dq\, q^{d-3}\right) + O(\lambda^4), \tag{8.38}$$

where K_d is a numerical constant that depends on the dimension of space. It may seem like we have achieved our goal, since we have established expressions of the interdependence between the KPZ parameters v, λ and D. However, an analysis of the convergence of higher order perturbations shows that we have done nothing of the sort. The expressions above are insufficient and it is necessary to renormalise by expressing the auto-affinity of the interface in the asymptotic regime.

8.4.1.2 Renormalisation Flow

We therefore proceed to an infinitesimal change of scale. We take a scale factor $b = 1 + \delta l$ close to 1, so we can write b^x as $(1 + x\delta l)$. Substituting this expression for b^x into (8.31), we obtain scaling relations for the two effective parameters for infinitesimal changes:

$$v'_{\text{eff}} = v_{\text{eff}}[1 + \delta l(z - 2)] \tag{8.39}$$

$$D'_{\text{eff}} = D_{\text{eff}}[1 + \delta l(z - d - 2\alpha)]. \tag{8.40}$$

Finally, by combining these with the results of the perturbation calculation (8.37) and (8.38), we can establish two evolution equations for the parameters v and D, to 4th order in λ and for an infinitesimal change in scale (the integral in (8.37) and (8.38) must be evaluated with limits $q = 1 - \delta l$ to $q = 1$):

$$\frac{dv}{dl} = v\left[z - 2 - \lambda^2 \frac{D}{v^3} K_d \frac{d-2}{d}\right] \tag{8.41}$$

$$\frac{dD}{dl} = D\left[z - d - 2\alpha + \lambda^2 \frac{D}{v^3} K_d\right]. \tag{8.42}$$

The parameter λ is only affected by the change in scale:

$$\frac{d\lambda}{dl} = \lambda\,[z + \alpha - 2]. \tag{8.43}$$

It is worth noting that the quantity $g = \lambda^2 D/v^3$, which we call a *coupling constant*, plays a special role in the interdependence of the parameters. It is possible to establish its independent evolution equation from (8.41) and (8.43):

$$\frac{dg}{dl} = \frac{2-d}{d}g + K_d \frac{2d-3}{d}g^3. \tag{8.44}$$

8.4.2   KPZ Regimes

We can now express the self-similarity of the interface, i.e. that all the parameters are invariant under a change in scale, in other words that the right hand sides in (8.41) to (8.44) are zero.

First of all, we recover the relation $z + \alpha = 2$, from (8.43).

8.4.2.1   KPZ in 1D

In one dimension (8.44) leads to two fixed points g_1^* and g_2^*:

$$g_1^* = 0 \quad \text{(unstable)} \qquad \text{and} \qquad g_2^* = (2/K_d)^{1/2} \quad \text{(stable)}. \tag{8.45}$$

Unlike for phase transitions, here it is the *stable* fixed point such that λ is non zero, which interests us. Linearising around it leads to the exact 1D values we have introduced:

$$\alpha = 1/2 \quad \beta = 1/3 \quad z = 3/2$$

8.4.2.2   KPZ in 2D

Two is the critical dimension of the Edwards–Wilkinson equation. The reader can actually check that the flow (8.44) for g only gives the trivial fixed point $g^* = 0$ and the perturbation calculation is insufficient to lead to a satisfactory fixed point. Only a calculation with strong coupling, i.e. one that does not assume λ is small, can evaluate the fixed point parameters ν, λ and D and derive their exponents in two dimensions.

8.4.2.3   KPZ in $d > 2$: A Phase Transition

In greater than two dimensions ($d > 2$), there is a non trivial critical point:

$$g_2^* = \left[\frac{d(d-2)}{2K_d(2d-3)} \right]^{1/2}. \tag{8.46}$$

The significant difference from the one dimensional ($d = 1$) case is that this fixed point is *unstable*. This has the following consequences:

- If g is small ($g < g_2^*$), i.e. if the nonlinear contribution in the equation is small, then the system will converge towards the trivial fixed point $g_1^* = 0$. The system will have an asymptotic behaviour identical to that of the Edwards–Wilkinson equation, i.e. that of a flat interface.

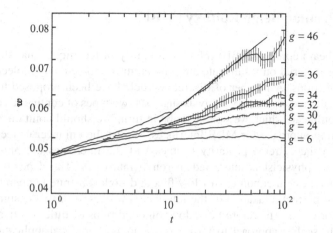

Fig. 8.9 Roughness as a function of coupling g obtained by numerically solving the KPZ equation in three dimensions (after [12])

- If g is large $(g > g_2^*)$, i.e. if the nonlinear contribution in the equation is large, g will tend to diverge under the effect of renormalisation. The effect of the nonlinear term becomes relevant, but the perturbation calculation cannot evaluate the exponents. In this case, the KPZ equation belongs to a new universality class.

Experiments and numerical simulations seem to show this transition, as shown in Fig. 8.9.

How do the "strong coupling KPZ" exponents behave as a function of dimension? Different conjectures have been suggested to explain the values of exponents observed by numerical simulation in the case of ballistic deposition or its variants (Eden model, "solid on solid" model, etc). As an example, we describe the conjecture proposed by Kim and Kosterlitz [9] for a discrete "solid on solid" model. This model is a version of random deposition modified by limiting the *local gradient* of the interface to a maximal value. A site is chosen at random and an atom is deposited provided that the height difference between the site and its neighbours does not exceed a certain height of N atoms. Kim and Kosterlitz [9] proposed the following relations:

$$\alpha = \frac{2}{d+3} \qquad \beta = \frac{1}{d+2} \qquad z = 2\frac{d+2}{d+3}. \tag{8.47}$$

Let us now see how growth behaves in a more complex situation, that of *molecular beam epitaxy* commonly called MBE. In this case, atoms arrive in isolation, adsorb, diffuse, aggregate in the preferred sites, or possibly desorb. We will discover that, despite its complexity, this growth mechanism can often be well described by the KPZ equation.

8.5   Molecular Beam Epitaxy (MBE)

Molecular beam epitaxy (MBE) refers to a family of techniques that share a way of deposing material on a substrate atom by atom or molecule by molecule. It has many applications. A number of discrete models have been proposed to describe the growth method. Symmetry analysis suggests two types of continuous equations, linear or nonlinear, which we will describe in turn. We should point out a paradox in this scaling approach for MBE interfaces. Whilst the engineer or technician is interested in the extreme planarity (flatness) of the interfaces that this methods produces, the physicist is interested in rough interfaces. The deposition of low energy molecules[3] in conditions of low flux and high substrate temperature, leads to an almost perfect relaxation of the interface. Most experiments designed to test scaling theories have therefore been done on conditions of little practical interest. However, the scaling approach to MBE has two areas of practical application. Firstly it guides the search for conditions of *two dimensional* growth, that is to say which lead to a flat interface. Secondly, it described well the formation of two dimensional aggregates during construction of an atomic monolayer.

Compared to the previously discussed cases, here two new mechanisms are considered: *diffusion* of atoms on the interface by displacement from site to site and their possible *desorption*, that is to say their return to the gas phase.

In addition, the mobility of atoms here can depend on their state of contact with the interface. In the classic terrace ledge kink (TLK) model represented in Fig. 8.9, it is assumed, for example, that the only mobile atoms are those on *terrace* sites (T) which have a single contact with the interface, or those on *ledge* sites (L) which have two connections. Atoms on *kink* sites (K) which have three contacts are assumed to be immobile. Unlike all the models of *numerical* growth we have mentioned up to now, the TLK model is a model of *realistic* microscopic growth in which physical characteristics can in principle be calculated if we know the nature of the atoms and contacts. Predictions of this model can therefore be tested by physical measurements.

8.5.1   Linear MBE Equation

We will address, in turn, the two new mechanisms that have to be taken into account, desorption and then diffusion. We will then construct the equation for h and discuss the relevance of the corresponding extra terms.

---

[3]In the following, we no longer mention the case of *molecules*. For simplicity we use the word *atoms*. This does not restrict the generality of the system as long as it does not produce chemical reactions such as dissociation.

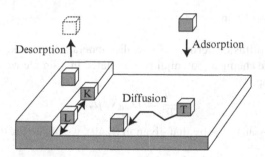

Fig. 8.10 Microscopic mechanism involved in molecular beam epitaxy (MBE). Compared to previous cases, it involves two new mechanisms: diffusion of atoms on the interface by displacement from site to site and their possible desorption i.e. return to the gas phase. In the illustrated terrace ledge kink (TLK) model, three classes of sites are distinguished: terrace sites (T) with a single contact with the substrate, *ledge* sites (L) with two connections and *kink* sites (K) which have three connections

8.5.1.1 Desorption Term

In the microscopic description illustrated in Fig. 8.10, the atom's interaction energy determines the probability of desorption. We assume that desorption can be characterised by an activation energy nE_1, where n is the number of contacts of the adsorbed atom and E_1 is the interaction energy of one contact. The corresponding term in the evolution equation is therefore of the form:

$$\left.\frac{\partial h}{\partial t}\right|_{\text{desorption}} = -B \exp\left(-\frac{nE_1}{kT}\right) \tag{8.48}$$

In a continuous description of the interface, the number of contacts n is replaced by the local curvature $\nabla^2 h$. If the atom is in a trough or valley, it is less easily desorbed than if it is at a peak. If the interaction energy nE_1 is of the order or less than kT, the exponential can be linearised (note also that n and $\nabla^2 h$ correspond to the opposite of the local chemical potential):

$$\exp\left(-\frac{nE_1}{kT}\right) \approx 1 - \frac{nE_1}{kT} \approx 1 - a\nabla^2 h \approx 1 + b\mu(r,t). \tag{8.49}$$

To first order, up to a constant which will just add the deposition flux, the desorption term is written

$$\left.\frac{\partial h}{\partial t}\right|_{\text{desorption}} = B\,\nabla^2 h \tag{8.50}$$

and behaves like the relaxation term in the Edwards–Wilkinson equation.

8.5.1.2 Diffusion Term

To evaluate the diffusion term, we use the general expression of the diffusion equation when the chemical potential is not related in a simple way to the diffusing quantity (see Chap. 4):

$$\frac{\partial h}{\partial t} = D_{\text{diff}} \nabla^2 \mu. \tag{8.51}$$

By replacing the value of μ by that given in (8.49), we obtain the diffusion term:

$$\frac{\partial h}{\partial t}\bigg|_{\text{diffusion}} = -K\nabla^4 h. \tag{8.52}$$

If we add the noise term η and the flux F of incident atoms, we obtain an equation for growth with diffusion, but without a relaxation term:

$$\frac{\partial h}{\partial t} = -K\nabla^4 h + F + \eta. \tag{8.53}$$

Since this equation is linear, we can solve it by expressing its scale invariance, as in Sect. 8.2.2. We obtain the following expressions for the exponents:

$$\alpha = \frac{4-d}{2} \qquad \beta = \frac{4-d}{8} \qquad z = 4. \tag{8.54}$$

Note that the value of the roughness exponent α is high. High roughness implies local gradients can be large, which seems to contradict our starting linear equation that assumed small variations in h. In practice, if the roughness is large, we probably should not neglect the relaxation and/or desorption terms.

8.5.1.3 Linear Equation

Adding the relaxation term to (8.53), we obtain a linear expression including all the mechanisms present in epitaxy:

$$\frac{\partial h}{\partial t} = \nu \nabla^2 h - K\nabla^4 h + F + \eta. \tag{8.55}$$

Let us summarise the meaning of each term:

- $\nu\nabla^2 h$ corresponds to the *relaxation* in the Edwards–Wilkinson equation, as well as the *desorption*. These two mechanisms tend to smooth out the interface whose curvature is $\nabla^2 h$. The parameter ν is the analogue of a surface tension.
- $K\nabla^4 h$ is the expression of the *diffusion* of atoms at the interface. It is fourth order due partly to the local chemical potential μ, which is itself proportional to

the local curvature, and partly to the diffusion mechanism itself characterised by $-\nabla^2\mu$. The parameter K plays the role of a diffusion coefficient.

- F accounts for the exchange with the gas phase. It measures the *incident flux* minus the average *desorption flux*.
- η is the noise term. Physically it comes from for example the stochastic adsorption and desorption.

Changing the scale by a of factor of b, we obtain:

$$\frac{\partial h}{\partial t} = vb^{z-2}\nabla^2 h - Kb^{z-4}\nabla^4 h + F + b^{(z-d-2\alpha)/2}\eta. \tag{8.56}$$

From the first two terms, we can define a length L_{MBE} characterising their relative weights:

$$L_{\text{MBE}} = \left(\frac{K}{v}\right)^{1/2}. \tag{8.57}$$

The behaviour of these two terms when the scale is changed identifies two regimes:

- At small scales ($b \to 0$), i.e. at short times, the term b^{z-4} dominates:

Diffusion regime for $\xi(t) < L_{\text{MBE}}$ i.e. diffusive behaviour (8.54).

- At large scales ($b \to \infty$), i.e. long times, the term b^{z-2} dominates:

Desorption regime for $\xi(t) > L_{\text{MBE}}$ i.e. Edwards–Wilkinson type behaviour.

Our custom being to be interested only in asymptotic behaviours, we should simply ignore the diffusion regime and note that the corresponding term is not relevant. However, this attitude, although conforming with our starting philosophy, is not always in agreement with the practical situation. The length L_{MBE} of the change in regime should be compared to the characteristic size L of the real system under consideration. Each of these two mechanisms in competition, desorption and diffusion, is *thermally activated*, that is to say is controlled by energy barriers characterised by an activation energy. The characteristic length L_{MBE} is:

$$L_{\text{MBE}} \sim \exp\left(\frac{E_{\text{des}} - E_{\text{diff}}}{2kT}\right). \tag{8.58}$$

The thermal dependence of this length is related to the difference between the characteristic activation energies E_{des} of desorption and E_{diff} of diffusion. A simple reasoning saying that it is inherently easier to overcome a diffusion barrier than a desorption barrier (otherwise the atoms desorb before being able to diffuse) would imply a positive difference and therefore an increase of L_{MBE} with temperature.

However, to compare the corresponding fluxes, we must take into account the nature of the sites that dominate desorption and diffusion respectively (these sites may not be the same).

8.5.2 Nonlinear MBE Equation

We already know a general nonlinear term, that of the KPZ equation (8.27). Remember that this term expresses the irreversible nature of deposition in that the interface is not invariant under exchange of the two media. It can describe the nonlinear nature of growth mechanisms involved in MBE, except that of diffusion. As we saw in the previous subsection, diffusion is the only new mechanism we have to account for in MBE and we should therefore identify its associated nonlinear terms.

8.5.2.1　Nonlinear Diffusion Term

The nonlinear terms to be taken into account must respect conservation of mass, so they can be considered as the divergence of a current. However, the term $(\nabla h)^2$ does not fulfil this condition. To fourth order in the differential we can identify two possible contributions:

$$\nabla^2 (\nabla h)^2 \quad \text{and} \quad \nabla (\nabla h)^3. \tag{8.59}$$

The effects of the first term have been studied by several authors, which is not the case for the second. Actually, no physical system seems to correspond to this second term which is relevant at short times. Therefore we will take into account the first term to form the following nonlinear equation:

$$\frac{\partial h}{\partial t} = -K\nabla^4 h + \lambda_1 \nabla^2 (\nabla h)^2 + F + \eta. \tag{8.60}$$

This equation must be solved by the dynamic renormalisation method, just like the KPZ equation in Sect. 8.4.

8.5.2.2　Solution by Dynamic Renormalisation

By following the same approach as for the KPZ equation (Sect. 8.4), we obtain the following equations for the renormalisation flow:

Table 8.1 Comparison of exponents of the diffusive regime predicted by the linear and nonlinear MBE theories, for a two dimensional interface

Exponent	α	β	z
Linear MBE equation (diffusive regime)	1	1/4	4
Nonlinear MBE equation (diffusive regime)	2/3	1/5	10/3

$$\frac{dK}{dl} = K \left[z - 4 + K_d \lambda_1^2 \frac{D}{K^3} \frac{6-d}{d} \right]$$

$$\frac{dD}{dl} = D \left[z - 2\alpha - d \right] \tag{8.61}$$

$$\frac{d\lambda_1}{dl} = \lambda_1 \left[z + \alpha - 4 \right],$$

where K_d is the same numerical constant as in Sect. 8.4.1. These equations determine a single non trivial fixed point, which leads to values for the exponents as follows:

$$\alpha = \frac{4-d}{3} \qquad \beta = \frac{4-d}{8+d} \qquad z = \frac{8+d}{3}. \tag{8.62}$$

8.5.2.3 Comparison with the Linear Diffusive Regime

In practice interfaces often have two dimensions. In this case, Table 8.1 shows the values of exponents of the diffusive regime ($\xi(t) < L_{\text{MBE}}$), obtained by the linear and nonlinear models.

The results of the theory including the nonlinear term are more convincing than that of the linear theory. For example the exponent α is smaller, as is the resulting characteristic slope of the roughness. Equation (8.60) describes well the experimental reality of MBE growth without desorption. We can actually obtain it from discrete models such as the TLK model illustrated in Fig. 8.10. In conclusion, we have learnt that:

- If desorption dominates ($\xi > L_{\text{MBE}}$), the Edwards–Wilkinson equation (8.17) describes MBE well.
- If diffusion dominates ($\xi < L_{\text{MBE}}$), (8.60) describes the growth well.

8.6 Roughening Transition

What is commonly called a *roughening transition* refers to the roughness of a surface of a three dimensional solid, at *equilibrium*. In this way it is a bit of a digression from the other sections in this chapter covering nonequilibrium mechanisms. The topic is of great practical interest as we will see later, however its description requires a special treatment due to the *two dimensional* aspect of the interface and the *discrete* nature of its fluctuations. Since it acts as an equilibrium phase transition, we can examine its universality class.

As h is a natural choice of order parameter, we could think of following the example of the Ising model since it corresponds to a scalar order parameter. However, as we have already pointed out, the energetic term driving growth is related to the local *curvature* of the surface, linear in h. In practice, as we will show below, the combination of a *continuously varying* order parameter and the discrete nature of fluctuations produced by a *periodic potential*, leads to a system of XY type, which has a two component order parameter. An analogy can be established with the model for two dimensional melting described by Nelson and Halperin [13] or with the magnetic 2D XY model, which we described in Sect. 3.6.5.

In this description we showed the microscopic excitations leading to the transition are vortex/antivortex pairs. Table 8.2 compares different mechanisms of excitation responsible for the KT transition in different physical systems where the XY model in two dimensions can be applied [14, 16]. A direct physical analogy with two dimensional melting can be made with the aid of pairs of *screw dislocations* (Fig. 8.11). A screw dislocation in a solid is a helical defect corresponding to a shift of one crystalline plane as we go round one turn of the defect axis.

8.6.1   Continuous Model

We want to describe the interfacial energy of a solid that has reached an asymptotic growth regime. As in the previous descriptions, we aim to transform the discrete nature of the solid into a continuous model. We assume that the energy of the interface is simply proportional to its surface area, with the proportionality coefficient being the surface tension v. In two dimensions, the area element is

Table 8.2 Comparison between different excitation mechanisms of Nelson–Halperin type leading to a Kosterlitz–Thouless transition

Physical system	Excitations leading to KT transition
Magnetism	Vortex/antivortex pairs
Superconductivity/superfluidity	Vortex/antivortex pairs
Melting of a 2D solid	Dislocation/anti-dislocation pairs
Roughening transition	Screw dislocation/anti-screw dislocation pairs

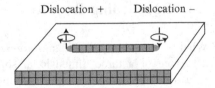

Dislocation +   Dislocation –

Fig. 8.11 Screw dislocation/anti-screw dislocation pair

$\left[1 + (\nabla h)^2\right]^{1/2} dxdy$, so the energy has the form:

$$E = v \iint dxdy \left[1 + (\nabla h)^2\right]^{1/2}. \tag{8.63}$$

If the gradient remains small, to first order in $(\nabla h)^2$ we obtain:

$$E = vL^2 + \frac{v}{2} \iint dxdy \, (\nabla h)^2. \tag{8.64}$$

This is formally identical to (3.60), simply relating this energy to the stationary state of the Edwards–Wilkinson equation, (8.17), under the effect of surface tension. It predicts that there is no transition: roughness exists at all temperatures. Since $d = 2$ is the critical dimension of the Edwards–Wilkinson equation, the growth of roughness as a function of size is marginal (that is to say logarithmic, $\log(L)$, and not a power law).

This equation does not explain numerical simulations that clearly show that a transition does exist. The main difference between the simulations and (8.64) is the *discrete* nature of the solid. In general this difference is not relevant (see Chap. 3) for transitions belonging to a universality class. Here, we expect a *non universal* XY type transition in two dimensions. In this case the microscopic features, such as the periodicity of the lattice, are crucial for the transition. We therefore need to add a periodic term to the energy. We will also ignore the constant term as it plays no role in the transition:

$$E = \iint dxdy \left[\frac{v}{2} (\nabla h)^2 - V \cos\left(\frac{2\pi h}{a}\right)\right]. \tag{8.65}$$

8.6.2 Renormalisation

The renormalisation procedure is the same as for equilibrium phase transitions. A differential change in scale leads to the flow (8.67), introducing two reduced variables:

$$x = \frac{2av}{\pi kT} \qquad y = \frac{4\pi V}{kT\Lambda} \tag{8.66}$$

$$\frac{dx}{dl} = \frac{y^2}{2x} A\left(\frac{2}{x}\right)$$

$$\frac{dy}{dl} = 2y \frac{x - 1}{x}, \tag{8.67}$$

where Λ is a characteristic length used in the renormalisation. The complex function $A(u)$ is only used in the region near $u = 2$ in which its value is $A(2) = 0.398$.

These flow equations were established by Kosterlitz and Thouless for the 2D XY model (see Sect. 3.4.5 for the magnetic homologue). We have good evidence that this model corresponds well to the physical system being studied. The results show the following:

- A transition, the *roughening transition*, occurs at temperature

$$T_R = \frac{2av}{\pi k}. \tag{8.68}$$

- Far below this temperature, the interface is flat, whereas near T_R the roughness diverges, independently of the system size L, as:

$$w(T \to T_R) \sim (T_R - T)^{-1/4}. \tag{8.69}$$

- For temperatures $T < T_R$, the correlation length $\xi(T < T_R)$ remains infinite. This feature of the XY model means that it remains *critical* as long as the temperature does not reach the transition temperature. Remember that in the 2D XY model, there is never long range order at finite temperature. The transition switches the system from a state of finite order containing few defects to a totally disordered state.
- When the temperature is very close to the transition, the roughness saturates at a constant value which it keeps beyond the transition temperature. The system then recovers the logarithmic behaviour predicted by the Edwards–Wilkinson equation (8.17):

$$w(T > T_R) \sim \log(L). \tag{8.70}$$

- In this region ($T \sim T_R$), the correlation length diverges in the usual way for the KT transition:

$$\xi(T > T_R) \sim \exp\left(\frac{B\,T_R^{1/2}}{(T - T_R)^{1/2}}\right) \quad \text{where } B \sim 1.5. \tag{8.71}$$

This complex behaviour is well reproduced by numerical simulations [17] as illustrated in Fig 8.12.

As well as numerical simulations, several experimental results confirm the existence of the roughening transition. Experiments cannot be carried out on all solids, because for many materials T_R is near to or above the melting temperature, making observation of the transition impossible. Examples of reliable observations include indium (110) at $T_R = 290$ K, lead (110) at $T_R = 415$ K and silver (110) at $T_R = 910$ K [10]. Various techniques are used, but they most often use reflection high-energy electron diffraction (RHEED). The properties, notably the transition temperature, depend on the orientation of the crystal observed, as there is no reason why the associated surface tension should be the same for different crystal faces.

Another physical system giving very interesting information is that of solid helium 4, detailed studies on which have been done [5], showing a roughening

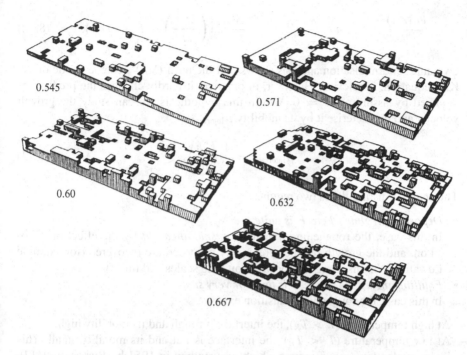

Fig. 8.12 Evolution of surface roughness as a function of temperature for a "solid on solid" simulation. Numbers show the reduced temperature at the characteristic interaction energy J. The transition (saturation of roughness) is expected at $T/J = 0.62$ (after [17])

transition at 1.28 K. The experiments measured the crystal growth velocity around this transition. The growth is very slow below 1.2 K, increases rapidly and then saturates above 1.3 K. When the interface is flat ($T < T_R$), the only way for incident atoms to irreversibly aggregate is nucleation, that is to say the rare statistical formation of a cluster which is large enough to be stable. Such cluster formation is a rare event and the process is slow. Also, when the interface is rough, incident atoms easily find adsorption sites where they are irreversibly fixed (for example kink site K in Fig. 8.9). The roughening transition also involves profound changes in the dynamics of the interface, which we will briefly describe in the next section.

8.6.3 Nonequilibrium Roughening Transition

Now that we know the relevance, in the *non* universal case of the transition, of the periodic potential term modelling the crystal periodicity, we can include it in an evolution equation such as KPZ equation (8.27):

$$\frac{\partial h(r,t)}{\partial t} = \nu \nabla^2 h + \frac{\lambda}{2} (\nabla h)^2 - \frac{2\pi V}{a} \sin \left(\frac{2\pi h}{a} \right) + F + \eta(r,t). \tag{8.72}$$

Dynamic renormalisation analysis shows that the term $(\nabla h)^2$, characteristic of the KPZ equation, is relevant. However, in practice, it hardly changes the predictions obtained by assuming $\lambda = 0$. Within this hypothesis we can study the growth velocity and characterise it by its mobility $\mu_{\text{interface}}$:

$$\mu_{\text{interface}} = \frac{1}{F} \left\langle \frac{\partial h}{\partial t} \right\rangle. \tag{8.73}$$

The analysis distinguishes two regimes:

- *Deposition regime where F is finite.*
  In this case, the roughening transition is *smoothed out* by the effect of deposition, and the growth regimes previously proposed are recovered (for example Edwards–Wilkinson regime for MBE on long scales and times).
- *Equilibrium regime where F is finite or very small.*
  In this case, the roughening transition exists:

- At high temperature $(T > T_R)$, the interface it rough and its mobility high.
- At low temperature $(T < T_R)$, the interface is flat and its mobility small. This is the *nucleation growth* regime which was studied in 1951 by Burton et al. [2]. The, very nonlinear, nucleation regime has been quantitatively described by many authors [15].

8.7   Universality Classes of Growth

In this section we summarise results obtained from continuous equations describing growth mechanisms. The equations are grouped into universality classes according to their symmetries. For mode details on this classification, we refer the reader to the excellent work of Barabasi and Stanley [1].

The general form of equations of growth,

$$\frac{\partial h}{\partial t} = G + \eta \tag{8.74}$$

contains two terms: the *deterministic* term G and the *stochastic* term η. These terms have two types of characteristic properties:

- The deterministic term G is linear (L) or nonlinear (N). We have met two types of linear terms: the Edwards–Wilkinson linear term $\nabla^2 h$ which we call (L2) and the linear term $\nabla^4 h$ related to the diffusion which we call (L4). We have also met

Table 8.3 Main universality classes of growth. In the case of KPZ, the exponents are evaluated numerically for $d > 1$ (see Table 8.1)

Symmetries		Name	Equation	Exponents		
G	η			α	β	z
–	D	Random	$\dfrac{\partial h}{\partial t} = \eta$	–	$1/2$	–
L2C	D	EW (Edwards–Wilkinson)	$\dfrac{\partial h}{\partial t} = \nu\nabla^2 h + \eta$	$\dfrac{2-d}{2}$	$\dfrac{2-d}{4}$	2
N2D	D	KPZ(Kardar Parisi Zhang)	$\dfrac{\partial h}{\partial t}=\nu\nabla^2 h +\dfrac{\lambda}{2}(\nabla h)^2 + \eta$	$1/2\,(d{=}1)$	$1/3\,(d{=}1)$	$3/2\,(d{=}1)$
L4C	D	MBE 1 linear diffusion with deposition	$\dfrac{\partial h}{\partial t} = -K\nabla^4 h + \eta$	$\dfrac{4-d}{2}$	$\dfrac{4-d}{8}$	4
L2C	C	EW with diffusive noise	$\dfrac{\partial h}{\partial t} = \nu\nabla^2 h + \eta_d$	$-\dfrac{d}{2}$	$-\dfrac{d}{4}$	2
L4C	C	linear diffusion without deposition	$\dfrac{\partial h}{\partial t} = -K\nabla^4 h + \eta_d$	$\dfrac{2-d}{2}$	$\dfrac{2-d}{8}$	4
N4C	D	MBE 2 nonlinear with deposition	$\dfrac{\partial h}{\partial t} = -K\nabla^4 h +\lambda_1\nabla^2\left[(\nabla h)^2\right] + \eta$	$\dfrac{4-d}{3}$	$\dfrac{4-d}{8+d}$	$\dfrac{8+d}{3}$
N4C	C	MBE 3 nonlinear diffusion without deposition	$\dfrac{\partial h}{\partial t}=-K\nabla^4 h +\lambda_1\nabla^2\left[(\nabla h)^2\right] + \eta_d$	$1/3(d{=}1)$ $\dfrac{2-d}{2}\,(d>1)$	$1/11\,(d{=}1)$ $\dfrac{2-d}{8}(d>1)$	$11/3\,(d{=}1)$ $4(d>1)$

two nonlinear terms: $(\nabla h)^2$ in the KPZ equation, denoted (N2), and $\nabla^2\left[(\nabla h)^2\right]$ for diffusion, denoted (N4).

- The number of atoms at the interface can be conserved (C) or not (D) for each of the terms G and η. For G, only the nonlinear KPZ term, $(\nabla h)^2$, does not conserve the number of atoms. Therefore there is no LD type universality class for a linear equation that does not conserve atoms. Regarding the noise, we have met two sources of noise: the noise, η, related to the atomic deposition process (non conservative, D) and the noise, η_d, related to diffusion (conservative, C).

The main universality classes can be inferred as follows in Table 8.3.

The values of the exponents deduced, for the equations we have studied in this chapter, are collected in Table 8.4.

In conclusion, we emphasise that a two dimensional interface is both the most common case in practice and the most difficult theoretically. Two effects combine to make the job difficult for theorists. One is the fact that $d = 2$ is the critical dimension of the basic (Edwards–Wilkinson) equation (see Table 8.3), and the other is that

Table 8.4 Numerical values of exponents for $d = 1$, 2 and 3 for the main equations

Equation	$d = 1$			$d = 2$			$d = 3$		
	α	β	z	α	β	z	α	β	z
EW	1/2	1/4	2	0	0	2	−1/2	−1/4	2
KPZ	1/2	1/3	3/2	0.38	0.24	1.58	0.30	0.18	1.66
MBE 2 with deposition (N4CD)	1	1/3	3	2/3	1/5	10/3	1/3	1/11	11/3
MBE 3 with deposition (N4CC)	1/3	1/11	11/3	0	0	4	−1/2	−1/8	4

the roughening transition occurs specifically in two dimensions, with its complex non universal features. This leads to a wide variety of possible behaviours [1] as confirmed by experiments and numerical simulations. When the interface has a single dimension, the exponents α and β are large and the interface is usually rough. If we imagine a three dimensional interface in four dimensional space, the exponents are small or negative, leading to a flat interface. Hopefully we have convinced the reader that our three dimensional space provides the largest variety of growth mechanisms. Readers will also realise that this domain has many avenues to be explored and groundwork to be done, maybe by themselves ... we wish them success!

References

1. A.L. Barabasi, H.E. Stanley, *Fractal Concepts in Surface Growth* (Cambridge University Press, Cambridge, 1995)
2. W.K. Burton, N. Cabrera, F.C. Frank, The growth of crystals and the equilibrium structure of their surfaces. Philos. Tr. Roy. Soc. S-A **243**, 299 (1951)
3. S.F. Edwards, D.R. Wilkinson, The surface statistics of a granular aggregate. P. Roy. Soc. Lond. A Math. **381**, 17 (1982)
4. D. Forster, D.R. Nelson, M.J. Stephen, Large-distance and long-time properties of a randomly stirred fluid. Phys. Rev. A **16**, 732 (1977)
5. F. Gallet, S. Balibar, E. Rolley, The roughening transition of crystal-surfaces. 2. Experiments on static and dynamic properties near the first roughening transition of HCP He-4. J. Phys.-Paris **48**, 369 (1987); see also S. Balibar, F. Gallet, E. Rolley, P.E. Wolf, La transition rugueuse. La Recherche **194**, 1452 (1987)
6. S. Havlin, S.V. Buldyrev, A. Bunde, A.L. Goldberger, P.Ch. Ivanov, C.-K. Peng, H.E. Stanley, Scaling in nature: from DNA through heartbeats to weather. Physica A **273**, 46 (1999)
7. P.C. Hohenberg, B.I. Halperin, Theory of dynamic critical phenomena. Rev. Mod. Phys. **49**, 435 (1977)
8. M. Kardar, G. Parisi, Y.C. Zhang, Dynamic scaling of growing interfaces. Phys. Rev. Lett. **56**, 889 (1986)
9. J.M. Kim, J.M. Kosterlitz, Growth in a restricted solid-on-solid model. Phys. Rev. Lett. **62**, 2289 (1989)
10. J. Lapujoulade, The roughening of metal-surfaces. Surf. Sci. Rep. **20**, 191 (1994)
11. R.N. Mantegna, S.V. Buldyrev, A.L. Goldberger, S. Havlin, C.-K. Peng, M. Simons, H.E. Stanley, Linguistic features of noncoding DNA-sequences. Phys. Rev. Lett. **73**, 3169 (1994)

12. K. Moser, J. Kertesz, D.E. Wolf, Numerical solution of the Kardar-Parisi-Zhang equation in 1, 2 and 3 dimensions. Physica A **178**, 215 (1991)

13. D.R. Nelson, B.I. Halperin, Dislocation-mediated melting in 2 dimensions. Phys. Rev. B **19**, 2457 (1979)

14. B.I. Halperin, D.R. Nelson, Resistive transition in superconducting films. J. Low Temp. Phys. **36**, 599 (1979)

15. P. Nozières, F. Gallet, The roughening transition of crystal-surfaces. 1. Static and dynamic renormalization theory, crystal shape and facet growth. J. Phys.-Paris **48**, 353 (1987)

16. G.H. Gilmer, J.D. Weeks, Statistical properties of steps on crystal-surfaces. J. Chem. Phys. **68**(3), 950 (1978)

17. J.D. Weeks, G.H. Gilmer, Thermodynamic properties of surface steps. J. Cryst. Growth **43**(3), 385 (1978)

Chapter 9
Dynamical Systems, Chaos and Turbulence

We continue our exploration of systems without characteristic scales and specific methods into the field of dynamical systems, with the analysis of chaotic and turbulent behaviours. Due to the sheer magnitude of this field, our presentation will be deliberately selective, focusing only on critical aspects and behaviours and associated scaling laws. A brief introduction to the theory of dynamical systems takes us first of all to the important concept of *bifurcation*, a qualitative change in the asymptotic dynamics. An anecdotal but familiar example is that of a "dripping tap" whereby as we turn the tap off water flow passes abruptly from a continuous stream to a periodic regime of drops. This concept of bifurcation is closely connected to that of *instability*, which we will illustrate in a very rich experimental system, *Rayleigh–Bénard convection*, shown in Fig. 9.1. We will show that bifurcations are analogous to phase transitions, in terms of scaling or universality properties (Sect. 9.1). We will then go through the concept of chaos, an extraordinary type of dynamics that is *perfectly deterministic but nevertheless unpredictable* on the long term. We will see that chaotic dynamical systems are *temporal analogues of critical systems*, since the smallest of perturbations end up having repercussions at all scales. This sensitivity removes all usefulness from the concept of a trajectory, which must be replaced by a statistical description in terms of *invariant measures*. So analysis does not give details of a particular trajectory but *global* properties of the flow. Characteristic scenarios have been used to describe the transition to chaos, for example in Rayleigh–Bénard convection. As in the study of phase transitions, the relevant questions are considered to be to look for universal properties, determine universality classes and in the same way structural stability analysis of the models (Sect. 9.2). We have seen in Sect. 4.4.1, how the description of diffusion was justified by the chaotic properties of the movement of diffusing particles when viewed at the microscopic scale. More generally, chaos at the molecular scale validates the foundations of equilibrium statistical mechanics and is also behind that recently proposed to develop statistical mechanics of systems far from equilibrium (Sect. 9.3). We will also talk about a particular chaotic behaviour, *intermittency*, which has remarkable scaling properties (Sect. 9.4). We will end this chapter with a brief presentation of turbulence and its hierarchical structure – a cascade of

A. Lesne and M. Laguës, *Scale Invariance*, DOI 10.1007/978-3-642-15123-1_9,   293
© Springer-Verlag Berlin Heidelberg 2012

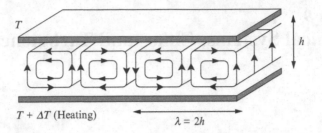

Fig. 9.1 Rayleigh–Bénard experiment. By heating from below a liquid film confined between two horizontal glass plates, we maintain an *adjustable* vertical temperature gradient. The existence of an instability is intuitively understood from the fact that hotter, so less dense, lower regions of liquid tend to rise, whereas relatively heavy upper regions of liquid tend to sink. This mechanism is dampened by thermal diffusion and viscous friction such that convective motion only appears if the temperature gradient is large enough. Quantitatively this *instability threshold* is expressed by the value of a dimensionless parameter, the *Rayleigh number* $Ra = g\alpha h^3 \Delta T / \kappa \nu$, depending on the gravitational acceleration g; temperature difference ΔT and separation h between the two plates; kinematic viscosity ν, isobaric thermal expansion coefficient α and thermal diffusivity κ of the liquid (with $\kappa = \chi / C_p$ where χ is the thermal conductivity and C_p the heat capacity per unit volume) [16,32,55]. For $Ra > Ra_c \approx 1,700$, circular convection currents appear whose size is fixed by the fluid film thickness h. The liquid is all moving, but in an organised way, forming a *steady state* structure. If ΔT (and so Ra), is increased further, *secondary instabilities* appear. Waves of period τ start moving along the convection cells deforming their linear sides into sine waves and then subharmonic waves of period 2τ appear. If the experimentalist is very careful a few further period doublings can be observed. The succession of these instabilities leads to a chaotic regime (Sect. 9.2.3). We describe this *transition to chaos* in Sect. 9.2.4. If we continue to increase ΔT, a *fully developed turbulent* regime is observed. This is qualitatively analogous to hydrodynamic turbulence which we will address in Sect. 9.5, but is relatively simpler because the temperature and density of the fluid are constant.

The case where the upper surface is open to the air is different, because surface tension effects are important, even dominant, and another dimensionless number (*Marangoni number*, Mg) controls the fluid dynamics

eddies – leading to ideal scaling laws, namely Kolmogorov theory and, more recently, the multifractal model (Sect. 9.5).

9.1 A Different View of the Dynamics of a System

9.1.1 A "Geometry" of Dynamics

The natural approach in studying the dynamics of a physical system is to solve the equations of motion as a function of the initial conditions, *exactly* following the trajectory[1] of the system over a *finite* duration. This approach is not always practical, for the simple reason that it is often not possible to analytically integrate the

---

[1]Let us emphasize at the start that throughout this chapter, "trajectories" are trajectories *in phase space* not in real space (unless the two spaces coincide).

differential equations governing the movement. Poincaré showed that a Hamiltonian system composed of N interacting bodies (for example the solar system where $N=10$) is in general not integrable if $N \geq 3$. It is even more difficult if we are interested in the system behaviour as $t \to \infty$ (which we called the *asymptotic regime*) and its dependence on the initial conditions. Perturbative methods are only valid on short times. Numerical integration possible today does not remove this difficulty because it also can only follow the dynamics over a limited duration. This is because it relies on discretisation schemes which may introduce uncontrolled deviations and numerical errors due to the finite precision of the machine ("numerical noise") can be amplified by the nonlinear dynamics and grow exponentially fast with the integration time. Other difficulties are added when we return to the physical reality, due to the fact that, since the initial condition is never known with infinite precision, the dynamics of a *bundle of trajectories* must be followed. In addition, the dynamical model itself is only a simplification of the real dynamics, since influences judged as secondary are always left out, but this judgement, although justified for short times, can turn out to be incorrect at long times. We should therefore think of a *set of evolution laws*, derived by perturbing the initial model. In conclusion, determination of individual trajectories can not form predictions of long term behaviour of a physical system. These were the problems facing Poincaré when he was trying to answer the question of the stability of the solar system. He therefore adopted a *geometrical* and *global* point of view, the basis of the modern theory of dynamical systems [69].

In summary, the defining characteristic of the theory of dynamical systems, compared to mathematical analysis of ordinary differential equations (the approach mentioned at the beginning of this section), is to address the following questions, of great interest to physicists trying to make reliable and robust predictions:

- *What is the long term behaviour of the system?*
- *How is it changed if we slightly alter the initial condition?*
- *What does it become if the evolution law itself is perturbed?*

9.1.1.1   Fundamental Concepts of Dynamical Systems

The first step in modelling a physical system consists of describing its instantaneous state by a set of variables, which we denote in shorthand as x, *not to be confused with a position in physical space*. The set $\mathscr{X}$ of these points x, each representing a possible configuration of the system is called the *phase space*. Note that this set will depend on the particular modelling and therefore the observation scale.

A *continuous dynamical system* is therefore a deterministic model of evolution of the form $\dot{x}(t) = V[x(t)]$, where the dot means the derivative with respect to time t. For example, the evolution equation $m\ddot{x} + \gamma\dot{x} + U'(x) = 0$ following from the principle of Newtonian dynamics applied to a damped oscillator of mass m, friction coefficient γ and potential energy $U(x)$, takes the form of a dynamical system:

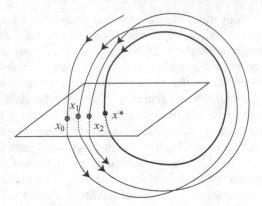

Fig. 9.2 Discretisation by "Poincaré section". We keep only the trace of a continuous time evolution in a section of phase space transverse to a periodic trajectory (a planetary orbit in Poincaré's original work), here in bold. Instead of studying a trajectory $x(t)$, we study the sequence of points of intersection $(x_n)_{n \geq 0}$ with this section. Note that this discretisation is *intrinsic to the dynamics* (the "first passage time" taken to pass from x_n to x_{n+1} depends on n and the trajectory under consideration). The presence of a periodic trajectory is not an absolutely necessary condition: it is invoked to ensure that neighbouring trajectories will actually recross the chosen section

$[\dot{x} = v, \ \dot{v} = -U'(x)/m - \gamma v/m]$ of two variables x and v. Another example is that of the system of equations obtained in the field of chemical kinetics [47].

One of the fundamental ideas of the theory of dynamical systems is to describe, not a particular trajectory over a finite time, but the *set of trajectories* (in phase space) which we call the *flow*. It is this global knowledge that will enable prediction, at least qualitatively, of the asymptotic behaviour of the system.

We also use[2] discrete models of evolution: $x_{n+1} = f(x_n)$. For example, models of population dynamics (n is then the number of generations) and discrete dynamical systems obtained by the method called *Poincaré section* illustrated in Fig. 9.2.

9.1.1.2 Conservative and Dissipative Systems

An important distinction is that separating conservative and dissipative systems. Conservative systems are Hamiltonian systems with conserved total energy and dissipative systems are all the rest, in which trajectories evolve in phase space to

---

[2]There is an important difference to understand intuitively between continuous and discrete dynamical systems. Trajectories of an autonomous (i.e. where V does not depend explicitly on time) continuous dynamical system can only cross or meet themselves at fixed points, which they only reach asymptotically. This topological constraint does not exist for discrete trajectories generated by a transformation. Given a discrete dynamical system in d dimensions, continuous dynamical systems generating dynamics whose trace obeys this discrete dynamical system take place in a dimension *strictly larger than d*, as explicitly shown in Fig. 9.2.

particular subsets called *attractors*.[3] They reduce to equilibrium states (fixed points) if the system is isolated (i.e. not driven by any input of energy or matter). More generally, dissipative systems are systems in which the dynamics is accompanied by a reduction in the natural volume of phase space, whereas this volume is invariant in conservative systems (Liouville's theorem). The evolution of a dissipative system can nevertheless lead to a non trivial asymptotic state if it is driven, that is if the dissipation is compensated for by an injection of energy or matter. For example, in the case of a chemical system, dissipation associated with the progress of a reaction must be compensated for by the injection of reactants to observe a non trivial steady state.

9.1.1.3   Fixed Points and Linear Stability Analysis

A *fixed point* of the dynamics (in continuous time) is a point $x^* \in \mathcal{X}$ such that $V(x^*) = 0$, which corresponds to an *equilibrium state*. To determine the *stability properties* of this equilibrium state, we study the evolution of a small perturbation y_0 of the equilibrium state, small enough that we can use the linearised evolution equation: $\dot{y} = DV(x^*) \, y$, where $y = x - x^*$ is the deviation from the equilibrium point. Integrating this directly gives: $y(t) = e^{tDV(x^*)} \, y_0$. Trajectories will approach x^* in directions of eigenvectors of $DV(x^*)$ that are associated with eigenvalues with strictly negative real parts, called the stable directions. On the other hand, trajectories will move away from x^* in the unstable directions, associated with eigenvalues with strictly positive real parts. If all the eigenvalues have non zero real parts, it can be demonstrated that the flow is equivalent to the linearised flow and analysis of the matrix $DV(x^*)$ will therefore be sufficient to determine the behaviour of trajectories in the vicinity of x^*. On the other hand, the marginal situation, in which one or more of the eigenvalues has a zero real part, is singular and associated with bifurcation which will be discussed in Sect. 9.1.2.

In the case of a system in discrete time, x^* is a fixed point if it satisfies $f(x^*)=x^*$. The linear evolution is written $y_{n+1} = Df(x^*) \, y_n$ where $y_n = x_n - x^*$ and where $Df(x^*)$ is the Jacobian matrix for the transformation f of the fixed point x^*. Therefore, the stability of the fixed point in the direction of an eigenvector of $Df(x^*)$ is obtained by comparing the modulus of the associated eigenvalue λ with 1: stable directions are those for which $|\lambda| < 1$. In this discrete case, the condition for which the flow is equivalent to the linearised flow is that no eigenvalues have modulus 1.

---

[3] The intuitive concept of an attractor $\mathscr{A} \subset \mathscr{X}$ has appeared in various mathematical formulations, more or less strict depending on the context and the authors. They share the condition of invariance ($f(\mathscr{A}) = \mathscr{A}$ in discrete time) and the fact that $\mathscr{A}$ "attracts" (strictly it is the limit ensemble) all or some of the trajectories passing nearby. This means that the study of asymptotic behaviour of trajectories becomes the study of that of the dynamics restrained to $\mathscr{A}$. See [59] for an in depth discussion of the concept of attractor and the different definitions proposed.

9.1.1.4 Limit Cycles and Attractors

A limit cycle is a periodic trajectory asymptotically attracting all trajectories passing near it. At long times, all trajectories starting from a point near a limit cycle will therefore have an oscillation behaviour with the same frequency and form as that of the limit cycle. A limit cycle is the simplest case of an asymptotic regime that does not reduce to an equilibrium state. A typical example is the following, given in Cartesian and polar coordinates respectively:

$$\begin{cases} \dot{x} = ax(r_0 - \sqrt{x^2 + y^2}) - \omega y \\ \dot{y} = ay(r_0 - \sqrt{x^2 + y^2}) + \omega x \end{cases} \Longleftrightarrow \begin{cases} \dot{r} = ar(r_0 - r) \\ \dot{\theta} = \omega. \end{cases} \tag{9.1}$$

If $a < 0$, the fixed point $(0,0)$ (i.e. $r = 0$) is stable and the cycle $r = r_0$ is unstable, i.e. a perturbation δr is amplified over time. If $a > 0$, the fixed point is unstable, but the cycle $r = r_0$ (with angular velocity ω) has become stable. It can be explicitly verified in this example that $a = 0$ is a special case, which we will return to in Sect. 9.1.2 under the name of Hopf bifurcation. The eigenvalues are complex conjugates ($\pm i\omega$) and their real parts cancel.

Note that a limit cycle, due to its status as an attractor, corresponds to a robust oscillation behaviour: its period and amplitude are not permanently affected by a perturbation, the influence of which eventually ends up as just a phase shift.

More complex asymptotic regimes exist than fixed points and limit cycles. In conservative systems, *quasiperiodic* regimes are often seen, which have several incommensurable frequencies: $x(t) = \varphi(\omega_1 t, \ldots, \omega_n t)$. Amongst dissipative systems, where the asymptotic dynamics is mainly governed by dynamics constrained to attractors, the most notable example is that of strange attractors associated with chaotic dynamics (Sect. 9.2).

9.1.2 Bifurcations

The global "geometric" approach of asymptotic dynamics taken by the theory of dynamical systems, lead to the key concept of *bifurcation*. By this term we mean all *qualitative* changes in the asymptotic dynamics observed when a parameter μ of the dynamics is varied. The value $\mu = \mu_0$ where the change is produced is called the *bifurcation point*. This concept has been particularly exploited in the case of dissipative systems, where a bifurcation corresponds to a qualitative change in the attractor (*bifurcation theory*). This change can be visualised on a *bifurcation diagram* representing the attractor as a function of the parameter μ.

9.1.2.1 Experimental Realisation

The parameter controlling the asymptotic dynamics (which here we call *control parameter*) is related to the amplitude of the nonlinear amplification terms, rate of injection of energy, amplitude of the nonlinear saturation terms and size of dissipation. These first two terms have a destabilising effect, whereas the other two counterbalance these. The qualitative analysis of the phenomenon is enough to identify the different mechanisms involved in the dynamics, but a step of modelling is necessary to elucidate the dimensionless control parameter encapsulating the result of their competition. In this way, we have met the Péclet number in Sect. 4.1.6, Rayleigh number in Fig. 9.1 and we will see the Reynolds number in Sect. 9.5.2. In practice, it is usually the experimental device which prescribes the adjustable parameter μ (for example the temperature difference ΔT in the Rayleigh–Bénard experiment). The parameter μ is varied slowly enough that the system has the time to stabilise to its asymptotic regime for each value of μ.

An initial example is that of secondary instabilities observed in the Rayleigh–Bénard experiment (Fig. 9.1). If we parametrise the shape of the circulating convection currents and study the *temporal* variation of the parameters, each instability appears as a bifurcation during which a limit cycle is destabilised and replaced by a cycle with double the period. Bifurcations can also be observed in the continuous flow of liquid between concentric cylinders rotating at different velocities (Couette–Taylor problem) [18]. Another famous example is that of a "dripping tap" where, on decreasing the water output, we pass from a regime where the jet is continuous to periodically falling drops [57, 67, 78]. We also mention the bifurcation observed when we tilt a sheet of corrugated iron on which a bead has been placed. At a critical incline, the stable and unstable positions join and there is no longer any stationary positions because the incline is too steep and the bead rolls down the sheet. Many examples of bifurcations are found in chemistry, for example the Belousov–Zhabotinski reaction (Sect. 9.2.1) can start spontaneously oscillating [47, 79]. We also find bifurcations in biology, in population dynamics [61] as well as at the cellular scale, for example in the glucose cycle, enzymatic reactions or in the activity of certain neurons [34].

9.1.2.2 Mathematical Analysis

Dynamical systems describing real dynamics usually take place in phase spaces of dimension greater than 1. However, amazingly the dynamics near a bifurcation is dominated by modes that become unstable, enabling the dynamics to be reduced to a system in 1 or 2 dimensions (see Sect. 9.1.4 for details). In the case of spatiotemporal evolutions, as we have seen in the case of Rayleigh–Bénard convection, the problem must be reduced to purely temporal dynamics of amplitudes or other parameters of spatiotemporal solutions of the given parametrised form [55]. So explicit treatment of concrete examples often starts with a step reducing the dynamics. Here we assume that this step has already been made.

The simplest entirely identified case is that where the attractor before bifurcation, (for $\mu < \mu_0$), is a stable fixed point $x_0(\mu)$, so the eigenvalues of the stability matrix $DV[x_0(\mu)]$ all have strictly negative real parts. Destabilisation occurs at $\mu = \mu_0$, when (at least) one eigenvalue crosses the imaginary axis. In the case of a discrete dynamical system, a bifurcation is produced when the largest modulus eigenvalue(s) cross the unit circle. $\lambda_1(\mu)$ being the eigenvalue with the maximum real part (or maximum modulus in the discrete case), we can summarise a bifurcation situation as follows:

- In continuous time: $\mathrm{Re}\,\lambda_1(\mu) < 0$ if $\mu < \mu_0$, $\mathrm{Re}\,\lambda_1(\mu_0) = 0$.
- In discrete time: $|\lambda_1(\mu)| < 1$ if $\mu < \mu_0$, $|\lambda_1(\mu_0)| = 1$.

The references [71] and [17] give complete presentation of bifurcation theory. Here we will limit ourselves to giving a few mathematical examples which are among the most representative and that are involved in the transition to chaos, as we will see in Sect. 9.2.4. The universality and therefore the scope of these models will be discussed in Sect. 9.1.4.

9.1.2.3 A Few Typical Bifurcations

We first mention a bifurcation that we will return to in Sect. 9.4.1 in the context of intermittency: the *saddle-node* bifurcation (also sometimes called tangential or fold bifurcation). It corresponds to the collision of two fixed points, one stable and one unstable, followed by their "annihilation". This is the bifurcation observed in the example of a bead on an inclined sheet of corrugated iron. A typical model is presented in Fig. 9.3.

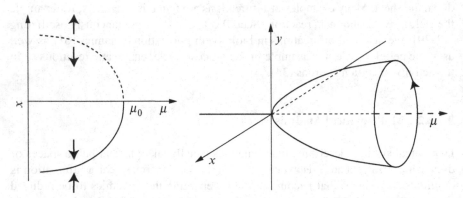

Fig. 9.3 *Left*: Saddle-node bifurcation observable in the evolution $\dot{x} = \mu - \mu_0 + x^2$. For $\mu < \mu_0$, we have two fixed points, $\sqrt{\mu_0 - \mu}$ (unstable) and $-\sqrt{\mu_0 - \mu}$ (stable). They collide at μ_0 and disappear for $\mu > \mu_0$. *Arrows* indicate the instability of the upper branch and stability of the lower branch. *Right*: Hopf bifurcation observable in the evolution $\dot{r} = r(\mu - r^2)$, $\dot{\theta} = \omega$. The branch $r \equiv 0$ is stable for $\mu < 0$ and destabilises at $\mu_0 = 0$ being replaced by a limit cycle $r(\mu) = \sqrt{\mu}$ (as attractor)

Fig. 9.4 Period doubling bifurcation (*pitchfork bifurcation*) in discrete time. It is for example observed in the evolution $x_{n+1} = 1 - \mu x_n^2$ at $\mu_0 = 3/4$. Here the fixed point $x_0(\mu)$ is replaced by a cycle of period 2: $x^{\pm}(\mu) = f_\mu[x^{\mp}(\mu)]$. *Arrows* indicate the stable branches

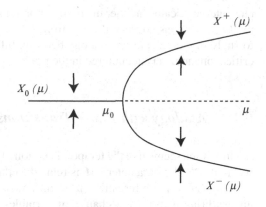

A second example is the onset of an oscillating regime corresponding to the destabilisation of a fixed point in favour of a limit cycle. This is the bifurcation observed in, for example, the dripping tap. We have seen a theoretical example in Sect. 9.1.1 with the system $\dot{r} = \mu r(r_0 - r)$, $\dot{\theta} = \omega$. In this case, called *subcritical Hopf bifurcation*, finite amplitude oscillations ($r = r_0$) suddenly appear at the bifurcation point, at $\mu_0 = 0$. Before bifurcation the cycle is present but unstable. In the example shown in Fig. 9.3, the amplitude of the cycle is zero at the bifurcation point (no cycle exists for $\mu < 0$) and the oscillating regime therefore continuously connects with the equilibrium state. This case, shown in Fig. 9.3, is called a *supercritical Hopf bifurcation*. In both cases, transfer of stability from a fixed point to a limit cycle is produced when *two* complex conjugate eigenvalues cross the real axis (or the unit circle for a discrete dynamical system). Their values $\pm i\omega$ ($e^{\pm i\omega}$ in the discrete case) at the bifurcation point determine the period $2\pi/\omega$ of the oscillating regime at the moment it appears. In the discrete case, the value $e^{\pm i\omega} = -1$ must be treated separately. Since its value is real, the destabilisation only involves one eigenvalue. We therefore observe a fixed point being replaced by a cycle of period 2 (Fig. 9.4). If the discrete dynamical system is obtained by a Poincaré section of a continuous dynamical system, this will show a *period doubling*, during which a limit cycle is destabilised leaving a limit cycle with double the period. This bifurcation is shown, for example, by the secondary instabilities in Rayleigh–Bénard convection.

9.1.2.4   Sensitivity of Dynamics at Bifurcation Points

Bifurcations are critical points in the sense that they are very sensitive to perturbations and noise. Consider the case of a bifurcation corresponding to an exchange of stability between two attractors (opposite case from saddle-node bifurcation treated in Sect. 9.4.1). If we apply a perturbation, even very small, to a system at the bifurcation point, it will interfere with the competition between the two marginally stable attractors on offer to the system. This will result in a response without characteristic scales (neither in duration nor amplitude), which is very different from the linear response observed away from bifurcation points. This property leads us to study in more detail the analogy between bifurcations and phase transitions and critical phenomena encountered in the previous chapters.

9.1.3   Analogy with Phase Transitions

A bifurcation seems like the temporal version of a thermodynamic phase transition,[4] where the "order parameter" M is related to the asymptotic state of the dynamical system. Before the bifurcation ($\mu < \mu_0$), $x_0(\mu)$ is a stable fixed point and we can always bring it to 0 by a change of variables such that $M(\mu) = x_0(\mu) = 0$ if $\mu < \mu_0$. Above the bifurcation, the asymptotic state can be a fixed point $x_1(\mu)$, in which case $M(\mu) = x_1(\mu) \neq 0$, or even a cycle of amplitude $r(\mu)$, in which case $M(\mu) = r(\mu) > 0$. In particular, in one dimension, the evolution $\dot{x} = V(x)$ can be written $\dot{x} = -\partial F/\partial x$ and the fixed points are therefore obtained by minimising $F(x)$, in exactly the same way as the value of the order parameter is obtained by minimising the free energy.

9.1.3.1   Asymptotic Regime and Thermodynamic Limit

The asymptotic limit $t \to \infty$ is analogous to the thermodynamic limit. It also acts as an idealisation corresponding to a limiting case which is never strictly reached, but that we hope is a correct approximation to what happens at long but finite times. Since in taking the limit we keep only the dominant terms, the description is both simpler and more universal. It helps draw out the relevant concepts, for example bifurcation, just like taking the thermodynamic limit is necessary for the system to display clear-cut phase transitions.

In the Ising model, the thermodynamic limit does not commute with the limit $H \to 0$; we have to take the magnetic field H to 0 after the thermodynamic quantities have been determined to access the actual behaviour of the system. One analogous difficulty is encountered here when the dynamical system depends on a small parameter ϵ in a singular way, in the sense that the system for $\epsilon = 0$ corresponds to qualitatively different dynamics (for example if ϵ appears as a factor of the time derivative). In this case, the limits $t \to \infty$ and $\epsilon \to 0$ do not commute. Therefore the bifurcation diagram obtained for $\epsilon = 0$ will be qualitatively different from the limiting behaviour obtained by considering the attractor for values of ϵ tending to 0 [45]. Ensuring a correct treatment of such singular systems with

---

[4]Note that the analogy is with phase transitions in classical thermodynamics, described in a "mean field theory" in the sense of neglecting fluctuations and where only the average order parameter is written: the "critical exponents" of a bifurcation are always rational.

several limits being involved together is one of the successes of renormalisation methods [48].

9.1.3.2 Critical Exponents of a Bifurcation

As in all phase transitions where the system is poised between two regimes, a bifurcation is the result of temporal critical properties. Consider an evolution $\dot{x} = V(\mu, x)$ in one dimension, with $x_0 = 0$ as a stable fixed point if $\mu < \mu_0$ and with a bifurcation at $\mu = \mu_0$. For $\mu < \mu_0$, the derivative $\partial V/\partial x$ taken at $x_0 = 0$ is negative and can be written $-1/\tau(\mu)$. The linearised system is written $\dot{x} = -x/\tau(\mu)$ and has exponential solutions $x(0) \exp[-t/\tau(\mu)]$, depending on the initial condition $x(0)$. The characteristic time $\tau(\mu)$ is the time required to reach the fixed point. It is also the response time after which the system stabilises in a new equilibrium state after a (small) stimulation. In other words, $\tau(\mu)$ is the relaxation time of the system. At the bifurcation point, the derivative $\partial V/\partial x(\mu_0, 0)$ is zero, corresponding to the divergence of this characteristic time: $\tau(\mu_0) = \infty$. This *divergence of the time to reach the attractor* and, more generally, the characteristic time of the dynamics is a general signature of bifurcation points. In our one dimensional example, the system is written to lowest order $\dot{x} = -cx^2$ where c is equal to $(-1/2)\, \partial^2 V/\partial x^2(\mu_0, 0)$. Integrating gives $x = x_0/[1 + x_0 ct]$, which behaves as $1/ct$ at long times. We therefore have an asymptotic regime that is scale invariant and independent of the initial condition. This regime is characterised by the exponent 1. Summarising we have:

$$\mu < \mu_0 : \frac{dV}{dx}(\mu, 0) = -\frac{1}{\tau(\mu)} < 0 \text{ and } x(t) \sim e^{-t/\tau(\mu)}$$

(9.2)

$$\mu = \mu_0 : \frac{dV}{dx}(\mu_0, 0) = 0, \quad \tau(\mu_0) = \infty \text{ and } x(t) \sim 1/t.$$

It could happen that all the derivatives $\partial^k V/\partial x^k(\mu_0, 0)$ are zero for all $k \leq n$. It can be shown that in this case we have a different exponent, equal to $1/n$ and asymptotically the approach to the fixed point behaves as $x(t) \sim t^{-1/n}$.

9.1.3.3 Finite Size Effects

If trajectories are observed over a too short *finite duration* T, they may not yet be perfectly stabilised on the attractor but only locally in its neighbourhood. This deviation from the asymptotic behaviour will be more marked the closer a system is to a bifurcation point, because the time taken to reach the attractor diverges at bifurcation. So the bifurcation diagram obtained has distortions similar to those observed on the phase diagram of a sample of *finite size* N. The sharp features of the diagram, for example jumps or vertical tangents, disappear leaving instead a continuous line without kinks. This is expected as trajectories of finite duration T are regular with respect to variations in the control parameter μ which is not the case in the limit $T \to \infty$ where singularities appear at bifurcation points.

9.1.3.4 Summary of the Analogy

$$
\begin{array}{rcl}
\text{bifurcation} & \longleftrightarrow & \text{phase transition} \\
t & \longleftrightarrow & r \\
\text{duration } T & \longleftrightarrow & \text{size } L \\
\text{asymptotic regime} & \longleftrightarrow & \text{thermodynamic limit} \\
\text{divergence of characteristic time} & \longleftrightarrow & \text{critical slowing down}
\end{array}
$$

To conclude, we insist on the fact that this analogy is so profound it could be seen as an identity: *a phase transition is nothing other than a bifurcation in the underlying microscopic dynamics*, which will explore different regions of phase space both sides of the transition. Bifurcations are thus the underlying dynamics of phase transitions. A phase will be an invariant region in configuration space. It is only exactly invariant in the thermodynamic limit. For finite size, we still could have a few transitions between different regions, corresponding to a finite free energy barrier and a blurred separation between phases.

9.1.4 Normal Forms and Structural Stability

The important result is that *close to bifurcation*, components of the evolution in directions becoming unstable at the bifurcation point dominate the behaviour [37]. This makes sense intuitively since in stable directions the dynamics relaxes exponentially fast to zero. We therefore start by reducing the description to only these marginally stable degrees of freedom, leading, for example, to a dynamical system in one dimension if only one eigenvalue becomes unstable. A second result is that the reduced system can then be reduced by conjugacy to a polynomial evolution law, which we call the *normal form* of the bifurcation. It is a sort of common denominator of all the dynamics that have this bifurcation and is actually the simplest dynamical system in which it can be seen. Once the fixed point has been transformed to $x_0 = 0$ and the bifurcation point to $\mu_0 = 0$, the normal forms of two typical bifurcations[5] of continuous systems are (Fig. 9.3):

- Saddle-node bifurcation: $\dot{x} = \mu - x^2$.
- Hopf bifurcation: $\dot{r} = \mu r - r^3$, $\dot{\theta} = 1 + ar^2$.

to which we add (Fig. 9.4):

---

[5]Other notable normal forms exist, but are associated with non generic bifurcations (the associated bifurcation theorems involve equalities not inequalities):
– *transcritical* bifurcation: $\dot{x} = \mu x - x^2$;
– *pitchfork* bifurcation: $\dot{x} = \mu x - x^3$ for the supercritical case or $\dot{x} = \mu x + x^3 - x^5$ for the subcritical case (Fig. 9.5).

- Period doubling bifurcation, unique to discrete dynamical systems: $f(x) = 1 - \mu x^2$ with here $\mu_0 = 3/4$ and $x_0(\mu_0) = 2/3$.

These bifurcations are *generic*[6] in the sense that the bifurcation theorems describing the systems where they occur only involve strict inequalities (involving the derivatives of $V(\mu, x)$ or $f_\mu(x)$ with respect to x and μ, taken at the point $(\mu_0, x_0(\mu_0))$) [2, 3]. Therefore these remain true if we slightly change the evolution law for V or f.

9.1.4.1   Normal Forms and Universality

Although proofs leading to normal forms and associated bifurcation theorems are mathematical, their conclusion is most interesting for physicists: models with the same normal form have qualitatively identical bifurcations. To understand the implications of this result, we need to remember that a physical model is an approximation, often quite rough, to reality. To build a model, a physicist voluntarily puts on a partial blindfold in choosing to describe only a small number of quantities and only consider a limited number of parameters. In doing so, everything that happens outside the system or at different scales is taken into account in an average (or negligible) manner. With this theory of normal forms, we see the appearance of *universality* of properties envisaged by the theory of dynamical systems. In this way we justify the claim that analysis of very simple, even simplistic, dynamical systems can shed light on the behaviour of real systems.

We should nevertheless remember that the validity of normal forms is limited to the vicinity of bifurcation points: a normal form describes the generic mechanisms whereby the instability of the dynamics at $(\mu = \mu_0,\ x = x_0(\mu_0))$ develops, giving way to another branch of steady states. It does not accurately reflect the mechanisms that will control these steady states beyond the bifurcation point $(\mu \gg \mu_0)$.

9.1.4.2   Structural Stability and Modelling

This discussion of the theory of normal forms and its scope brings us to the more general concept of *structural stability* or *robustness*. It proves to be essential in modelling a physical phenomenon, since the phenomenon predicted by the model should not significantly change if we slightly change the model. In other words, the results of the model should be robust with respect to negligible influences and small fluctuations of the parameters. A structurally stable model (with respect to a given type of perturbations) is equivalent to all the perturbed models. Therefore its robust

---

[6] Take a model depending on a real parameter a and a phase space $\mathscr{X}$. The statement "for $a < a_0$, the solution ..." is generic whereas the statements "for $a = a_0$, the solution ..." or "for $a \leq a_0$, the solution ..." are not generic since a small variation in a changes the hypothesis to another hypothesis for which the statement is no longer true.

predictions have a chance of being reproduced in the observed reality. Note that structural stability is implicit in all properties of universality, which is more general in that any universal phenomenon is structurally stable, but a universality class can contain systems whose "operating rules" can not be deduced from each other by a small perturbation.

9.1.4.3 Bifurcation Theory and Catastrophe Theory

The concept of bifurcation is reminiscent of that of *catastrophe*, introduced and developed (earlier) by Thom [82]. On one hand, bifurcation theory is more general because it is not constrained to the dynamics $\dot{x} = -\nabla V(x)$, produced by a potential $V(x)$, with only fixed points at which the catastrophe theory applies. Qualitative changes identified in bifurcation theory are not limited to changes in stability of one or more fixed points, but also include the appearance of limit cycles and even chaos. On the other hand, catastrophe theory is more universal, because the space in which Thom [82] carried out his classification is the product space of phase space and parameter space $\{\mu, \nu\}$. It is also a general classification of the singularities of a surface, applied in particular to the surface representing the fixed points as a function of the parameters. Catastrophe theory will therefore describe the way in which bifurcations appear and succeed one another as we move in the $\{\mu, \nu\}$ plane of control parameters. One result is the demonstration of hysteresis: the bifurcations actually experienced depends on the path followed in parameter space and on the starting point of the system in phase space, in other words the whole history of the system (Fig. 9.5).

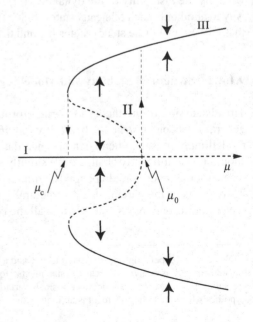

Fig. 9.5 Hysteresis associated with a subcritical bifurcation, as the control parameter μ is varied. When μ is increased, the passage from branch *I* to branch *III* occurs at μ_0; but when μ is decreased, the return from branch *III* to branch *I* occurs at a value $\mu_c < \mu_0$

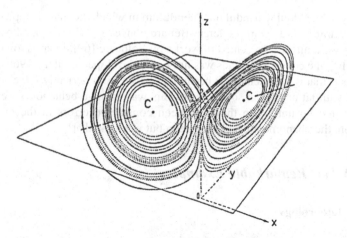

Fig. 9.6 Lorenz attractor, for $\sigma = 10$, $r = 28$ and $b = 8/3$ (reprinted with permission from [47]). It is an example of a *strange attractor*: its structure is fractal and the dynamics on it is chaotic

9.1.4.4 Bifurcations and Instabilities

Bifurcation is similar to the more general concept of *instability*. The term bifurcation is traditionally reserved for temporal evolutions, in a phase space of generally low dimension, whereas the term instability is used for spatiotemporal evolutions. However these two concepts cover the same idea of *qualitative* change of the asymptotic regime. It is often possible to bring one to the other, in other words to describe an instability (of an extended system) as a bifurcation of a purely temporal evolution. The example of secondary instabilities of Rayleigh–Bénard convection cells can clarify the general idea, which we will return to in Sect. 9.2 to justify the Lorenz model (Fig. 9.6). This studies the *variation of the parameters* of a spatial or spatiotemporal function describing the state of the system before it is destabilised.[7] The instability threshold transition therefore corresponds to a bifurcation in the purely temporal dynamics of these parameters.

9.2 Deterministic Chaos

The term chaos means *perfectly deterministic* dynamics which is *nevertheless unpredictable* in the long term. Familiar examples, which we will come back to in detail in Sect. 9.2.1, include; atmospheric dynamics from which we try to predict the weather (Fig. 9.6), making flaky pastry, balls moving on a pool table (Fig. 4.8),

---

[7]This boils down to introducing a collective variable that destabilises at the instability threshold and whose behaviour therefore dominates the dynamics (general result the details of which we will give in Sect. 9.2.4).

the solar system, a double pendulum[8] (pendulum in which the mass is replaced by a second pendulum) and again Rayleigh–Bénard convection (Fig. 9.1). Although the main ideas were already presented in works by Poincaré [69], Lyapounov [53] and Birkhoff [10], the concept of chaos was not really developed until the 1960s. One of the reasons is that it was then that it became possible to numerically solve equations of motion without analytical solutions. In this way strange behaviours were seen, so strange that up until then they had been considered errors in the method or observation: the phenomenon had been seen but not looked at!

9.2.1 A Few Remarkable Examples

9.2.1.1 Meteorology

Decades of research in the first half of the twentieth century, achieved what was hoped to be a conclusive step to weather forecasting: a complete scheme of mathematical equations describing the motion of the atmosphere was established and numerical methods to solve them were developed. However, scientific progress was not made in the direction initially hoped in that long term weather forecasts remain mediocre and are not reliable. However profound understanding of the reasons for this failure lead to the key idea of *sensitivity to initial conditions* and the intrinsic unpredictability that follows. Initial conditions arbitrarily close to each other lead to trajectories ending up having no relation to each other [64].

In 1963 Lorenz proposed a purely temporal model in three dimensions obtained by reducing the spatiotemporal equations describing atmospheric convection[9]:

$$\begin{cases} dX(t)/dt = \sigma(Y - X) \\ dY(t)/dt = rX - Y - XZ \\ dZ(t)/dt = XY - bZ. \end{cases} \tag{9.3}$$

where σ, r and b are constant parameters related to hydrodynamic properties of the atmosphere [52]. Solving this reveals an asymptotic object, since baptised *Lorenz attractor* and represented on Fig. 9.6. It is a typical example of a *strange attractor*. Not only is its structure fractal but also the dynamics restricted to the attractor is very complicated. In particular, trajectories pass from one "wing" to the other in an unpredictable way. By denoting the left wing 0 and right wing 1, the sequence of 0 and 1 associated with a typical trajectory follows the same statistics as a sequence of heads and tails generated by tossing a coin. In this sense a chaotic motion appears

---

[8]The oscillations and/or their coupling must be nonlinear; two ideal springs (harmonic oscillators) in series are equivalent to a single spring whose behaviour is perfectly predictably that of a harmonic oscillator.

[9]The derivation of this system of differential equations from the (spatiotemporal) hydrodynamic equations describing the evolution of the atmosphere can be found in [6].

random when observed over a long time, although it is deterministic and therefore perfectly predictable on short times.

9.2.1.2 Chemical Systems

The first example of *spontaneous* chemical oscillations[10] was observed in the reaction now associated with the name of its "inventors" Belousov (1958) and then Zhabotinski (1964) who disseminated and confirmed by an experimental study the work of Belousov [79]. This reaction is quite complicated (it involves around 15 chemical components) but the origin of the oscillating behaviour can be understood by a simplistic description. A first, slow, reaction consumes a species A which blocks a second reaction. This second reaction is auto-catalytic and faster than the first but produces species A which inhibits it. The second reaction occurs from the moment when the first reaction has consumed enough A. After a certain delay, the amount of A produced by this second reaction becomes high enough again to block it therefore leaving the first reaction to "clean up" A bringing the cycle back to the start and restarting it. In an open reactor with high enough concentrations of reagents,[11] this scheme leads to chaotic oscillations, appearing random even though the phenomenon is perfectly described by a system of kinetic equations [47]. Starting from a steady state and very slowly increasing the concentration of reagents (in other words increasing the rate at which they are injected into the reactor), a whole series of bifurcations can be seen: first of all a Hopf bifurcation, corresponding to the appearance of oscillations, then a succession of period doubling bifurcations, leading to chaos (the situation is in reality much richer and more complex). Theoretical aspects of this scenario are addressed in Sect. 9.2.4.

A biological example which is very similar is the coupling in series of two enzymatic reactions that are self-amplifying,[12] showing complex asymptotic behaviours, including chaos [34].

9.2.1.3 The Solar System

Since Poincaré we know that the solar system, as all systems composed of N interacting bodies for $N \geq 3$, is not integrable. In response to the question of the

---

[10]They are spontaneous in the sense that observed temporal variations are not simply the reflection of external temporal variations (for example, a periodically varying rate of injection of reagents).

[11]Let us emphasise that the reactor here is *continuously fed* (to maintain the concentrations of reagents constant) and *agitated* to ensure spatial homogeneity and avoid the formation of structures (incidentally very interesting and also much studied).

[12]By this we mean that the product of the reaction activates the enzyme and increases the reaction rate and therefore the product formation and so on, if there is no other mechanism capable of consuming the product.

stability of the solar system it seemed that the state of the solar system cannot be predicted beyond a certain duration T_0, although the only forces involved come from the perfectly deterministic law of universal gravitation.[13] This unpredictability is an example of chaos in a conservative system. Observable manifestations are:

- Irregular variations in the angle of Mars' axis, up to 60°.
- Saturn's rings' complex and very heterogeneous structure.
- Complex and very heterogeneous structure ("Kirkwood gaps") in the asteroid belt between Mars and Jupiter's orbits.
- Hyperon's (a satellite of Saturn) very rapidly and strongly fluctuating speed of rotation about itself.
- This satellite's irregular shape, due to ejection along chaotic trajectories of fragments detached on collisions with meteorites (without chaos these fragments would remain close to the satellite and end up joining it restoring its initial spherical shape).
- Halley's comet's trajectory: a deviation of about 5 years between the date observed for one of its passages (-1403) and the date calculated by numerically "integrating back" the equations of motion.[14]
- Deviation $D(t)$ between possible future trajectories of Earth: taking into account the uncertainty of the current state, this deviation behaves as $D(t_0+\Delta) = 3D(t_0)$, with Δ of the order of 5 million years; for every 5 million years added to the time of prediction, the uncertainty in Earth's trajectory around the sun triples. The same exponential growth of uncertainties is true for the trajectory of Pluto, with Δ of the order of 20 million years. We deduce from this that the state of the solar system is totally unpredictable beyond 100 million years.

Observations show that the motion of celestial bodies cannot be considered as a model of regularity and perfection. So astronomy has the same limitations as other domains of physics: for *intrinsic* reasons the predictability of phenomena in the future is limited [46,51].

9.2.1.4 Billiards (Pool) and Lorentz Gas

Figure 4.8 showed the Lorentz gas model accounting for molecular chaos. The chaotic motion observed in a game of pool (billiards) has the same origin: defocusing collisions (illustrated in Fig. 9.7) reinforced by the confinement of trajectories in a bounded domain. There is a risk of confusion: here the trajectories are in real space (the table surface) however the example illustrates in a convincing way the

---

[13]It is not relativistic effects that explain this unpredictability. Nevertheless, consequences of relativistic effects, as well as influences of bodies outside of the solar system, can be amplified due to the chaotic nature of the motion and the associated sensitivity to perturbations.

[14]These are reversible so they can be used to "go back" in time.

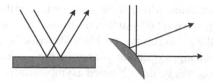

Fig. 9.7 Sensitivity to initial conditions. The phase space is here real space (the surface of a billiard or pool table) but the idea generalises to trajectories in an abstract phase space. *Left*: bouncing off a flat surface conserves the difference between the incident trajectories. *Right*: bouncing off a convex surface amplifies the difference between incident trajectories by a factor increasing with the curvature of the obstacle

concept of sensitivity to initial conditions which we will return to in more detail, quantitatively, for more abstract trajectories, in an arbitrary phase space.

Other examples and an accessible presentation of chaos and its range of applications can be found in [12, 23, 33, 48].

9.2.2 Statistical Description and Ergodicity

Due to the sensitivity and unpredictability of trajectories in chaotic dynamics, the only relevant description is a statistical one. Two points of view are then possible:

1. We could try to describe the *frequency of visits* to a given region of phase space $\mathscr{X}$. This is a priori[15] an observable quantity, obtained by constructing a histogram from the saved trajectory.
2. We could also describe the *existence probability* of the system in the phase space at a given instant. In mathematical terms this weighting of different regions of $\mathscr{X}$ is called a *measure* (on $\mathscr{X}$) [38]. If we are only interested in the steady state regime observed at long times, disregarding transient regimes, we will study the measures that are invariant with respect to the motion: m is invariant under the action of ϕ if for all parts A of $\mathscr{X}$, A and its reciprocal images $\phi_t^{-1}(A)$ have the same measure: $\forall t, m[\phi_t^{-1}(A)] = m(A)$. Such a measure is *adapted to the dynamics*, in the sense that the associated weighting does not change over time, and it will describe in a global manner, *one* steady state regime of the system. Generally several invariant measures exist. The problem for the

---

[15]This qualification is linked to the fact that the recorded signal $Z(t)$ is generally scalar, whereas the phase space $\mathscr{X}$ can be of dimension higher than 1. Therefore a procedure has to be used to reconstruct the trajectory $z(t) \in \mathscr{X}$ from which the signal is derived ($Z(t) = \phi[z(t)]$ where ϕ is the measure function). The underlying idea is that each variable is affected by the set of other variables and so contains information about the global dynamics of the system. The most common procedure used is the *method of delays*, where the trajectory is reconstructed in $n + 1$ components: $[z(t), z(t - \tau), \ldots, z(t - n\tau)]$. A discussion of this procedure, in particular of the choice of parameters τ and n is in [1, 21] and the original article [80].

physicist is then to determine which is *the* measure that will describe the observed asymptotic regime, given a set of initial conditions. The statistical description of the dynamics gives the answer: it is possible to describe how the initial existence probability (in $\mathscr{X}$) changes over the course of the evolution and the relevant invariant measure m_∞ will be that obtained asymptotically.

The link between these two points of view is formed by the *Birkhoff's ergodic theorem* [11]. This states[16] the equality of the temporal average and the statistical average with respect to the measure m_∞ in $\mathscr{X}$, on condition that this measure is invariant and ergodic with respect to the motion. This theorem therefore requires an additional property, *ergodicity* of the measure m_∞. We say that an invariant measure m_∞ is *ergodic*[17] (with respect to the flow ϕ_t) if each invariant set A (that is such that $\phi_t(A) \subset A$ for all t) is of measure zero ($m_\infty(A) = 0$) or full measure ($m_\infty(\mathscr{X} - A) = 0$) [39].

More qualitatively, ergodicity means that $\mathscr{X}$ cannot be decomposed into two invariant disjoint parts of strictly positive measure, in other words that *there exists no two sets of states which evolve separately, without ever communicating*. Almost all trajectories have the same temporal statistical properties, which can be obtained as ensemble averages with respect to the invariant measure m_∞. Inversely, knowing one typical trajectory is enough to reconstruct the invariant measure. Such a trajectory is therefore representative of any other typical trajectory.

9.2.3   The Basic Ingredients

In practice the details of chaos are quite different depending on whether the system is *conservative* or *dissipative*. The first case is the field of Hamiltonian systems, which can show chaotic behaviours if they are non integrable (for example $N \geq 3$ interacting bodies). The second case reduces to the study of the attractor of the system, called strange when the system is chaotic due to the complexity of its structure and the specific properties of the dynamics restricted to the attractor.

9.2.3.1   Baker's Transformation

One of the simplest models for understanding which are the essential ingredients of deterministic chaos is the *baker's transformation*, so called because it reproduces

---

[16]More precisely it is stated: For each trajectory starting from a point x_0 belonging to a subset of $\mathscr{X}_0$ of $\mathscr{X}$ of full measure (that is to say such that $m_\infty(\mathscr{X} - \mathscr{X}_0) = 0$) and for each observable F, we have: $\lim_{t \to \infty} \frac{1}{t} \int_0^t F(\phi_s(x_0))ds = \int_{\mathscr{X}} F(x)dm_\infty(x)$.

[17]We emphasize that ergodicity is a property of the pair formed by the motion ϕ_t and the invariant measure m_∞. Nevertheless, we often use the short hand "ergodic measure" or "ergodic motion" when there is no ambiguity over the partner.

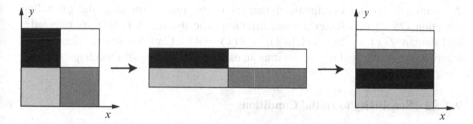

Fig. 9.8 Baker's transformation illustrating mechanisms underlying chaos. The dynamics dilates distances by a factor $b > 1$ along x, contracts them by a factor $a < 1$ along y, and "folds" the results. The dynamics conserves areas if $ab = 1$ (here $a = 1/2$ and $b = 2$). If $ab < 1$, the dynamics is dissipative and the attractor is fractal (Cantorian type) in the y direction and regular in the x direction

schematically the topological transformation undergone by bread dough when kneaded:

$$(x, y) \xrightarrow{B} \begin{cases} (2x, y/2) & \text{if } x \le 1/2 \\ (2x - 1, (y + 1)/2) & \text{if } x > 1/2. \end{cases} \tag{9.4}$$

It can be seen straight away that B is bijective within the square $[0, 1[\times [0, 1[$, which corresponds to reversible dynamics. The way in which it transforms the different regions of this square is depicted in Fig. 9.8. It has three unusual properties, which prove to be characteristics of all chaotic dynamical systems:

1. It has a dilating (unstable) direction, the x axis, along which the dynamics doubles the distances. It also has a contracting direction, the y axis, along which the dynamics decreases the distances by a factor of 2; this direction Oy becomes the dilating one if we reverse the direction of time.
2. It mixes points. Look, for example, at the evolution of points $M = (2^{-n} + \epsilon, 2^{-k} + \eta)$ and $N = (2^{-n} - \epsilon, 2^{-k} - \eta)$, for ϵ and η arbitrarily small. After a long enough time, $\log(1/\epsilon)$, independent of η, the trajectories of M and N are completely separated. This mixing property is also used in practice to mix granular materials.[18]
3. It has an infinite number of periodic orbits,[19] of arbitrarily long periods (and therefore on infinity of arbitrarily long characteristic times).

---

[18]A remarkable property of mixtures of different types of grains is that any vibration or rotation motion leads to segregation of the different types of grain. The homogeneity of the mixture cannot be improved by shaking, as we do for a suspension or emulsion. The idea is to include the grains to be mixed in a neutral paste, which is subjected to the baker's transformation. The paste is then removed or a paste that does not interfere with the subsequent use of the granular mixture is used.

[19]We can always write points (x, y) in the square $[0, 1] \times [0, 1]$ in the following form (dyadic expansion): $x = \sum_{n=0}^{\infty} 2^{-(n+1)} \sigma_n$ and $y = \sum_{n=1}^{\infty} 2^{-n} \sigma_{-n}$ where $\sigma_n = 0$ or 1. It can be shown that B acting on (x, y) ends up shifting the indices of the set $[\sigma]$. The set $[\sigma']$ associated with $B(x, y)$ is given by $\sigma'_n = \sigma_{n+1}$ (shift operator). It then follows that the points (x, y) for which $\sigma_{n+N} = \sigma_n$ for all integer ratios of n and N (arbitrarily fixed), will have a periodic trajectory of period N under the action of B.

A "minimal" model of chaotic dynamics is the projection onto the extending direction Ox of this Baker's transformation. The dynamics is thereby reduced[20] and written $f(x) = 2x$ (modulo 1), or $f(x) = \text{Frac}(2x)$ where Frac denotes the fractional part. We will often take it as an example in the rest of this chapter.

9.2.3.2 Sensitivity to Initial Conditions

The examples mentioned in Sect. 9.2.1 show that one of the reasons certain *deterministic* dynamics are unpredictable in the *long term* is a sensitivity to initial conditions, in other words, the dynamics exponentially amplify errors or perturbations. An uncertainty ϵ_0 in the initial state becomes equal to $\epsilon_t = \epsilon_0 e^{\gamma t}$ after a time t. The real positive exponent γ is a global characteristic of the dynamics called an *Lyapounov exponent*.[21] This exponent gives a characteristic evolution time: $1/\gamma$ is the time scale at which two initially neighbouring trajectories separate (their distance is multiplied by $e \approx 2.718$ every $\Delta t = 1/\gamma$) [53]. For example we mentioned above, without calling them that, the Lyapounov exponent $(\gamma = (\log 3)/\Delta)$ of the solar system. The Lyapounov exponents of the baker's transformation are $\log b > 0$ and $\log a < 0$. Let us also look again at the example of weather forecasting. Expected improvements in instrumentation for meteorological observations will change the precision with which we can measure the instantaneous state of the atmosphere by a factor $\alpha < 1$. Consequently this will increase the duration over which we can make reliable predictions. This duration becomes T_0' where $e^{\gamma T_0} = \alpha \, e^{\gamma T_0'}$, i.e. $T_0' - T_0 = \gamma^{-1} \log(1/\alpha)$ where γ is the maximum Lyapounov exponent of the atmospheric dynamics. For example, if the number of quadrangles of the earth's surface in which meteorological satellites measure atmospheric parameters is multiplied by 100, i.e. the resolution of initial conditions is improved by a factor of 10, the reliability of prediction will only be increased over a duration $(\log 10)/\gamma$.

We have just attributed the unpredictability to the imprecision in *perception* of the initial conditions. This appears to reflect only a weakness in our ability to detect the phenomenon in all its details and not to bring into question Laplace's old dream to "calculate the world". However this is forgetting that as well as the uncertainty in measuring the *finite* set of *chosen* observables to describe the state of the system, at each time step perturbations due to the influence of all the degrees of freedom not taken into account in the description are added. This "noise" is

---

[20]Note that this projection in the unstable direction is the only one leading to a reduced dynamics that is deterministic and closed. The projection operation nevertheless transforms the underlying reversible dynamics into an irreversible reduced dynamics [19].

[21]This relation $\epsilon_t = \epsilon_0 e^{\gamma t}$ is approximate: it is not valid at short times, due to the influence of the transient regime and the local characteristics of the dynamics; nor is it valid at long times if the phase space (or the attractor if there is one) is bounded which "folds" the trajectories. It cannot therefore be considered as the definition of a Lyapounov exponent, which is a global and asymptotic quantity, but only as a simple and intuitive interpretation of this quantity.

small but it is amplified by the chaotic dynamics and contributes significantly to the unpredictability. Fed by (at least) this source of uncertainty, chaotic dynamics is seen as an uncontrollable source of stochasticity.

In addition, if the system has several attractors, even their statistical (asymptotic) properties are unpredictable. In order to make any prediction, we would have to precisely know the attraction basin (points from which the trajectory asymptotically joins the attractor) of each attractor, and for each observation the basin to which the initial condition belongs.

Definition of Lyapounov exponents

First of all, let us imagine an evolution law f. For every point $x_0 \in \mathcal{X}$, we construct:

$$\gamma(f, x_0) = \liminf_{n \to \infty} \log\left(|(f^n)'(x_0)|^{\frac{1}{n}}\right) = \liminf_{n \to \infty} \frac{1}{n} \sum_{0 \le j < n} \log|f'(f^j(x_0))|.$$

$$(9.5)$$

The scale invariance $\gamma(f, x_0) = \gamma[f, f(x_0)]$ ensures that $\gamma(f, .)$ is m-almost everywhere constant, equal to $\gamma(f, m)$ for all invariant measures m that are ergodic under the action of f. Birkhoff's ergodic theorem (Sect 9.2.2) then proves the existence of the limit and gives its value:

$$\gamma(f, m) = \int_{\mathcal{X}} \log|f'(x)| dm(x).$$

$$(9.6)$$

We will have as many exponents as invariant ergodic measures under the action of f. In general, $\gamma(f, m)$ cannot be expressed simply from f, due to the fact that $\gamma(f, m)$ is not a characteristic of f but a *global* characteristic of the flow it generates. The dynamics is chaotic if $\gamma(f, m) > 0$.

For a continuous flow $\phi_t(x)$ in $\mathcal{X} \subset \mathbf{R}$, the Lyapounov exponent is defined by: $\gamma(\phi, m) = \lim_{T \to \infty} \log|\phi'_T(x)|$, for all $x \in \mathcal{X}_m$, where m is an invariant ergodic measure and $\mathcal{X}_m \subset \mathcal{X}$ is of full measure (i.e. $m(\mathcal{X} - \mathcal{X}_m) = 0$).

We briefly describe the generalisation to a discrete dynamical system of dimension $d > 1$. Osedelec's theorem ensures that $q \le d$ Lyapounov exponents $\gamma_1 \ge \ldots \ge \gamma_q$ exist[22] (in total d if we count the repeated ones), defined as the eigenvalues of the limit $\lim_{n \to \infty} [Df^n(x)^\dagger Df^n(x)]^{1/2n}$ where $\dagger$ indicates the transpose. This limit matrix is (in general) independent of x according to the ergodic theorem. The dynamics is chaotic if at least γ_1 is strictly positive. The situation can be described more precisely: for almost all

---

[22]We should add a technical condition requiring that the law of motion f is continuously differentiable and Hölder-continuous.

x, there exists a family $[E_i(x)]_{1 \leq i \leq q}$ of sub-spaces nested within each other $(E_{i+1} \subset E_i)$, such that [21, 36]:

$$\lim_{n \to \infty} \frac{1}{n} \log \|Df^n(x).u\| = \gamma_i \qquad \text{if} \quad u \in E_i(x) - E_{i+1}(x). \qquad (9.7)$$

E_1 is the space (of dimension d) that is tangent in x to the phase space. The take home message is that after n steps ($n \gg 1/\gamma_1$), the deviation initially equal to u (i.e. $x_0 - y_0 = u$) behaves *generically* as $e^{n\gamma_1}$ (i.e. $\|x_n - y_n\| \sim e^{n\gamma_1}$). It is only if $u \in E_2$ that we observe γ_2, and to observe the subsequent exponents more and more special initial deviations u are required. If we are only interested in generic behaviour, it is the maximal Lyapounov exponent γ_1, and only this one, which describes the property of sensitivity to the initial conditions.

9.2.3.3 Mixtures

The amplification of initial deviations and perturbations is not sufficient to generate chaos; a mixing mechanisms must be added, as we will see by comparing three discrete dynamical systems.

- The transformation $x \to 2x$ on $\mathbf{R}$ amplifies perturbations by a factor of 2 at each step, but there is no mechanism mixing the trajectories. The order of initial conditions is preserved at each subsequent time, as is the relative distances $(z_t - y_t)/(x_t - y_t) = (z_0 - y_0)/(x_0 - y_0)$. Therefore the dynamics remains perfectly predictable.
- On the other hand, the evolution generated by $x \to x + \alpha$ (modulus 1) has a mechanism of reinjection in $[0, 1]$ but does not amplify errors. Even though successive images of a point are mixed, the deviation between two trajectories remains unchanged over time: for all t, $x_t - y_t = x_0 - y_0$. This transformation, associated with a rotation of angle $2\pi\alpha$ on the unit circle, is therefore not chaotic.
- Finally let us consider the transformation $x \longrightarrow 2x$ (modulus 1), which can also be written $x \longrightarrow \text{Frac}(2x)$ where Frac means the fractional part. On the unit circle, this transformation corresponds to doubling the angle. It can also be seen as the projection of the baker's transformation on the expanding direction Ox. The evolution causes both a gain and a loss of information. On one hand, at each step, each trajectory merges with another since $f(x) = f(x + 1/2)$. On the other hand, at fixed resolution ϵ, neighbouring initial conditions are separated: x_0 and $x_0 + \epsilon$ are indistinguishable at $t = 0$, but are distinguishable at $t = 1$. The combination of these two opposing effects induces a mechanism mixing the trajectories and such resulting dynamics is chaotic.

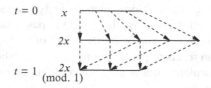

Fig. 9.9 Expansion by a factor of 2 and folding superimposing images of x and $x + 1/2$, which is the root of the chaotic nature of the dynamics generated by the transformation $x \rightarrow 2x$ (modulus 1)

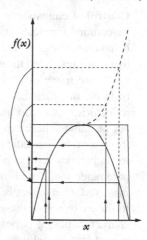

Fig. 9.10 Inhomogeneous expansion (by a factor $|a(1 - 2x)|$) and folding superimposing images of x and $1 - x$, at the root of the chaotic nature of the dynamics generated by the logistic map $ax(1 - x)$ for $a > a_c = 3.58$ (*thick line*)

To summarise, it is the combination of expansion in certain directions of phase space and a folding mechanism causing reinjection of trajectories in a bounded region of phase space which produces the mixing characteristic associated with chaotic dynamics and reconciles their deterministic nature and random features. This principle is illustrated in Figs. 9.9 and 9.10.

9.2.3.4 Strange Attractors and Unstable Periodic Orbits

In the case of dissipative systems, an attractor in which the dynamics has mixing properties and is sensitive to initial conditions is called a *strange attractor* [23] [65].

However to be comprehensive, we should mention a third ingredient necessary to obtain chaos: the dynamics must have an infinite number of *unstable periodic orbits*. They contribute to the chaotic nature by favouring the succession of regimes with very different characteristic times, the periods of trajectories within which being

---

[23]The exact definition of a strange attractor is a compact attractor containing a "homocline" trajectory (or orbit), that is to say a trajectory emerging from a point called a "homocline point" situated at the intersection of the stable manifold and the unstable manifold of a saddle type fixed point [36]. The dynamical complexity of such trajectories, in which it can be easily shown that all the points are homocline, had been already highlighted by Poincaré [23].

able to take arbitrarily large values. Consequently the dynamics has no characteristic scale: the smallest fluctuation can cause the system to pass from the neighbourhood of one unstable periodic orbit to that of another orbit of very different period. This explains why the system reacts to perturbations in an unpredictable way and at all time scales [66]. In particular, a strange attractor contains an infinite number of these unstable periodic trajectories.

Control of chaos:
In certain practical situations we might want to avoid having a chaotic behaviour. For dissipative systems in which the asymptotic dynamics is the dynamics restricted to an attractor, a method introduced by Grebogi et al. [35] under the name of "control of chaos" achieved this objective. Its principle rests on the fact that a strange attractor contains not only dense chaotic trajectories, but also an infinite number of closed trajectories, so periodic but unstable. These researchers showed that it is possible to stabilise any of these unstable periodic orbits, by applying a particular perturbation, adjusted to each time step (in practice, a succession of modifications of the control parameter, calculated from one time step to another). Once this technique is mastered, chaos, that is the existence of a strange attractor, is beneficial due to the existence of these periodic trajectories. Since they have very different characteristics, it is possible to choose the one having a desired shape and period and to stabilise it by the adapted perturbation. The control of chaos therefore not only enables a chaotic behaviour to be replaced by a periodic behaviour, but also to change the characteristics of the obtained periodic regime in a very flexible and fast way.

9.2.4   Transition to Chaos

A remarkable set of results shows that the onset of chaos occurs following *universal scenarios*. These results can be summarised by the fact that the transition from a regular behaviour to a chaotic one is produced following a well determined succession of *qualitative changes* (bifurcations in the case of dissipative systems) [9]. In this section we will present the transition to chaos in Hamiltonian systems (KAM theorem), Ruelle and Takens' scenario associated with the appearance of strange attractors and period doubling scenario. Intermittent behaviours and the temporal scaling laws they show will be described in Sect. 9.4.

9.2.4.1   Transition to Chaos in Hamiltonian Systems

Let us start with the case of conservative systems with Hamiltonian dynamics (therefore of even dimension $2n$). We start from a regular situation, with dynamics associated with an integrable Hamiltonian H_0, that is for which the system has n constants of motion $I_1 \ldots I_n$ ("first integrals"). Transferring to action-angle coordinates $(I_1, \ldots, I_n, \theta_1, \ldots, \theta_n)$ puts the equations of motion in the set $[\dot{I}_j = 0, \dot{\theta}_j = \omega_j(I_1, \ldots, I_n) = cte]$. This shows that in this case the motion is *quasiperiodic*: the trajectories are described by functions of the form $t \to \Phi(\omega_1 t + \varphi_1, \ldots, \omega_n t + \varphi_n)$. They lie on invariant tori of dimension n, parametrised by the invariants $I_1, \ldots, I_n$ and on which they "circulate" with angular velocities $\omega_1, \ldots, \omega_n$, in other words with periods $2\pi/\omega_1, \ldots, 2\pi/\omega_n$. We can study the appearance of more complex dynamics by adding a non integrable perturbation to the Hamiltonian H_0, i.e. $H = H_0 + \epsilon V$. A series of theorems by Kolmogorov, Arnold and Moser rigorously describe what we observe when ϵ is increased: the invariant tori will be deformed and little by little disappear, *in a well established universal order*, depending on the arithmetic properties of the ratios $\omega_1/\omega_n, \ldots, \omega_{n-1}/\omega_n$ but independent of the form of the perturbation V [36]. The tori for which these ratios are rational disappear as soon as $\epsilon > 0$. The more irrational[24] these ratio are, the more the associated invariant torus persists (although deformed) for large values of ϵ. In this way a remarkable link exists between the *dynamic* stability properties of the motion and the *arithmetic* properties of its eigen frequencies. If $n = 2$, the final invariant surface to disappear is that for which $\omega_1/\omega_2 = \sigma = (\sqrt{5}-1)/2$, one of the properties of the golden ratio σ being it is the "most irrational" real number [48]. The invariant tori behave like barriers that cannot pass over other trajectories and they will therefore partition the phase space. Their progressive disappearance is accompanied by the appearance of complex trajectories, which are able to erratically explore larger and larger regions of phase space and are characterised by a positive Lyapounov exponent. For small ϵ, these chaotic regions are very localised because they are trapped between the invariant surfaces on which the movement remains quasiperiodic. These regions spread as ϵ increases, which increases the chaotic nature of the observed dynamics since the trajectories are less and less localised. The region becomes totally chaotic once the final invariant torus has disappeared. This type of "Hamiltonian chaos" is that observed in the solar system. Even though it has the same sensitivity to initial conditions and mixing properties, it is very different from the chaos observed in dissipative systems from the point of view of its localisation and geometry in phase space. In dissipative systems, describing the asymptotic regime boils down to describing the dynamics restricted to the attractor and it is in the geometry of this attractor that the chaotic nature of the motion is reflected. In contrast, *the concept of*

---

[24]The *degree of irrationality* of a real number $r \in [0, 1]$ can be quantified by studying its rational approximations. For each integer q, we denote by $p_{q,r}/q$ the best approximation of r by a rational number of denominator q. We can then define the subsets $\mathcal{F}_\alpha$ of $[0, 1]$, containing the real numbers r such that $|r - p_{q,r}/q| \leq q^{-\alpha}$ for an infinite number of integers q. It can be shown (Dirichlet's theorem) that $\mathcal{F}_2 = [0, 1]$ and that for all $\alpha > 2$, $\mathcal{F}_\alpha$ is a fractal of dimension $2/\alpha$ (Jarnik's theorem). These sets are nested: $\mathcal{F}_{\alpha_2} \subset \mathcal{F}_{\alpha_1}$ if $\alpha_1 < \alpha_2$. The larger α, the more the elements of $\mathcal{F}_\alpha$ are "well approximated" by the rational numbers. By determining which sets $\mathcal{F}_\alpha$ the ratios $\omega_1/\omega_n, \ldots, \omega_{n-1}/\omega_n$ belong to, we determine the order in which the associated invariant tori will disappear [24].

an attractor does not exist in conservative systems and the invariant measure remains the natural volume of phase space. The picture to remember for the transition to chaos in conservative systems is that of a *growth of regions of phase space* where trajectories have chaotic behaviour.

9.2.4.2 Ruelle and Takens' Scenario

This scenario concerns dissipative dynamical systems and it describes the conditions under which a strange attractor appears. Ruelle and Takens showed that a general mechanism is a succession of three Hopf bifurcations with incommensurable frequencies (irrational ratios of each other). At the first bifurcation the system passes from an equilibrium state to a limit cycle, at the second a quasiperiodic regime appears at two frequencies and after the third the asymptotic regime is *in general*[25] a strange attractor. More precisely, Ruelle and Takens' result reveals that a quasiperiodic regime with three incommensurable frequencies is not a generic situation in the sense that the smallest perturbation destabilises it. In contrast, the existence of a strange attractor is a robust (structurally stable) property in the sense that it is not destroyed by adding an additional term, in other words by the introduction of a new influence in the law of motion [63, 77]. We retain that it is sufficient that a system passes three Hopf bifurcations when its control parameter is increased for it to *be able* to have chaotic behaviour associated with a strange attractor. This scenario can be shown experimentally in experiments of convection in confined geometries [55] or in the Belousov–Zhabotinski reaction [4]. Characterisation can be made by a spectral analysis of the phenomenon: the first Hopf bifurcation is seen by a peak in ω_1 (and smaller peaks corresponding to the harmonics); the second Hopf bifurcation is manifested by the appearance of a peak in ω_0 (and the harmonics) and the third Hopf bifurcation, leading to chaos, will be seen by the transformation of the line spectrum to a chaotic spectrum, which is a broad band spectrum stretching to the lowest frequencies and no longer showing any significant peaks.

This scenario, even though not quantitative, was a major conceptual breakthrough because it shattered the picture we had of turbulent regimes. The old view, due to Landau, consisted of the destabilisation of infinitely many modes such that the motion looked erratic and was impossible to predict on the long term. Consequently it was thought that chaos could only appear in systems with an infinite number of degrees of freedom. Ruelle and Takens' scenario shattered this dogma by showing

---

[25]The mathematical difficulty of the theorem is to define the term "in general" which requires considering a space of dynamical systems and endow it with a topology [20]. The structural stability of a quasiperiodic motion $t \rightarrow \Phi(\omega_1 t + \varphi_1, \ldots, \omega_n t + \varphi_n)$ with n periods depends on the perturbations considered. There is a structural instability from $n = 3$ if the perturbations are only constrained to be continually differentiable twice (class $\mathscr{C}^2$), whereas the instability appears from $n = 4$, if only perturbations that are infinitely differentiable are allowed (class $\mathscr{C}^\infty$). The quasiperiodic regime *can* therefore be replaced by a strange attractor.

that the nonlinear coupling of three incommensurable modes is sufficient to obtain such behaviour [77]. Deterministic chaos, in this way, was proposed by these authors as a possible explanation of "weakly turbulent" regimes observed just after destabilisation of the laminar regime. Today we talk of *chaos* when the system is of low dimension (or the essential dynamics brings the system to low dimension) and of *turbulence* (fully developed) when the system has a large number of fundamental unstable modes (Sect 9.5).

> **Chaos, three dimensions and Ruelle and Takens' theorem**
> Ruelle and Takens' result is often formulated incorrectly by invoking the dimension of phase space d. Its exact (but simplified) statement is: in a dissipative system, 3 (incommensurable) unstable *modes* generically lead to chaos (i.e. existence of a strange attractor). The example of the Lorenz attractor shows *in addition* that we can observe chaos in a phase space of *dimension* $d = 3$. On the other hand, in a phase space of dimension $d = 2$ the whole dynamics is predictable, because each trajectory behaves as an impenetrable barrier for the others. A dimension of $d \geq 3$ is therefore a necessary condition for observing chaos, but this condition has nothing to do with Ruelle and Takens' result. No condition in the dimension d is required for discrete dynamical systems: the logistical map shows chaos from $d = 1$. This is not incoherent if we remember that a discrete dynamical system is typically obtained by Poincaré section of a continuous dynamical system in a larger dimension. The order in which successive images of $x \rightarrow 1 - \mu x^2$ ($\mu > \mu_c$) are placed on $[-1, 1]$ shows that they cannot come from a planar trajectory (since it would intersect itself, which is prohibited) and hence any associated continuous dynamical systems have a dimension at least equal to three.

9.2.4.3 Period Doubling Scenario

This period doubling scenario, also called subharmonic cascade, is the most impressive due to its universality and therefore its predictive power. It is widely treated in the literature, so we will mention just the main ideas [48, 55, 66].

In this scenario, the transition from a situation in which the physical system is stabilised in an equilibrium state (stable fixed point) to a situation in which the asymptotic regime is chaotic is made by a succession of *"period doublings"* (period doubling bifurcations). As we increase the control parameter μ, at $\mu = \mu_0$ the stable fixed point will give way to a cycle with a certain period T (the fixed point still exists for $\mu > \mu_0$ but it is unstable). Then, at $\mu = \mu_1$, this cycle will in its turn destabilise and be replaced by another stable cycle of period $2T$. The sequence therefore follows in this way: the stable cycle of period $2^{j-1}T$ observed for $\mu < \mu_j$ will destabilise at $\mu = \mu_j$ and simultaneously a stable cycle of double the period, $2^j T$, will appear which becomes the attractor for $\mu > \mu_j$. The first point to note is the fact that the sequence of period doublings continues indefinitely. The monotonously increasing sequence of bifurcation points $(\mu_j)_{j \geq 0}$ tends to a limit μ_c. At $\mu = \mu_c$, the onset of a chaotic regime is observed [58]. The second point, even more remarkable, is the *universality* of this scenario. The values of the bifurcation points $(\mu_j)_{j \geq 0}$ are specific to the system considered, as well as their limit μ_c. However, the accumulation of these values up to μ_c follows a geometric progression:

$$\lim_{j \to \infty} \frac{\mu_{j+1} - \mu_j}{\mu_{j+2} - \mu_{j+1}} = \delta \quad \Longleftrightarrow \quad \mu_c - \mu_j \sim \delta^{-j}$$

where δ is a *universal number*: $\delta = 4.66920\ldots$ [25]. This indicates that it is identical for all systems in which this accumulation of period doublings leading to chaos is observed. Here a *qualitative* analogy between the behaviours involves a *quantitative* analogy. This scenario is particularly popular for the following reasons:

- **It is easily observed numerically**, for example in the discrete dynamical system of evolution law $f_\mu(x) = 1 - \mu x^2$ or the equivalent system $g_a(x) = ax(1-x)$ [44]. The corresponding bifurcation diagram is represented in Fig. 9.11.
- **It is also called** *subharmonic cascade* because in frequency space, each bifurcation corresponds to the appearance of a peak at half frequency (subharmonic). It is also the simplest and most reliable experimental criterion to implement. For $\mu_j < \mu < \mu_{j+1}$, the spectrum is composed of peaks at $\omega = \omega_0$, $\omega_1 = \omega_0/2, \ldots, \omega_j = \omega_0/2^j$. At $\mu = \mu_{j+1}$, spectral lines $\omega_{j+1} = \omega_j/2$

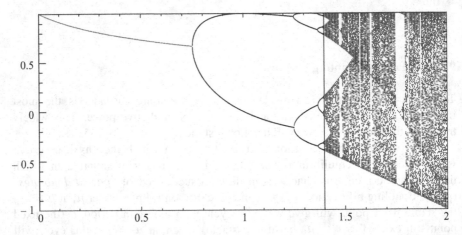

Fig. 9.11 Bifurcation diagram of the logistic map $f(x) = 1 - \mu x^2$. The parameter μ is placed on the abscissa and the attractor on the ordinate. We clearly see the accumulation of period doublings leading to chaos at $\mu_c \approx 1.4011550$. This structure is often called "Feigenbaum tree"

reflecting the period doubling of the asymptotic regime appear. The spectrum observed at the chaos threshold, at $\mu = \mu_c$, is a broad spectrum reflecting the random nature of chaotic dynamics.

- **It is observed in many experimental systems** (for example in Rayleigh–Bénard [50] or the Belousov–Zhabotinski reaction [47]) and measurements tend to confirm the universality of the scenario and the value of δ.
- **It is completely described and understood analytically.** The characterisation by a parameter $(f_\mu)_\mu$ of the families of dynamical systems which will show an accumulation of period doublings is known and the universality of the exponent δ can be *shown* by an renormalisation method and its value *calculated*. These families form the *universality class* of the scenario [15, 83]. Other universality class can be shown, determined by the regularity of the family at the critical point $x = 0$: if the behaviour is $|x|^{1+\epsilon}$ in the vicinity of 0 (and the family satisfies the conditions necessary for the observation of an accumulation of period doublings), an exponent δ_ϵ is observed, which we know how to determine perturbatively for small enough ϵ.

If we replace the spatial variables by the temporal variable, we can see a fruitful analogy between the transition to chaos and critical phase transitions, leading to the application of renormalisation methods to access envisaged scale invariance properties [48].

9.2.5 Range and Limits of the Concept of Chaos

9.2.5.1 Abundance of Chaotic Behaviours

The ingredients of chaos are present in all dynamics in which the underlying action is composed of an stretching in certain direction and a folding (Figs. 9.9 and 9.10). This qualitative argument, completed in the dissipative case by the structural stability of strange attractors demonstrated by Ruelle and Takens (Sect. 9.2.4), explains the omnipresence of chaotic behaviours. In addition to the physical examples that we have mentioned in this section, we can cite other examples in the domain of biology [49]:

- A historical example is that formed by logistic maps and resulting more refined models in population dynamics [58].
- The normal heart rate is chaotic and the disappearance of chaos is pathological. More generally, this led to the introduction of the concept of "dynamical disease" [54].
- Enzyme reactions and resulting biochemical oscillations can show chaotic components (e.g. glycolytic oscillations, calcium oscillations) [34].

Note however that chaotic behaviour may be less common in reality than in models. Actually, while chaos is robust to weak deterministic perturbations, it is more sensitive to stochastic perturbations. Noise, by destroying the structure of flow in phase space, destroys the chaotic properties.

9.2.5.2 Analysis of Chaotic Signals

Possibly the most fruitful consequence of chaos theory is the set of "nonlinear signal analysis" methods, exploiting the concepts introduced to describe chaos, in order to obtain quantitative information about observed dynamics [42]. The indices resulting from chaos theory (for example Lyapounov exponents or the fractal dimension of the attractor) reveal changes in the underlying dynamics. They are for example used in the analysis of electrocardiograms (ECGs) and electroencephalograms (EEGs) [5].

However, exploiting temporal data with the help of this nonlinear analysis to *interpret* the phenomenon and its origin, beyond a simple quantitative diagnosis, is more problematic. Implementing deterministic chaos in a real phenomenon is actually a sensitive question, often seeming to be a "ill posed problem" and it is difficult to give a clear response. For example, the question of discriminating between chaos and stochastic dynamics has the prerequisite of determining the scale at which the motion is to be considered: a deterministic chaos model and a stochastic model could easily coexist at different levels of description. For a deterministic description, we must be able to ensure that a low dimensional model is acceptable. To do so, we have to extract information about the *global* dynamics, of unknown dimension, from a temporal recording, most often scalar, which we call a *reconstruction* of the dynamics. Finally, we need a rigorous statistical analysis to estimate the reliability and accuracy of the values determined for the different chaos indices considered. We refer the reader to [1, 21] and [41] for the methodological aspects and to [72,73] for a discussion of the precautions to take in using the concept of chaos.

9.2.5.3 Chaos: An Internal Source of Randomness

A final, more conceptual, conclusion is that chaos provides an internal source of stochasticity. A deterministic motion can generate an identical trajectory, in terms of statistical properties, as that due to a stochastic process, but the resulting random behaviour is contained in the law of motion itself. In this way stochastic behaviour is taken into account without the need to invoke an external cause. Chaos involves a fast temporal decorrelation along each trajectory, which leads to the use of a statistical description in terms of invariant measure. In modern terms, this is the idea that encompasses Boltzmann's hypothesis of molecular chaos, which we will develop below in Sect. 9.3.

9.3 Chaos as the Foundation of Statistical Mechanics

9.3.1 Boltzmann's Ergodic Hypothesis

Even though the concepts of ergodicity and chaos were formulated long after Boltzmann's work, they nevertheless play an essential role in the approach he used, and afterwards that of Gibbs, to build the foundations of statistical mechanics. Remember that the aim of statistical mechanics is to constructively connect theories and knowledge available at about the mechanisms at microscopic scales (usually molecular), and macroscopic phenomena, that is the behaviour of the same system observed at much larger scales. The idea that we will develop here is the following: *chaos present at the microscopic scale leads to reproducible, regular and robust behaviours observed at our scale*. We refer the reader to the work of Dorfman [19] and [13] for a deeper discussion.

For example, let us imagine a system of particles of low enough density to be in the gaseous state.[26] We can get a qualitative idea of its microscopic dynamics by considering a gas of hard spheres in a box, subjected to elastic collisions between each other and the walls of the container. As shown schematically in Fig. 4.8, such dynamics is very sensitive to initial conditions and to perturbations and the presence of walls makes the dynamics mixing [44]. The chaotic nature,[27] whose experimental reality is known today [31], insures a fast decorrelation: the molecular environments encountered by a particle at times t and $t + dt$ can be effectively considered independent. It is this property of decorrelation that was introduced by Boltzmann under the name of *molecular chaos*. He then used it to justify a more directly exploitable mathematical hypothesis, the *ergodic hypothesis*.

In the case of an *isolated* system, with fixed total energy E, this ergodic hypothesis boils down to assuming that all the microscopic configurations of the system of energy E will be visited with the same frequency over the microscopic evolution. This particular case is also called the *microcanonical hypothesis*. It is then generalised to systems in thermal equilibrium. The reasoning rests on the fact that the combination of the system and the thermostat is isolated, validating the microcanonical ergodic hypothesis for this combination. We then exploit the fact that the evolution of the state of the system affects the thermostat only infinitesimally, by definition of a thermostat. Therefore the ergodic hypothesis is

---

[26]The arguments remain qualitatively valid for simple liquids but their technical implementation will be different, since the approximations allowed by the low density of gases could no longer be made.

[27]Note however that it behaves with a type of chaos a bit different from that presented in Sect. 9.2: the number of degrees of freedom is very large, whereas a characteristic of deterministic chaos is that it takes place in systems in low dimension. However this characteristic is not exclusive and molecular chaos seems to be an example of spatiotemporal extension of low dimensional chaos, involving the same ingredients: sensitivity to initial conditions, mixing and the existence of periodic trajectories of all periods [30].

reformulated by saying that the configurations $[s]$ of the system will be visited with a frequency equal to their probability at equilibrium $P([s]) \sim e^{-\beta E([s])}$ where $\beta = 1/kT$ (Boltzmann distribution), which identifies temporal averages (averages along the evolution of a configuration of the system) with statistical averages (averages over all possible instantaneous configurations of the system, weighted by the probability distribution at equilibrium).[28]

In this way Boltzmann's hypothesis seems to be an ergodicity hypothesis of the microscopic dynamics governing the evolution of the system with respect to its invariant measure (natural volume for a conservative system, Boltzmann distribution for a system at thermal equilibrium). This precise mathematical formulation of a qualitative, and even sometimes approximate, property[29] makes it functional. In what now appears to be a simple application of Birkhoff's ergodic theorem (Sect. 9.2.2), it enables the observable quantities to be expressed in the form of statistical averages and the relationships between these quantities to be deduced. The implicit idea is that these relationships are *robust* and remain valid beyond the restricted framework in which they were obtained, in other words, even if the ergodic hypothesis is not exactly valid.

9.3.2 Chaotic Hypothesis and Nonequilibrium Statistical Mechanics

An approach was recently proposed by Cohen, Gallavotti and Ruelle as a foundation of a statistical mechanics of systems far from equilibrium[30] [29, 74–76]. Called *chaotic hypothesis*, it can be summarised as follows:

---

[28]Another way of formulating the same point involves the concept of *statistical ensemble*: the temporal average is equal to the average over a large number of independent systems, constructed identically to the original system. This formulation, introduced by Gibbs, corresponds exactly to the actual concept of statistical sampling. In practice, what we call a "statistical ensemble" is an ensemble of microscopic configurations weighted by a probability distribution such that the statistical averages $\langle A \rangle$ correctly estimate the observed quantities A_{obs} in the system under consideration. The situations presented above correspond to the microcanonical ensemble and the canonical ensemble respectively.

[29]This hypothesis is rarely explicitly justified for two reasons: it leads to theoretical results in agreement with experiments and it is not known how to prove this ergodicity in general.

[30]The term "out of equilibrium" is ambiguous and we should distinguish:

- Systems *relaxing* towards their equilibrium state, sometimes slowly and in a complex way if metastable states exist.
- Systems that have reached a *steady* state that is out of equilibrium, in the sense that there non zero flows (of material, energy etc) across them. We use the term "far from equilibrium" for these systems to distinguish them from the previous case.

1. There exists fundamental mechanisms, still not well understood or even ignored, that produce mixing properties at the microscopic scale (chaotic nature of microscopic motion) *and* an irreversibility at the macroscopic scale.
2. A class of mathematical models, of *hyperbolic* dynamical systems, possesses these mixing properties. By definition (in discrete time for simplicity), these dynamical systems possess an invariant compact set, such that at each point the stable and unstable directions are transverse, depending continuously on the point. In addition, in these models, the rates of contraction (in the stable directions) are bounded above by $a < 1$ and the rates of expansion (in the unstable directions) are bounded below by $b > 1$. The nonequilibrium steady state is described by an invariant measure having special properties, in particular of mixing due to the hyperbolicity of the dynamics (*SRB measures*, from the names Sinai, Ruelle and Bowen).
3. We know how to describe the asymptotic behaviour of this ideal model, in particular quantifying its macroscopic irreversibility, for example the rate of entropy production, as a function of the indices quantifying its chaotic properties (typically the Lyapounov exponents).
4. We suppose that these expressions describing the macroscopic irreversibility as a function of the microscopic chaos are *universal*, robust and that they reflect the fundamental mechanism (1) and not the particular models (2), which they go far beyond. The class of models introduced in (2) is only a mathematical intermediary having the advantage of proposing an illustrative case where calculations are possible. This approach therefore seems to be the nonequilibrium analogue of Boltzmann's ergodic hypothesis.

It should be noted that we have asymptotic irreversibility and dissipation (reflected for example in the production of entropy) in both temporal directions. For $t \to +\infty$, it is the unstable directions and associated Lyapounov exponents $\gamma \geq \log b > 0$ that control the dynamics. For $t \to -\infty$, it is the stable directions and associated Lyapounov exponents $\gamma \leq \log a < 0$ that determine the dominant behaviour, which we immediately see after a time reversal $t \to -t$. However, it must be noted that irreversibility persists after this reversal (see e.g. the discussion in [13]).

A simplistic example, that nevertheless clarifies the fundamental ideas, is that of the baker's transformation B. The long term behaviour is dominated by what happens in the unstable direction Ox since the distances along Oy are contracted by a factor of 2 at each time step. The component along Ox is therefore interpreted as a macroscopic observable of the system. If this *reversible* evolution generated by B is projected on the unstable direction Ox, the obtained dynamics, associated with the transformation $x \to 2x$ (modulus 1) becomes *irreversible*: at each time step, the trajectory merges with another (given x_n, there are therefore 2^n possible starting points x_0). In reverse, the evolution is described by B^{-1} and it is its projection on Oy that becomes irreversible.

The irreversibility associated with hyperbolic evolutions is expressed in the same way: over time information about the initial state of the system along the stable directions becomes more and more inaccessible at macroscopic scales (due to the contraction of distances in these directions) and the only knowledge, that of the unstable component, is not enough to return to the initial state.

9.3.3   Chaos and Transport Phenomena

In a uniting of chaos theory and nonequilibrium statistical mechanics, current research seeks to obtain empirical transport laws and associated coefficients (for example coefficient of diffusion, thermal and electrical conductivity) from simplified but realistic deterministic microscopic models [14]. We have already mentioned a deterministic model, in Sect. 4.4.1, called *Lorentz gas*, in which a particle moves at constant velocity in a lattice of obstacles with which it is subjected to elastic collisions (Fig. 4.8). Analysis of this model fixes diffusion in the equations of motion of the molecules (we can therefore relate the coefficient of diffusion to the chaotic characteristics of molecular motion) and reconciles explicitly its determinism and reversibility with the irreversibility and stochasticity of diffusion [30]. Another example is that of a one dimensional chain of anharmonic oscillators coupled nonlinearly and in contact at its extremities with two thermostats at different temperatures. In this case we recover Fourier's law and associated linear temperature profile, explaining the origin of the familiar irreversibility of this system at our scale [13, 19, 22].

   In these examples, we see the reappearance of real space, absent in the theory of (low dimensional) dynamical systems presented in this chapter. The justification for this absence is that in the situations considered, for example in the vicinity of a bifurcation point, the evolution is dominated by the dynamics of a few modes, the others playing no role asymptotically [37]. The reduction is not always valid and we must therefore turn to the theory of extended dynamical systems and clarify the concept of spatiotemporal chaos.

9.4   Intermittency

Dynamic behaviours covered by the term *intermittency* are characterised by brief phases of intense and irregular activity alternating randomly with rest phases of varying length, also called "laminar phases". Different mechanisms are invoked according to the context and the statistical properties of the intermittent signal observed. Far from entering into a review of the different models of dynamic intermittency (for that we recommend [20] or [9]), we will limit ourselves to describing what connects intermittency to this book, namely the *existence of temporal scaling laws describing the laminar phases*.

9.4.1 Intermittency After a Saddle-Node Bifurcation

First identified theoretically by Manneville and Pomeau [56] then experimentally in Rayleigh–Bénard convection [8], one mechanism is associated with *saddle-node bifurcations* (Sect. 9.3.1, Fig. 9.3). To simplify the expressions, we will consider the case of an evolution g_ν in discrete time, typically obtained by Poincaré section of a continuous flow (Fig. 9.2) and dependent on a control parameter ν. A necessary condition for observing a saddle-node bifurcation when ν passes a certain value ν_c is that g_{ν_c} has a fixed point X^* where the stability matrix $D_x g_{\nu_c}(X^*)$ has one eigenvalue equal to 1, all the other eigenvalues being of modulus strictly less than 1. We will not specify the other two, more technical, necessary conditions,[31] but we will highlight that they only involve strict inequalities, which therefore remain satisfied if we slightly change the transformation g_ν. The saddle-node bifurcation is therefore a *generic* bifurcation and the associated intermittency mechanism will be *robust* (it will not be destroyed by a small perturbation of g_ν) and consequently *universal* (it will not depend on the details of g_ν, provided that the conditions for bifurcation are satisfied).

A significant simplification in the analysis is made by reducing the transformation g_ν to the normal form of the bifurcation (Sect. 9.1.4) [37]: by conjugacy,[32] transformation of the control parameter ν (which becomes $\mu = \mu(\nu)$) and then projection on the direction that becomes unstable at the bifurcation point, it is possible to reduce a family of transformations $(g_\nu)_\nu$ satisfying the above conditions[33] to the family $f_\mu(x) = -\mu + x - Ax^2$, in one dimension, $A > 0$ being a fixed parameter.

Plotting the graph of the transformation $x \to -\mu + x - Ax^2$ gives an intuitive understanding of the origin of the intermittent behaviour (Fig. 9.12): beyond the bifurcation point ($\mu > \mu_c$), there is no longer a fixed point but the dynamics remains slowed down in the vicinity of x^*. It can be shown that the average duration of laminar phases varies as $\tau(\mu) \sim \mu^{-1/2}$. This result, obtained by renormalisation, guarantees its universality. The idea is to consider the renormalised transformation $Rf(x) = \lambda^{-1} f \circ f(\lambda x)$ where λ is adjusted to make Rf as close to f as possible, here by putting the monomial x coefficient as 1 in $Rf(x)$. The exact proof follows by determining the fixed points of R and the eigenvalues of the linearised operator in the vicinity of these fixed points, following the standard procedure presented in Chap. 3 [48]. The argument is well understood for the typical family $(-\mu + x - Ax^2)_\mu$: the optimal choice is $\lambda = 1/2$, leading to the approximate relation $Rf_\mu \sim f_{4\mu}$. Moreover, by construction of R, the number of steps taken in the "ditch" separating the graph of f_μ from the bisector, in other words the

---

[31] If w is the linear form associated with the eigenvalue 1 of the stability matrix, it is necessary that $w[D_\nu g_{\nu_c}(X^*)] > 0$ and $w[D_x^2 g_{\nu_c}(X^*)] > 0$.

[32] A conjugacy is the replacement of g_ν by $\phi \circ g_\nu \circ \phi^{-1}$ and x by $\phi(x)$ where ϕ is an adequate diffeomorphism.

[33] It is even the key to the demonstration of bifurcation theory.

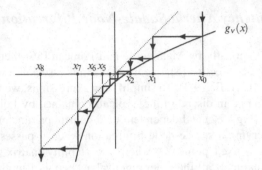

Fig. 9.12 Graph (*bold line*) of a map g_ν parametrised by ν and conjugated to the normal form $f_\mu(x) = -\mu + x - x^2$. The figure explains how slow and monotonous the discrete evolution $x_{n+1} = g_\nu(x_n)$ is if ν is slightly above a value ν_c associated with a saddle-note bifurcation at a fixed point x^* (here placed at the origin) characterised by $g'_{\nu_c}(x^*) = +1$

duration of the laminar phase (see Fig. latexfig), obeys $N(Rf_\mu) = N(f_\mu)/2$ since renormalisation amounts to taking the steps two at a time. From this we extract:

$$N(f_\mu) \sim \tau(\mu) \sim \mu^{-1/2}. \qquad (9.8)$$

In conclusion, all dynamics with a saddle-node bifurcation with respect to one of the parameters will have a behaviour not only qualitatively but also quantitatively (same exponent $-1/2$) identical to the family $(-\mu + x - Ax^2)_\mu$. As we anticipated, intermittent behaviour is universal,[34] and the family $(-\mu + x - Ax^2)_\mu$ is seen as the typical representative of the associated universality class. Note that we can say nothing about the chaotic interludes because the ingredients of the chaotic dynamics, in particular the mechanisms of mixing and reinjection in the bounded region containing the fixed point X^*, are no longer described correctly after the *local* reduction to the normal form.

9.4.2 On-Off Intermittency

A second intermittency situation is encountered for example[35] in ecology, when considering the population dynamics of a species called "resident" in competition

---

[34]The exponent $-1/2$ is associated with the quadratic form of the evolution law. The family $(-\mu + x - Ax^{1+\epsilon})_\mu$ has a similar behaviour, with in this case $\tau(\mu) \sim \mu^{-\epsilon/(1+\epsilon)}$. It represents another universality class (where $\epsilon = 1$ corresponds to the class of the classic saddle-node bifurcation).

[35]On-off intermittency is also encountered in a system of two coupled chaotic oscillators passing in an irregular manner from a synchronous motion, where the difference $x = X_1 - X_2$ between the states of the two oscillators cancels out, to an asynchronous motion where $x \neq 0$ [28,68]. Here we observe self-similarity of the signal $x(t)$ and the behaviour $P(\tau) \sim \tau^{-3/2}$ of the distribution

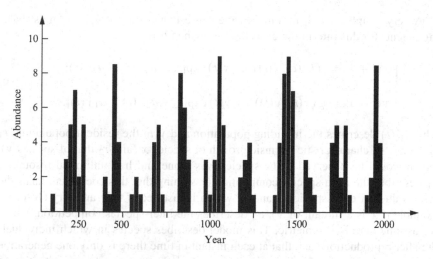

Fig. 9.13 Randomness and scale invariance of alternating phases of rarity and abundance of Pacific sardines *Sardinops sagax* (after [26]), from analysis of the density of scales in stratified marine sediments. A similar histogram is observed over a shorter duration (data from trade during the last century) or a longer duration (throughout the Holocene, roughly 12,000 years)

with an invading closely related species (a mutant form for example). Sometimes for long periods the population of the intruding species remains at very low levels, close to extinction. These "scarce phases" are punctuated by periods, of irregular duration and frequency, where this species is abundant. This phenomenon is observed, at least qualitatively, for many very different species, for example fish,[36] insects and viruses (epidemics). It is important from an ecological point of view because it shows that the rarity of a species is not necessarily followed by its extinction and can on the contrary form part of its survival strategy. We can retrospectively observe this type of dynamics in marine sediments by measuring the density of scales of a species of sardine along a sample core, that is to say over time (Fig. 9.13) Early studies showed that this spontaneous intermittency of rarity did not result from a variable change imposed by the environment but rather was the outcome of the population dynamics of the ecosystem and competition between species. The intermittent rarity is observed experimentally, on different time scales (century, millennium, era). We are particularly interested in the scale invariance shown by the experimental data. The probability distribution of the duration τ of phases of rarity is a power law: $P(\tau) \sim \tau^{-a}$ with $a \approx 3/2$. To specify the value of this exponent, study its possible universality properties and understand the origin of scale invariance, it was necessary to resort to modelling.

---

of the duration τ of the synchronous phases (state "off" $x = 0$, analogue of the scarce phases) characterising this type of intermittency.

[36] As for the historical example of the Lotka–Volterra model, observation of fish in particular could be made for "practical" reasons: fishing by trawling provides representative sampling and accurate data are available over long periods of time from auction sale records.

A very simple model, minimal in the sense that it only keeps the most basic ingredients for this intermittency, is the following [26]:

$$\begin{cases} x(t+1) = f(x(t), y(t)) \equiv x(t)\exp[r_1 - a_1 x(t) - a_2 y(t)] \\ \\ y(t+1) = g(x(t), y(t)) \equiv y(t)\exp[r_2 - a_2 x(t) - a_1 y(t)], \end{cases} \tag{9.9}$$

where $x(t)$ describes the intruding population and $y(t)$ the resident population. r_1 is the rate of characteristic intrinsic growth of species x (and r_2 that of species y), which would be observed if the species was alone and had unlimited resources; a_1 describes the intra-species competition (assuming that this coefficient takes the same value for each species x and y, which is justified if they are very close, for example if x is a mutant of y); a_2 describes the inter-species competition, which we assume here is symmetric. This model describes species in which individuals die after reproduction such that at each instant in time there is only one generation present, as is the case for insects, for example.[37] Let us start from a situation in which x is very weak (very few invaders, resident species by far the majority). We study the stability of this asymmetric state using a linear analysis, i.e. we study the dynamics of the population x keeping only the lowest order terms. In other words, we adopt a perturbative approach with respect to the dynamics $y_0(t)$ of the resident species alone, which is written $y_0(t+1) = g(0, y_0(t))$. The stability index is then the *invasion exponent*:

$$\chi = \lim_{T \to \infty} \frac{1}{T} \sum_{t=0}^{T-1} \log |\partial f / \partial x(0, y_0(t))|. \tag{9.10}$$

It consists of the temporal average of the instantaneous invasion exponents:

$$\chi(t) = \log |\partial f / \partial x(0, y_0(t))|,$$

describing the "ease" with which the intruding species can expand (starting from the state in which it is absent). This quantity χ is therefore a global index, analogue to a Lyapounov exponent[38] (Sect 9.2.3).

---

[37]To take into account the case where individuals survive after reproduction, the model can be modified by introducing survival rates s_1 and s_2 of the two species:

$$\begin{cases} x(t+1) = x(t)(s_1 + \exp[r_1 - a_1 x(t) - a_2 y(t)]) \\ y(t+1) = y(t)(s_2 + \exp[r_2 - a_2 x(t) - a_1 y(t)]). \end{cases}$$

The results obtained with this second model prove to be qualitatively and even quantitatively (same exponent $-3/2$) identical to those presented for the first model.

[38]It consists exactly of a *transverse Lyapounov exponent* describing the stability within the global attractor of the solution $y_0(t)$ corresponding to the resident species alone with respect to a transverse perturbation.

Studying the dynamics shows that this invasion exponent χ provides an invasibility criterion. If $\chi < 0$, the intrusive species cannot establish itself. If $\chi > 0$ large enough, it rapidly overpowers the resident species (and linear analysis soon becomes meaningless). Intermittent rarity is produced when y has random dynamics, due for example to chaotic dynamics, with $\chi > 0$ very small. The instantaneous invasion exponent, fluctuating around its average weakly positive value, will alternately take negative values, during which time the intruding population cannot expand (scarce phases), and positive values, during which time the exponential growth of x means the intruding population will reach a high level. This will then oscillate between phases of quasi-extinction and phases of prosperity, in a manner essentially governed by the dynamics of the resident species. In the limit where $\chi \to 0^+$, a scaling law appears in the statistics of the duration τ of the quasi-extinction phases. Its distribution $P(\tau)$ behaves as: $P(\tau) \sim \tau^{-3/2}$ where the exponent $-3/2$ is characteristic of this mechanism of intermittency, called *on-off intermittency* [40]. This exponent is in particular independent of the threshold chosen to define the phases of rarity and the duration over which the phenomenon is observed, which simply reflects the scale invariance of the phenomenon. From the moment when the phenomenon is manifested, it is universal, in the sense that it does not depend on the parameters of the dynamics nor the ecosystem considered. The existence of a scaling law signifies that the dynamics does not have a characteristic time scale and that we have universal critical dynamics. This is confirmed experimentally by the similarity (after normalisation) of data recorded over very different periods of time: century, millennium, era.

One correction to the power law predicted for $P(\tau)$ comes from noise due to the external environment. During rarity phases, this could accidentally induce extinction of the intrusive population (the weak value of its level makes it more vulnerable to environmental fluctuations), with a larger probability the longer the duration of the rarity phase. We therefore see a truncation of the power law at large values of τ.

As we move away from the critical point $\chi = 0^+$, that is to say as χ increases, the critical character disappears and we recover an exponential decay:

$$P(\tau) \sim \tau^{-3/2}\, e^{-\tau/\tau_0} \qquad \text{with} \qquad \tau_0 \sim \frac{1}{\chi^2} < \infty. \qquad (9.11)$$

We also recover the usual deviations from critical behaviour when the size of the ecosystem is finite. The model where the variables x and y take continuous values is only a valid approximation in the limit of an infinite medium; when the size L of the system is taken into account (the exact definition depends on the way in which limitations due to the finite size of the environment and populations will change the continuous dynamical model), it can be shown that τ_0 follows a scaling law $\tau_0 \sim L^a$.

9.5   Fully Developed Turbulence

A very complex but nevertheless familiar spatiotemporal behaviour is hydrodynamic or atmospheric turbulence. To have an intuitive idea of the phenomenon and its richness, just look at the movement of water in a river in flood around the pillars of a bridge. This phenomenon captures our attention here because it shows remarkable spatiotemporal scaling properties. They can be compared to dynamical critical phenomena (relaxation or response properties at the critical point, Chap. 4, or certain growth phenomena, Chap. 8). However we will see that the scale invariance here is much more complex, involving a *continuum of critical exponents*, which we will talk of as *multifractal* invariance [60].

9.5.1   Scale Invariance of Hydrodynamic Equations

The spatiotemporal behaviour of a fluid is described by the Navier–Stokes equation, governing the dynamics of its velocity field $v(r, t)$:

$$\partial_t v + (v.\nabla)v = -\rho^{-1} \nabla P + \nu \Delta v + g, \qquad (9.12)$$

where ν is the kinetic viscosity of the fluid (in m²/s), ρ its density, P its pressure and g the resultant of the external accelerations (force field per unit mass). The equation must be satisfied at every point in the domain accessible to the fluid or delimited by the observer (for example banks, obstacles and limits of the observed section in the case of a river). It is therefore completed by *boundary conditions* describing the behaviour of the velocity field at the limits of this domain. To this the equation for conservation of fluid is also added ($\partial_t \rho + \nabla(\rho v) = 0$), which takes the following form, of the "incompressibility equation", when the density ρ can be considered constant:

$$\nabla \cdot v = 0. \qquad (9.13)$$

These two (vector) equations are invariant by the scaling transformation with two parameters $\Lambda > 0$ and $\lambda > 0$:

$$
\begin{aligned}
r &\to \lambda r & P &\to \lambda^{2+d} \Lambda^{-2} P \\
t &\to \Lambda t & \rho &\to \lambda^{-d} \rho \\
v &\to \lambda \Lambda^{-1} v & g &\to \lambda \Lambda^{-2} g \\
\nu &\to \lambda^2 \Lambda^{-1} \nu.
\end{aligned}
\qquad (9.14)
$$

Let us introduce the typical dimension L of the system an the typical velocity V of the fluid, usually prescribed by the boundary conditions, for example the river flow or the speed of blades stirring the fluid. The choice $\lambda = 1/L$ and $\Lambda = V/L$ leads to dimensionless variables. This reduction shows that the quantity $\lambda^2 \Lambda^{-1} \nu = \nu/VL$

is the only real parameter of the problem. This introduces a dimensionless number, the *Reynolds number*:

$$Re = \frac{LV}{\nu}. \tag{9.15}$$

So the dynamic behaviour of the fluid will be controlled by the value of Re following to almost universal criteria, only depending on the system via its geometry and boundary conditions (value of velocity field at the walls or at other boundaries of the system). This scale invariance is the principle behind wind tunnel testing, i.e. to experimentally test the behaviour of a system of size L when it is subjected to winds of velocity V, it is sufficient to study a smaller system of size L/k but keeping the same Reynolds number (and of course a similar geometry). This condition is realised by imposing reduced velocities kV on the system.

9.5.2 Turbulence Threshold

We have just seen that the Reynolds number is the only parameter controlling fluid dynamics. In addition, this number can be interpreted as the ratio of the nonlinear amplification term $(v.\nabla)v$, which tends to destabilise the movement amplifying local arbitrarily small fluctuations in velocity, over the viscous dissipation term $\nu \Delta v$, which tends to dampen these fluctuations. In other words, this number expresses the ratio between the kinetic energy of the fluid and the energy dissipated by viscous friction. From this we can intuitively understand what is experimentally confirmed: turbulence appears in a fluid flow above an *instability threshold* Re^*. At small Reynolds numbers ($Re \ll Re^*$), the nonlinearities play a negligible role and we observe a *laminar* regime, with all the symmetries of the Navier–Stokes equation. As Re increases, the nonlinear amplification of fluctuations in the velocity field dominate more and more over their damping and the fluid develops instabilities which deviate qualitatively from the laminar regime. More specifically, a succession of (bifurcation) thresholds is shown experimentally. On passing each threshold, a new symmetry of the Navier–Stokes equation is broken such that the velocity field no longer has this symmetry. Above the final threshold, the nonlinearities totally dominate the fluid behaviour. The exact value Re^* of this turbulence threshold depends on the geometry of the problem, the boundary conditions and the definition we take for the beginning of turbulence but it always stays around the same order of magnitude, of 100. So the Reynolds number quantifies the degree of turbulence of the system, admittedly roughly but on an absolute scale. The expression for Re shows that three factors control the behaviour of the fluid: the turbulent regime will be reached faster the smaller the viscosity, the larger the dimensions of the system or the higher the average macroscopic velocity of the fluid. These last two are seen for example in a river that is turbulent when in flood but laminar normally. Finally let us highlight that turbulence only takes place permanently in *open* systems that are given an energy supply. For example, the turbulent motion created by the movement of blades in a tank is quickly dampened by viscous dissipation if we stop stirring the fluid.

We will not describe further the sequence of events[39] marking the transition from a laminar to a turbulent regime (for this see for example [55] and [27]). Because it shows scaling properties, we are only interested in the "fully developed turbulence" regime observed way above the threshold Re^*, typically for $Re > 10\ Re^* \approx 1,000$ (for example, Re reaches about 10^{10} in the atmosphere). We should highlight, unlike deterministic chaos presented in Sect. 9.2 and sometimes called *weak turbulence*, fully developed turbulence (or *strong turbulence*) brings into play a large number of coupled unstable modes. Furthermore, only a statistical description, namely that of the fluid velocity field, will be meaningful. We will study average quantities, denoted $\langle\ \rangle$, in particular the static structure functions (or factors), defined as the moments of the variation in the longitudinal velocity:

$$S_p(l) \equiv \langle \delta v(l, r, u, t)^p \rangle \tag{9.16}$$

where

$$\delta v(l, r, u, t) = [v(r + lu, t) - v(r, t)].u \qquad (u \text{ unit vector}) \tag{9.17}$$

and the dynamic structure functions (or factors):

$$\Sigma_p(l, \tau) \equiv \langle [\delta v(l, r, u, t + \tau)\delta v(l, r, u, t)]^{p/2} \rangle. \tag{9.18}$$

These functions are the equivalents of the static and dynamic correlation functions introduced to describe critical phenomena and we will see that they show analogous scaling behaviours.

9.5.3 A Qualitative Picture: Richardson's Cascade

The first fundamental step was to relate turbulence to the transfer of kinetic energy across a vast range of scales. Energy is introduced at a large scale L. The fluid then develops eddies at all scales, as a cascade, each one feeding relative[40] eddies of smaller scale, until the scale l^* where viscous dissipation is effective.

---

[39] Such a sequence leading first to chaos then to fully developed turbulence is also observed, in an exemplary way, in Rayleigh–Bénard's experiment described in Fig. 9.1. However the phenomenon is different because in this case the density of the fluid varies as a function of its temperature. The dimensionless number controlling the fluid dynamics is no longer the Reynolds number but the Rayleigh number, involving the vertical gradient of temperature imposed on the system.

[40] Note that it consists of *relative* eddies, whose motion is described with respect to eddies of larger size in which they are embedded.

More specifically, we can define local Reynolds numbers. The number characterising an eddy of size l and relative velocity v (with respect to the motion of the fluid as a whole at a larger scale) will be $Re_{loc} = lv/\nu$. When this number is large compared to 1, the energy dissipated by viscous friction is negligible and the eddy kinetic energy will fuel relative eddies at smaller scales. On the other hand, when this number reaches values close to 1, the eddy becomes a dissipative structure, degrading the coherent kinetic energy in molecular mechanisms. l^* is the typical size of the first (largest) dissipative eddies.

This picture is called *Richardson's cascade* [70]. This cascade can be widely observed, for example in a river in flood where we see small eddies superimposed on swirling movements on a larger scale. Turbulence therefore comes from the necessity to reconcile the energy injection mechanism at the macroscopic scale (e.g. movement of blades or injection of fluid at a given velocity) and the energy dissipation mechanism at the molecular scale. The larger Re, the greater the difference between the scale L at which energy is injected and the scale l^* of viscous dissipation, so the more room for many interlocked levels of organisation. The constant energy transfer between successive levels, before dissipation effectively comes into play, suggests that the Richardson cascade is self-similar in the domain $l^* \leq l \leq L$. Experimental results supporting this hypothesis are the subject of the next section.

9.5.4 Empirical Scaling Laws

Experimental studies of fully developed turbulence have led to three *empirical* laws, valid a very large Reynolds number ($Re \gg Re^*$) and confirming quantitatively the qualitative self-similarity of Richardson's cascade:

- **If we let the viscosity ν tend to** 0, all other characteristics of the system (L, V, geometry, boundary conditions) remaining unchanged, the energy ε dissipated per unit mass and time tends to a finite value:

$$\varepsilon \sim \frac{V^3}{L} \qquad (\nu \to 0). \qquad (9.19)$$

- **The quadratic mean** $S_2(r, l, u, t) \equiv \langle \delta v(l, r, u, t)^2 \rangle$ is independent of the point r, the unit vector u and time t if the turbulence is homogeneous, isotropic and stationary; so we call it simply $S_2(l)$ or $\langle \delta v(l)^2 \rangle$. In a range of scales $l^* \leq l \leq L$ called the *inertial domain*, it obeys the scaling law (*2/3 law*):

$$S_2(l) \equiv \langle \delta v(l)^2 \rangle \sim l^{2/3}. \qquad (9.20)$$

This law is replaced by $\langle \delta v(l)^2 \rangle \sim l^2$ at very small scales ($l < l^*$), in agreement with the regularity of the velocity field. However the behaviour $l^{2/3}$ observed in the inertial domain shows that the spatial derivative of the velocity field is not uniformly bounded, otherwise we would have $\langle \delta v(l)^2 \rangle \sim l^2$ at all scales. Therefore this law reflects the existence of anomalies in the velocity field in the turbulent regime, numerous and large enough with respect to the motion of the fluid to affect the behaviour of the average scale. Note that the value l^* marking the transition between the two scaling behaviours of $\langle \delta v(l)^2 \rangle$, is at this stage empirical. We will see below its physical interpretation and how it can be related to L and Re.

• **We often describe the turbulent regime** by its *power spectrum $E(k)$*. It is a function of the modulus of the wavevector k such that $E(k)dk$ is equal to the energy of the modes with wavevector modulus dk around k [81]. Therefore, by definition, we have:

$$\int_0^\infty E(k)dk = \frac{1}{2} \langle v^2 \rangle. \tag{9.21}$$

It can then be shown[41] that if $E(k)$ follows a scaling law $E(k) \sim k^{-\alpha}$, then $\langle |v(r + lu) - v(r)|^2 \rangle \sim l^{\alpha-1}$ (in the statistically stationary, homogeneous and isotropic regime). Experimentally we observe the scaling law:

$$E(k) \sim k^{-5/3} \qquad\qquad (2\pi/L \le k \le 2\pi/l^*). \tag{9.22}$$

This law, the equivalent to the 2/3 law in conjugate space, is only observed in a certain window of wave vectors. Its upper limit is the size of the system and lower limit the viscous dissipation taking place at small spatial scales.

The scale invariance of the Navier–Stokes equation is actually quite trivial. We have seen in Sect. 9.5.1 that it arises from a simple dimensional analysis. The exponents obtained are all as natural as that for example in the "scaling law" $\mathcal{V} \sim a^3$ relating the volume $\mathcal{V}$ of a cube to the length a of its sides and expressing its "self-similarity". The scale invariance that appears in the context of turbulence and is expressed in particular in the laws $S_2(l) \sim l^{2/3}$ and $E(k) \sim k^{-5/3}$, is less trivial. It reflects the complex spatiotemporal organisation of a turbulent flow. It is to the invariance of the Navier–Stokes equation what the scale invariance of a fractal is to the cube we have just mentioned.

9.5.5 Kolmogorov's Theory (1941)

The first quantitative analysis, due to Kolmogorov, is today a standard theory of scaling, based on three *hypotheses* of scale invariance [43].

---

[41]An intermediate result, the *Wiener–Khinchine theorem*, relates the spectrum $E(k)$ to the spatial correlation function of the velocity field according to the following formula: $E(k) = \int_0^\infty kr \, \sin(kr) \, \langle v(r + r_0)v(r_0) \rangle \, dr/\pi$ where $\langle v(r + r_0)v(r_0) \rangle$ only depends on r through the statistical isotropy and homogeneity of turbulence.

- **The first hypothesis** is that the *transfer of energy is constant* all along the cascade. This hypothesis, coming from experiments (Sect. 9.5.3), is justified by the fact that the energy starts to dissipate only at small scales l^* where the viscous dissipation becomes effective. The quantity ε of energy per unit mass and time given by the eddies of scale l_i to the set of those at the smaller scale l_{i+1} is therefore independent of i. The cascade stops at eddies of size l^* and relative velocity v^* giving a Reynolds number around 1, i.e. $l^*v^*/v \approx 1$. The energy of these eddies is totally dissipated by viscosity and is therefore no longer available to fuel movements on a smaller scale. We obtain:

$$l^* \sim L(Re)^{-3/4}. \qquad (9.23)$$

- **The second hypothesis** is to assume that in the inertial domain $l^* \leq l \leq L$, statistical properties of turbulence are stationary, homogeneous, isotropic, independent of injection of the material or energy creating the turbulence (scale L) and the viscosity (which just fixes the lower bound l^*). We therefore assume that the symmetries of the Navier–Stokes equation, broken when Re crosses the turbulence threshold, are restored but *in a statistical sense* at high values of Re. Under this hypothesis, taking the statistical average $\langle \; \rangle$ eliminates the dependence on r, u and t, so the moments $S_p(l)$ only depend on l.
- **The third hypothesis** is to assume a universal form for $\langle \delta v(l)^2 \rangle$, which, by dimensional analysis and the fact that ε is constant, is written:

$$\langle \delta v(l)^2 \rangle \sim \varepsilon^{2/3} l^{2/3}. \qquad (9.24)$$

In the same way *Kolmogorov's law* is obtained:

$$E(k) \sim \varepsilon^{2/3} k^{-5/3}. \qquad (9.25)$$

The theory also predicts that:

$$S_3(l) \equiv \langle (\delta v(l)^3) \rangle = -\frac{4}{5} \varepsilon l \quad \text{et} \quad S_p(l) \equiv \langle (\delta v(l))^p \rangle \sim (\varepsilon l)^{p/3}. \qquad (9.26)$$

9.5.6 Multifractal Analysis

More advanced experimental studies effectively demonstrate scaling laws $S_p(l) \sim l^{z_p}$, but the dependence of z_p on p appearing here is nonlinear, contradicting the prediction $z_p = p/3$ of Kolmogorov's theory. The difference comes from the fact that turbulence is far from being homogeneous and isotropic, which limits the range of validity of Kolmogorov's theory and makes it not very useful to finely characterise or control a turbulent regime. It gives a global framework and illuminates the fundamental principles at work in fully developed turbulence but it is not sufficient to account for all the complexity of the phenomenon. In fact, local rare but acute singularities can have a crucial influence on the behaviour of the fluid. We will see that they are enough to induce anormal scaling laws $S_p(l) \sim l^{z_p}$ where $z_p \neq p/3$. A local singularity of the velocity field can be described by a local exponent $\delta v(r, l) \sim l^{\alpha(r)}$ (at small values $l \ll L$). More rigorously, it should be written:

$$\delta v(r, \lambda l, u) \sim \lambda^{\alpha(r)} \, \delta v(r, l, u), \tag{9.27}$$

where the symbol $\sim$ means equal distributions: both terms have the same moments and the same joint distributions (i.e. at several times and components). Kolmogorov's theory corresponds to the perfectly homogeneous case where $\alpha(r) \equiv 1/3$.

The first extension is a bifractal mode where $\alpha(r) = \alpha_1$ on a set of fractal dimension f_1 and $\alpha(r) = \alpha_2 < \alpha_1$ on a set of fractal dimension $f_2 < f_1$. The behaviours of the structure functions $S_p(l)$ result from the superposition of two scaling laws with exponents α_1 and α_2 respectively. It is found that $S_p(l) \sim l^{z_p}$ where $z_p = \inf(p\alpha_1 + 3 - f_1, p\alpha_2 + 3 - f_2)$. This model therefore predicts two different exponents depending on the order p of the structure function S_p. The interpretation of this is that, depending on the order p, it is not the same set of singularities that dominates and controls the behaviour of S_p. Consequently, the profile of $p \to z_p$ has a discontinuity in the gradient (a crossover) passing from a line of slope α_1 at small values of p to a line of slope α_2 at large values of p. However, when this model is compared with experimental data, we observe that this graph is actually a concave curve and not the expected broken line, let alone the 1/3 slope predicted by Kolmogorov's theory.

This nonlinear dependence on p of the exponent z_p of the structure functions $S_p(l)$ reflects the existence of a continuum of exponents $\alpha(r)$, which leads to a generalisation of the bifractal model to a multifractal model [7]. A method inspired by the fractal geometry, *multifractal analysis*, provides both a theoretical framework to describe the spatial distribution of exponents $\alpha(r)$ and singularities of the associated velocity field and a way of experimentally accessing characteristics of this distribution.

What is a multifractal structure? In short, it is a "doubly fractal" structure. It has, first of all, local singularities described by a local exponent $\alpha(r)$. In the case we are considering the exponent describes the local singularity of the velocity field. Then, the place of points r where the exponent $\alpha(r)$ takes a given value α is itself a fractal structure, of fractal dimension $f(\alpha)$. The curve $\alpha \to f(\alpha)$ is called the *multifractal spectrum* of the velocity field. Typically, the smaller α the stronger the associated singularities and the more likely they are to have significant consequences, but also the rarer the points where this value α is observed, or in quantitative terms, the lower $f(\alpha)$. By substituting these ingredients into the expression for the moments and using the saddle method to evaluate the integral $S_p(l) \sim \int l^{\alpha p} l^{3-f(\alpha)} d\alpha$ obtained, we obtain[42] the behaviour at small values of l [27]:

$$S_p(l) \sim l^{z_p} \qquad \text{with} \qquad z_p = \inf_\alpha[p\alpha + 3 - f(\alpha)]. \qquad (9.28)$$

9.5.7 An Open Field Today

Turbulence is a classic example of complex spatiotemporal dynamics. As well as that presented here of an incompressible fluid violently agitated or arriving at an obstacle at high velocity, it appears in many other systems for example, atmospheric turbulence (Lorenz model Fig. 9.6 describes the "weak" version of the strong turbulence described here) or the Rayleigh–Bénard experiment, to mention just two examples already encountered in this chapter. More ingredients, and therefore more parameters (notably temperature) could come into play, but the qualitative ideas remain the same.

Understanding fully developed turbulence is a arduous problem because it involves a large number of space and time scales. The phenomenon must be addressed globally, giving due weight to singular events that are spatially localised (as we did in Sect. 9.5.6) but also transient (loss of statistical stationarity). This *intermittence* of turbulence (not to be confused with intermittency described in Sect. 9.4) at small scales will break the scale invariance observed in the inertial domain and consequently invalidate all the models and tools that rely on this invariance, in particular Kolmogorov's theory [27].

The problem also occurs in obtaining and processing experimental data since it is difficult to design sensors covering the whole range of the phenomenon without disturbing it. We usually adopt Taylor's hypothesis according to which instantaneous spatial averages (and static averages) coincide with temporal averages calculated from the recorded signal. To eliminate noise the analysis methods used are generally spectral methods. It not so easy to extract information on local and transient spatiotemporal structures. New tools, such as wavelet transform have been developed for this (Sect. 11.3.2) [62]. Methods developed for turbulence, for example multifractal analysis, are already widespread. No doubt it will be possible to establish fruitful analogies and applications, in terms of concepts as well as tools, in other contexts.

---

[42]To be more rigorous, multifractal analysis should be done on the cumulative probability distribution: $\text{Prob}[\delta v(r, l, u) > l^\alpha]$. We can also develop a multifractal analysis of the dissipation $\epsilon_l(r)$, as it happens related to that of the velocity field because $\delta v(l) \sim (l\epsilon_l)^{1/3}$, so that $\epsilon_l(r) \sim l^{3\alpha(r)-1}$ [27].

References

1. H.D.I. Abarbanel, *Analysis of Observed Chaotic Data* (Springer, New York, 1996)
2. V.I. Arnold *Ordinary Differential Equations* (MIT Press, Cambridge MA, 1978)
3. V.I. Arnold *Geometrical Methods in the Theory of Ordinary Differential Equations* (Springer, Berlin, 1983)
4. F. Argoul, A. Arneodo, J.C. Roux, From quasiperiodicity to chaos in the Belousov–Zhabotinskii reaction. I. Experiment. J. Chem. Phys. **86**, 3325–3338 (1987)
5. A. Babloyantz, A. Destexhe, Low-dimensional chaos in an instance of epilepsy. Proc. Natl. Acad. Sci. USA **83**, 3513–3517 (1986)
6. R. Badii, A. Politi, *Complexity: Hierarchical Structures and Scaling in Physics* (Cambridge University Press, Cambridge, 1997)
7. R. Benzi, G. Paladin, G. Parisi, A. Vulpiani, On the multifractal nature of fully developed turbulence and chaotic systems. J. Phys. A **17**, 3521 (1984)
8. P. Bergé, M. Dubois, P. Manneville, Y. Pomeau, Intermittency in Rayleigh–Bénard convection. J. Physique-Lettres **41**, L341-L344 (1980)
9. P. Bergé, Y. Pomeau, C. Vidal, *Order Within Chaos* (Wiley-Interscience, New York, 1987)
10. G.D. Birkhoff, *Dynamical Systems* (AMS, Providence, 1927)
11. G.D. Birkhoff, Proof of the ergodic theorem. Proc. Natl. Acad. Sci. USA **17**, 656–660 (1931)
12. R. Bradbury, A sound of thunder, in *R is for Rocket* (Doubleday, New York, 1952)
13. P. Castiglione, M. Falcioni, A. Lesne, A. Vulpiani, *Chaos and Coarse-Graining in Statistical Mechanics* (Cambridge University Press, Cambridge, 2008)
14. E.G.D. Cohen, Transport coefficients and Lyapunov exponents. Physica A **213**, 293 (1995)
15. P. Collet, J.-P. Eckmann, *Iterated Maps of the Interval as Dynamical Systems* (Birkhäuser, Boston, 1980)
16. M.C. Cross, P.C. Hohenberg, Pattern formation outside of equilibrium. Rev. Mod. Phys. **65**, 851 (1993)
17. M. Demazure, *Bifurcations and Catastrophes* (Springer, Berlin, 1999)
18. R.J. Donnelly, Taylor-Couette flow: The early days. Phys. Today, 32–39 (1991)
19. J.R. Dorfman, *An Introduction to Chaos in Non Equilibrium Statistical Mechanics* (Cambridge University Press, Cambridge, 1999)
20. J.P. Eckmann, Roads to turbulence in dissipative dynamical systems. Rev. Mod. Phys. **53**, 643 (1981)
21. J.P. Eckmann, D. Ruelle, Ergodic theory of chaos and strange attractors. Rev. Mod. Phys. **57**, 617 (1985)
22. J.P. Eckmann, C.A. Pillet, L. Rey-Bellet, Non-equilibrium statistical mechanics of anharmonic chains coupled to two heat baths at different temperatures. Commun. Math. Phys. **201**, 657 (1999)
23. I. Ekeland, *Le calcul, l'imprévu*, Points Science (Le Seuil, Paris, 1984)
24. K. Falconer, *Fractal Geometry* (Wiley, New York, 1990)
25. M. Feigenbaum, Quantitative universality for a class of nonlinear transformations. J. Stat. Phys. **19**, 25 (1978)
26. R. Ferrière, B. Cazelles, Universal power laws govern intermittent rarity in communities of interacting species. Ecology **80**, 1505 (1999)
27. U. Frisch, *Turbulence* (Cambridge University Press, 1996)
28. H. Fujisaka, T. Yamada, A new intermittency in coupled dynamical systems. Prog. Theor. Phys. **74**, 918 (1985)
29. G. Gallavotti, E.G.D. Cohen, Dynamical ensembles in nonequilibrium statistical mechanics. Phys. Rev. Lett. **74**, 2694 (1995)
30. P. Gaspard, *Chaos, Scattering Theory and Statistical Mechanics* (Cambridge University Press, 1998)
31. P. Gaspard, M.E. Briggs, M.K. Francis, J.V. Sengers, R.W. Gammon, J.R. Dorfman, R.V. Calabrese, Experimental evidence for microscopic chaos. Nature **394**, 865 (1998)

32. A.V. Getling, *Rayleigh–Bénard Convection* (World Scientific, Singapore, 1998)
33. J. Gleick, *Chaos: Making a new science* (Penguin, New York, 1988)
34. A. Goldbeter, *Biochemical Oscillations and Cellular Rhythms; The Molecular Bases of Periodic and Chaotic Behavior* (Cambridge University Press, 1996)
35. C. Grebogi, E. Ott, J.A. Yorke, Crises, sudden changes in chaotic attractors and transient chaos. Physica D **7**, 181 (1983)
36. J. Guckenheimer, P. Holmes, *Nonlinear Oscillations, Dynamical Systems and Bifurcations of Vector Fields* (Springer, Berlin, 1983)
37. H. Haken, *Advanced Synergetics* (Springer, Berlin, 1983)
38. P. Halmos, *Measure Theory* (Chelsea, New York, 1958)
39. P. Halmos, *Lectures on Ergodic Theory* (Chelsea, New York, 1959)
40. J.F. Heagy, N. Platt, S.M. Hammel, Characterization of on-off intermittency. Phys. Rev. E **49**, 1140 (1994)
41. R.C. Hilborn, *Chaos and Nonlinear Dynamics* (Oxford University Press, Oxford, 1994)
42. H. Kantz, T. Schreiber, *Nonlinear Time Series Analysis* (Cambridge University Press, Cambridge, 1997)
43. A.N. Kolmogorov, The local structure of turbulence in incompressible viscous fluid for very large Reynolds numbers. C.R. Acad. Sci. USSR **30**, 301. English translation in *Turbulence classic papers on statistical theory*, ed. by S.K. Friedlander, L. Topper (Interscience, New York, 1961)
44. H.J. Korsch, H.J. Jodl, *Chaos: A Program Collection for the PC* (Springer, Berlin, Heidelberg, 1998)
45. H. Krivine, A. Lesne Mathematical puzzle in the analysis of a low-pitched filter. Am. J. Phys. **71**, 31 (2003)
46. J. Laskar, Large-scale chaos in the solar system. Astron. Astrophys. **287**, L9 (1994)
47. H. Lemarchand, C. Vidal, *La réaction créatrice: dynamique des systèmes chimiques* (Hermann, Paris, 1988)
48. A. Lesne, *Renormalization Methods* (Wiley, Chichester, 1998)
49. A. Lesne, Chaos in biology. Biol. Forum **99**, 413 (2006)
50. A. Libchaber, J. Maurer, Une expérience de Rayleigh–Bénard de géométrie réduite; multiplication, accrochage et démultiplication de fréquences. J. Phys. Coll. (Paris) **41**, C3-51 (1980)
51. J.J. Lissauer, Chaotic motion in the solar system. Rev. Mod. Phys. **71**, 835 (1999)
52. E.N. Lorenz, Deterministic non periodic flow. J. Atmospheric Science, **20**, 130 (1963)
53. A.M. Lyapounov, Sur la masse liquide homogène donnée d'un mouvement de rotation. Zap. Acad. Nauk, St. Petersbourg **1**, 1 (1906)
54. M.C. Mackey, L. Glass, Oscillation and chaos in physiological control systems. Science **197**, 287 (1977)
55. P. Manneville, *Dissipative Structures and Weak Turbulence* (Academic Press, New York, 1990)
56. P. Manneville, Y. Pomeau, Intermittency and the Lorenz model. Phys. Lett., **75A**, 1 (1979)
57. P. Martien, S.C. Pope, P.L. Scott, R.S. Shaw, The chaotic behavior of the leaky faucet. Phys. Lett. A **110**, 339 (1985)
58. R.M. May, Simple mathematical models with very complicated dynamics. Nature **261**, 554 (1976)
59. J. Milnor, On the concept of attractor. Commun. Math. Phys. **99**, 177 (1985)
60. D. Mitra, R. Pandit, Dynamic multiscaling in fluid turbulence: an overview. Physica A **318**, 179 (2003)
61. J.D. Murray, *Mathematical Biology*, 3rd edn. (Springer, Berlin, 2002)
62. J.F. Muzy, E. Bacry, A. Arneodo, Wavelets and multifractal formalism for singular signals: Application to turbulence data. Phys. Rev. Lett. **16**, 3515 (1991)
63. S. Newhouse, D. Ruelle, F. Takens, Occurrence of strange axiom-A attractors near quasi-periodic flow on $\mathbf{T}^m$, $m \geq 3$. Commun. Math. Phys. **64**, 35 (1978)
64. C. Nicolis, Chaotic dynamics, Markov processes and climate predictability. Tellus series A, **42**, 401 (1990)

65. E. Ott, Strange attractors and chaotic motions of dynamical systems. Rev. Mod. Phys. **53**, 655 (1981)
66. H.O. Peitgen, H. Jürgens, D. Saupe, *Chaos and Fractals* (Springer, Berlin, 1992)
67. R.D. Pinto, W.M. Gonalves, J.C. Sartorelli, M.J. Oliveira, Hopf bifurcation in a leaky faucet experiment. Phys. Rev. E **52**, 6896 (1995)
68. N. Platt, E.A. Spiegel, C. Tresser, On-off intermittency: a mechanism for bursting. Phys. Rev. Lett. **70**, 279 (1993)
69. H. Poincaré, *Les méthodes nouvelles de la mécanique céleste* (Gauthiers-Villars, Paris, 1892)
70. L.F. Richardson, *Weather Prediction by Numerical Process* (Cambridge University Press, Cambridge, 1922)
71. D. Ruelle, *Elements of Differentiable Dynamics and Bifurcation Theory* (Academic Press, New York, 1989)
72. D. Ruelle, Deterministic chaos: the science and the fiction. Proc. Roy. Soc. Lond. A **427**, 241 (1990)
73. D. Ruelle, *Chance and Chaos* (Penguin, London, 1993)
74. D. Ruelle, Positivity of entropy production in nonequilibrium statistical mechanics. J. Stat. Phys. **85**, 1 (1996)
75. D. Ruelle, Differentiation of SRB states. Commun. Math. Phys. **187**, 227 (1997)
76. D. Ruelle, Gaps and new ideas in our understanding of nonequilibrium. *Physica A* **263**, 540 (1999)
77. D. Ruelle, F. Takens, On the nature of turbulence. Commun. Math. Phys. **20**, 167; Commun. Math. Phys. **23**, 343 (1971)
78. R.A. Shaw, *The dripping faucet as a model chaotic system* (Aerial Press, Santa Cruz, CA, 1984)
79. J. Tabony in *Morphogenesis*, ed. by P. Bourgine, A. Lesne. Biological self-organisation by way of the dynamics of reactive processes (Springer, Berlin, 2010) p. 87
80. F. Takens, *Dynamical Systems and Turbulence*, ed. by D. Rand, L.S. Young (Springer, Berlin, 1981) p. 230
81. H. Tennekes, J.L. Lumley, *A First Course in Turbulence* (MIT Press, Cambridge MA, 1972)
82. R. Thom, *Structural Stability and Morphogenesis* (Benjamin, Reading MA, 1975)
83. C. Tresser, P. Coullet, Itérations d'endomorphismes et groupe de renormalisation. *C.R. Acad. Sci. Paris A* **287**, 577 (1978)

Chapter 10
Self-Organised Criticality

During this book we have encountered many examples of critical phenomenon and have highlighted their common characteristics of divergence of the range of correlations, absence of characteristic scales,[1] fluctuations of all sizes and anomalous response (at all scales in amplitude, spatial extent and duration) to even a tiny perturbation. These characteristics are reflected in many scaling laws, expressing quantitatively the scale invariance of the phenomena. Typically, criticality occurs for a particular value of the control parameter which is adjusted externally, for example the critical temperature in second order phase transitions, percolation threshold p_c or bifurcation point in a dynamic system. However, it turns out that certain systems, maintained in a *nonequilibrium* state by a continuous supply of material or energy, can evolve *spontaneously* to a critical state, without external regulation. This is the concept of *self-organised criticality*, which we will present in this chapter, discussing various examples.

10.1 A New Concept: Self-Organised Criticality

10.1.1 Sandpile

10.1.1.1 A Familiar Experience

One of the emblematic examples of self-organised criticality is that of a pile of sand. If we add new sand to the top of the pile, excess sand flows to the bottom of the pile. In this way it behaves as an *open* system. For the system to reach a fixed regime, it is necessary to remove sand. Despite being statistically in a steady state, this regime

---

[1]Other than the "trivial" scales of system size, duration of observation and, at the other extreme, the scales of the constitutive elements.

A. Lesne and M. Laguës, *Scale Invariance*, DOI 10.1007/978-3-642-15123-1_10,
© Springer-Verlag Berlin Heidelberg 2012

is an *out of equilibrium* state. Unlike in a true equilibrium state, the flow of sand along the slopes of the pile is not zero. The slope of the sides of the pile takes a value that is independent of the size of the pile. The actual situation is a bit more complicated: the slope oscillates between two values, θ_m and θ_M, a few degrees apart. θ_M is a *critical value* since all higher values correspond to an unstable pile shape and therefore cause an *avalanche*. This sudden and complex relaxation in general overshoots the strict condition required for static stability of the pile. More sand than necessary is removed, bringing the slope to a value of $\theta < \theta_M$, from which it starts to increase again as we continue to add sand. The rate of injection will be chosen to be small enough that avalanches are separated from each other in time. Once the slope reaches a value close to θ_M, the moment at which the avalanche occurs is random. The size of the avalanche (number of grains involved) as well as its lifetime are also random. Avalanches are seen *at all scales*, from very localised avalanches that only locally readjust the slope to global subsidence in which the whole side of the pile is renewed.

10.1.1.2 Numerical Simulations

This familiar experience was reproduced numerically by Bak et al. [3], using the model explained in Fig. 10.1. When we analyse the distribution of sizes A of these avalanches, we find that the histogram $N(A)$ follows a power law (Fig. 10.2):

$$N(A) \sim A^{-\mu}. \tag{10.1}$$

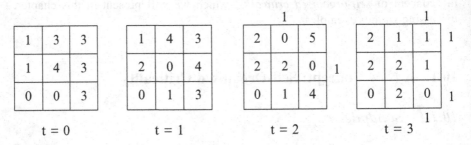

$$t = 0 \qquad\qquad t = 1 \qquad\qquad t = 2 \qquad\qquad t = 3$$

Fig. 10.1 Cellular automaton simulation of the critical behaviour of a sandpile. Space is divided horizontally into cells (or "sites"). Numbers indicate the local slope at each site (i.e. the slope in the sandpile region whose horizontal projection corresponds to that site). As soon as the slope at a particular site exceeds a threshold, here chosen to be 4, the state is updated such that the value of the slope at this point decreases by 4 and that of each of the 4 neighbouring sites increases by one. This modification may bring the slope of one of these sites above the threshold, in which case the network is rearranged again, and so on. The size of the avalanche is equal to the number of grains that leave the network (here 6) and its duration is the number of steps necessary to complete the rearrangement (here 3) such that each slope is below the threshold. The system is maintained out of equilibrium by a continuous supply of sand to the central site, the rate of which is very slow compared with the timescale of avalanches

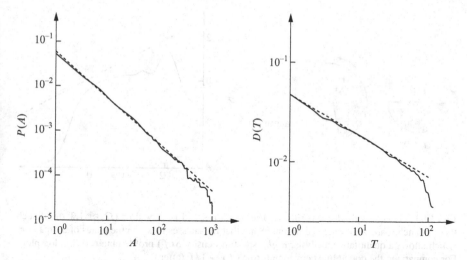

Fig. 10.2 Log-log plots showing the behaviour of a simulated sandpile (after [3]). *Left*: distribution of avalanches as a function of their size: $P(A) \sim A^{-\mu}$ with $\mu \approx 1.03$. *Right*: distribution of avalanches as a function of their lifetime weighted by the average response $\langle A(T) \rangle / T$ (see text): it scales as $D(T) \sim T^{-\mu'}$ with $\mu' \approx 0.43$, corresponding to a spectral density $S(f) \sim f^{-1.57}$

Numerical models of sandpiles give an exponent μ slightly larger than 1 ($\mu \approx 1.03$). The average time lapse $\tau(A)$ between two successive avalanches of size A grows with the size as $\tau(A) \sim A^2$. Finally, the distribution $P(T)$ of the lifetime T of avalanches also follows a power law. We actually plot (Fig. 10.2) the distribution $D(T)$ weighted by the average response $\langle A(T) \rangle / T$ (conditional average taken over avalanches with lifetime T):

$$D(T) \equiv P(T) \frac{\langle A(T) \rangle}{T} \sim T^{-\mu'}. \tag{10.2}$$

It can therefore be shown that the spectral density $S(f)$ (or power spectrum, equal to the Fourier transform of the temporal correlation function) is related to $P(T)$ according to the formula [2]:

$$S(f) = \int_0^{1/f} T D(T) dT \qquad \text{where} \qquad S(f) \sim f^{-2+\mu'} \sim f^{-\beta}. \tag{10.3}$$

10.1.1.3   Self-Organised Criticality and Marginal Stability

Bak et ak, [2] described the behaviour of their simulated sandpile as *self-organised criticality*. The word "self" reflects the intrinsic nature of θ_M, which is not fixed from outside but is "found" by the dynamics itself.

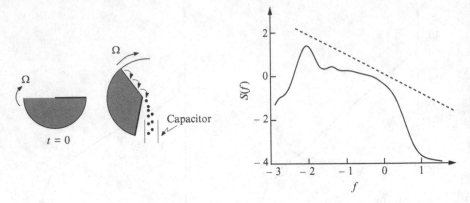

Fig. 10.3 Turning cylinder experiment. The rotation speed is very slow ($\Omega \approx 1.3°$/min) such that avalanches are well separated in time. Sand that falls passes between the plates of a capacitor, which allows a quantitative analysis. *Right*: spectral density $S(f)$ profile obtained (log-log plot). For comparison the dotted line corresponds to $S(f) \sim 1/f$ (after [21])

Starting from this numerical model and assuming that a single quantity X is enough to describe the local state of the system, a fundamental characteristic of self-organised criticality can be very systematically extracted: The system *spontaneously* tends to remain around an *intrinsic stability threshold* X_c. As long as $X < X_c$ and there is a supply of energy or material the system will evolve in such a way that X increases. As soon as X exceeds X_c, a relaxation phenomenon (an "avalanche") is suddenly produced bringing X back to values below X_c. In this way the value X_c corresponds to the *least stable* of the stable positions of the system, in the absence of external influences.[2] This is referred to as *marginal stability*, i.e. the smallest influence causing a growth δX is the "last straw".

10.1.1.4   Experimental Results

Quantitative experiments have been carried out by very slowly turning a filled half closed half cylinder (Fig. 10.3).

Here it is the rotation that maintains the system out of equilibrium. Avalanches are indeed seen, but the behaviour is far from the almost perfect power law observed numerically. In particular, a characteristic frequency appears (peak in the spectrum $S(f)$) and no convincing scale invariance emerges (Fig. 10.3). This can

---

[2]In this preliminary statement, we have deliberately omitted to take into account the spatial extension of the system. The quantity X is actually a function $X(r)$ and will therefore evolve differently at each spatial point r. Self-organised criticality appears when, in addition, spatial correlations develop as local values of X approach the threshold X_c. We will return to this point in Sect. 10.1.3.

be explained by the particular properties of real granular media: grains of sand exert sold friction forces on each other and the sand behaves both as a solid (if $\theta < \theta_M$) and a liquid (if $\theta > \theta_M$). In conclusion, although the analysis of simulated sandpiles brought forth the paradigm of self-organised criticality, it seems that real sand piles do not provide the best example of this type of behaviour, far from it.

10.1.2   Forest Fires

A second example, maybe more convincing, is the model called "forest fires". It better connects self-organised criticality to critical phenomena encountered in this book, namely percolation (Chap. 5).

The surface of the ground is discretised into cells, i.e. represented by a square lattice. The slow growth of trees is modelled by a random filling of empty squares, with a rate a_1. At time t, there are on average $a_1 t$ planted cells (equivalently we can fix a time interval $\Delta t_1 = 1/a_1$ between each replanting of a cell). The average density of trees, defined here as the average fraction of occupied cells, will therefore be a function $p(t)$ that grows over time. Since cells are filled randomly, filling will not be regular and homogeneous, but instead form clusters. The statistics of these clusters as time t is known, it is that of a percolation network of parameter $p(t)$. With a much lower frequency a_2, we light a fire in a lattice cell (the average time between two ignition events is therefore $\Delta t_2 = 1/a_2 \gg \Delta t_1$). If the sparks fall on a planted cell, the tree catches alight and the fire rapidly propagates to all neighbouring planted cells and so on until the whole cluster they are part of is ablaze. In one time step, the whole cluster burns and then becomes bare land. The range of the fire is simply measured by the number of burnt sites. We will therefore observe long periods of reforestation, during which the density $p(t)$ grows, separated by fires of comparatively short duration, during which the average density dramatically falls.

If $p(t)$ is low at the moment of ignition, far below the percolation threshold p_c, the cluster that burns will typically be of a small size $s(p)$, so that p will be very little affected by the fire and will quickly continue spreading. If on the other hand $p(t) > p_c$, a cell will have a probability $P[p(t)] > 0$ of belonging to the infinite cluster. The fire will therefore typically burn a sizable fraction of trees, so that $p(t)$ will significantly drop. If we let the system evolve, it will stabilise in an intermediate state between the two unstable extremes we have just described. It is numerically observed that p tends to p_c. So the system spontaneously self-organises in such a way that its average density is equal to the percolation threshold p_c. We therefore clearly have self-organised criticality situation. This affirmation can be specified quantitatively by showing that the size distribution $P(A)$ of fires follows a power law [4, 10]:

$$P(A) \sim A^{-\mu} \qquad \text{with} \quad \mu \approx 1.3. \qquad (10.4)$$

The density fluctuations are scale invariant, they have the form of a power law reflecting the absence of a characteristic size in this phenomenon (apart from, of course, effects related to the finite size of the simulation lattice, which will truncate the power laws). A variant of the model is to burn, in one time step, only cells that are nearest neighbours to inflamed cells. In this case, the lifetime T of the fire becomes another way to measure its magnitude and we similarly observe a power law distribution for the lifetimes. Here the simulation correctly reproduces real data. Analysis of data recorded for forest fires produced in large forests in America or Australia show a power law behaviour $P(A) \sim A^{-1.3}$ [30].

To conclude, we highlight two important characteristics of this model, which we will find again in other examples of self-organised criticality:

- The large *separation* between the reforestation timescale and the very fast relaxation time (fire lifetime).
- The *global stability of the critical point*, i.e. the global dynamics spontaneously leads the system to this point.

10.1.3   The Basic Ingredients

The behaviour of sandpiles and forest fires have been taken as reference models to address other problems such as avalanches and landslides, volcanic eruptions and earthquakes, ecosystems, financial and economic markets, traffic flow, diffusion and growth fronts. Before presenting these different situations and to be better prepared to discus their (potential) similarity, we can already summarise the main points of the concept of self-organised criticality introduced by Bak and collaborators. It applies to systems that are *out of equilibrium* and *spontaneously* evolving to a *critical state*, critical in that it does not have any characteristic scales but instead has a *self-similar* spatiotemporal organisation [3,7]. The critical nature of these systems is clearer in their response to small external perturbations: *a very small influence can have arbitrarily large consequences*, on arbitrarily long space and time scales. Quantitatively this criticality results in the *divergence of correlation lengths and times*. The zone and lifetime of the influence of a very localised event (e.g. addition of a grain of sand, ignition of a tree) are not bounded. The most characteristic signature is the $1/f$ *noise*. We mean by this that the spectrum $S(f)$ of correlations observed in these systems follows a power law $S(f) \sim 1/f^\beta$ with β close to 1 (different from the value $\beta = 0$ corresponding to white noise and the value $\beta = 2$ corresponding to the Wiener process). As we have mentioned in Sect. 10.1.1, it can be shown that such scaling behaviour comes from the existence of many different relaxation times following a power law distribution. We will see, in Sect. 11.3, that it reflects the existence of temporal correlations at all scales.

Self-organised criticality is a fundamentally collective phenomenon. Many simple (but nonlinear because they have a stability threshold) interactions gradually organise by developing long range correlations. The system then arrives in a state where many correlated regions are near to breaking point. It is therefore sensitive to respond to the smallest perturbation in an extremely amplified way. We observe an event whose magnitude is incomparable to that which triggered it. This critical state, which is marginally stable with respect to local perturbations, seems to be a stable state of the global dynamics, such that the system returns to it spontaneously, without being a result of regulation of a control parameter. Self-organised criticality then has no control parameter. For example, changing the injection rate of material or energy, which maintains the system far from equilibrium, does not at all affect the observed behaviours (as long as this rate remains small). The concept of self-organised criticality highlights that it is the same causes and mechanisms of initiation or nucleation underlying small events and major catastrophes, so that both types of event are equally unpredictable.

Self-organised critical systems are necessarily open and dissipative systems. However, the existence of a local stability threshold implies that the timescale of relaxation of the stored energy is very short compared to that of injection. The associated mechanism is schematically represented in Fig. 10.4. The system slowly accumulates energy in the form of stresses or strains and suddenly releases it with an uncontrolled amplitude once a tolerance threshold has been crossed. Note that this relaxation occurs randomly (and not as soon as the threshold is crossed) due to the large number of coupled degrees of freedom of the system under consideration. Of course, the more stress accumulated in the system above the threshold the higher the probability of a catastrophe releasing these stresses. Nevertheless, *the only prediction possible will be of a statistical nature.*

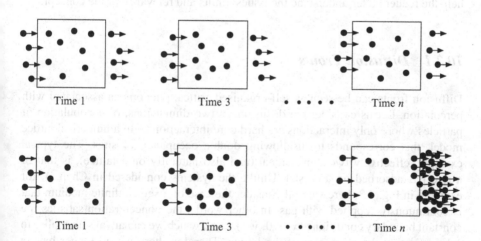

Fig. 10.4 Temporal asymmetry of flows entering and leaving in self-organised criticality (*lower section* of figure). Even though the average values of these flows on very large timescales are equal, there is a slow accumulation of energy or constraints, over $n \gg 1$ time steps, and very fast relaxation in a single time step. This mechanism, related to the existence of a stability threshold in the local dynamics, contrasts with the case of "dynamic equilibrium" (*upper section* of figure) where flows equilibrate in short timescales and produce a "normal" steady state

10.1.4   In Practice

The simplest way to show a self-organised criticality is to construct the histogram of the number of events as a function of their amplitude or lifetime. This gives the size distribution $P(A)$ of events. A signature of self-organised criticality is that this distribution follows a power law $P(A) \sim A^{-\mu}$. From experimental data, it is easier and more reliable to consider the probability $\mathcal{P}(A > A_0)$ that events have an amplitude larger than A_0, called "cumulative distribution". The signature of self-organised criticality is again a power law behaviour $\mathcal{P}(A > A_0) \sim A_0^{1-\mu}$ (with $\mu > 1$). We can also determine the distribution $P(T)$ of events of duration T, or the number $N(\tau)$ of events of duration τ. These also follow power laws in the case of a self-organised critical system. A more detailed analysis is obtained by determining the spatio-temporal correlations. The power spectrum $S(f)$ (Fourier transform of the temporal correlation function) is studied, whose form of $1/f^\beta$ with β close to 1 seems to be the clearest signature of self-organised criticality.

10.2   Examples

Having discussed the, now classic, examples of sandpiles and forest fires, we will next give a glimpse of different phenomena related (often in a debated and sometimes questionable way) to self-organised criticality. This presentation will help the reader better understand the issues, limits and relevance of the concept.

10.2.1   Diffusion Fronts

Diffusion fronts can be seen as self-organised critical phenomena associated with percolation. Let us consider the diffusion, in two dimensions, of a population of particles whose only interactions are hard core interactions.[3] In a numerical lattice model, this corresponds to disallowing double occupancy of sites. The typical example, effectively two dimensional (and also naturally on a lattice), is that of a rare gas adsorbed on a crystal. Unlike the systems considered in Chap. 4, for example in Fig. 4.3, here we will consider the case of a semi-infinite medium that is continuously supplied with gas. In other words, the concentration satisfies the constant boundary condition $c(x = 0, y, t) \equiv c_0$, which we can always normalise to 1 by adjusting the mesh size of the lattice. Therefore here we no longer have a relaxation phenomena but a system maintained out of equilibrium.

---

[3] We call *hard core interaction* a short ranged quasi-infinite repulsion, modelled by a hard core of radius equal to the range of this repulsive interaction.

The distribution of particles in this case is the same as that which would be obtained by *locally* filling the lattice like a percolation network, with probability $p(x,t) = c(x,t)$, which is the solution of the diffusion equation. This model, in which the probability varies in space is called *gradient percolation*. At the numerically accessible microscopic scale, the diffusion front is defined as the external border of the cloud of particles, which in two dimensions corresponds to the forwardmost connected path.[4] We therefore observe that at every moment in time the front is localised around $x_c(t)$ such that $p(x_c(t), t) = p_c$. In other words, the diffusion front localises in the region where the concentration is near the percolation threshold, the critical properties of which we have described in Chap. 5 [13].

10.2.2 Traffic Flow and Traffic Jams

Traffic jams and other annoying characteristics of traffic circulation are studied very scientifically under the name of "traffic theory" making use of different concepts developed in other domains, which at first glance seem nothing to do with the problem, e.g. granular media, phase transitions, kinetic theory of gases and self-organised critical phenomena. It is an example of a phenomenon at the interface of statistical mechanics and nonlinear physics [16]. The theoretical approach dates back to Nagel and Schreckenberg, in 1992, and their simulation of traffic flow using a cellular automaton model.[5]

Many studies of more realistic traffic models have been carried out [31] (see also Sect. 4.6). It can then be shown that traffic spontaneously evolves to the dynamic state in which transport is most efficient and that, unexpectedly, this optimal state is also critical from the point of view of transport. On one hand, it is more sensitive to obstacles, which can lead to massive and sudden congestion. On the other hand, it is also sensitive to the smallest internal fluctuations and spontaneously produces traffic jams of all sizes, without any external cause. The power spectrum of the intermittent dynamics associated with this critical regime is exactly $1/f$, indicating the existence of long range correlations and a broad distribution of characteristic times, in good agreement with both numerical simulations and observations of real traffic flow [19, 25]. More complex regimes are observed if several parallel routes

---

[4]In three dimensions a percolating path no longer makes a border and the front will extend over a whole range of concentrations. However, its foremost part will still be spontaneously localised in the region where $p = p_c$.

[5]A cellular automaton is a model in which time, space but also state variables take only discrete values. Such a model is therefore particularly suitable for numerical studies. Typically, particles are moved from site to site according to very simple probabilistic rules. Cellular automata are used to study many transport phenomena, for example reaction-diffusion phenomena or those encountered in hydrodynamics and population dynamics [8]. In the context of self-organised criticality, sandpiles, forest fires and traffic flow are numerically studied using this type of simulation [20].

are considered (e.g. motorway lanes) with transfer of vehicles from one to the other, entrance and exit sliproads and several types of vehicles.

10.2.3   Earthquakes

When the deformation of the earth's crust caused by slow movements of tectonic plates exceeds a certain threshold, a rupture is produced, in other words an earthquake. Energy is then released in the form of seismic waves. The magnitude M of an earthquake is measured on a logarithmic scale. By definition an earthquake of magnitude M is of amplitude $S(M) \sim 10^M$. The experimental Gutenberg–Richter law expresses the cumulative frequency $N(M > M_0)$ of earthquakes of magnitude M above M_0 as $N(M > M_0) \sim 10^{-bM_0}$. It is valid over a large range of scales, specifically for $2 \leq M \leq 6.5$, with a universal exponent b where $0.8 \leq b \leq 1.1$ [14, 15]. Using the frequency $N(S)$ derived from the cumulative frequency, it can be written as a function of S:

$$N(S) \sim S^{-\mu} \qquad \text{with} \quad \mu = 1 + b \approx 2. \qquad (10.5)$$

If we record the surface A affected by each earthquake in a given region over a period long enough to get many points and plot the results as a histogram $N(A)$, we also obtain a power law $N(A) \sim A^{-\mu'}$ with $\mu' \approx \mu \approx 2$. So it seems that earthquakes show self-organised criticality. Here we find a great disparity between mechanisms of injection of energy and its release since stresses accumulate over very long timescales, set by the movement of tectonic plates, while the events in which the stored energy is released occur over a range of very short timescales (although very variable from one event to another). Earthquakes depend on the organisation of the earth's surface which they change themselves. This feedback is the cause of the observed self-organised criticality.

A simple model was proposed to quantitatively study the consequences of this basic mechanism. Blocks of mass m are placed on a horizontal plate, on which they move with solid friction (coefficient of static friction F_s higher than the coefficient of dynamic friction F_d). They are connected together by springs (harmonic restoring force, stiffness k_c) and connected, also by a spring (of stiffness k_p), to a higher horizontal plate moving with uniform velocity v. The control parameters of this model are m, v, and ratios $\Phi = F_s/F_d$ and $a = k_c/k_p$ (see Fig. 10.5). For a small number of blocks, the purely deterministic system will show a chaotic behaviour. However, when there are $N \gg 1$ blocks, it shows self-organised criticality characterised by a scaling law $N(A) \sim A^{-\mu'}$ analogous to that observed in seismic data, but with $\mu'_{th} \approx 1.3$, smaller than the measured exponent $\mu' \approx 2$ [30].

Earthquakes are actually the manifestation of complex spatiotemporal dynamics and not isolated events. This is reflected in Omori's law describing the decay over time T of the frequency with which aftershocks are detected after an earthquake (where T is counted from the main earthquake):

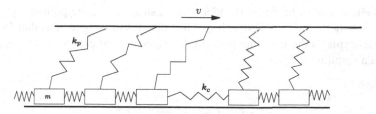

Fig. 10.5 Model of tectonic plates and resulting earthquakes. The figure shows a transverse cross-section of the two dimensional model: imagine rows of blocks each connected by elastic springs to its four neighbours

$$N(T) \sim T^{-\alpha} \qquad \alpha \approx 1. \qquad (10.6)$$

It is also observed that the distribution of epicentres is fractal, with fractal dimension $d_f = 1.2$. A global law, unifying the Gutenberg–Richter law, Omori's law and the fractal nature of the seismically active regions has been recently proposed [9]. It describes the distribution of intervals T separating earthquakes with amplitude larger than S (i.e. magnitude larger than $\log_{10} S$) in a region of linear size L:

$$P_{S,L}(t) \sim T^{-\alpha} \, f(T \, L^{d_f} \, S^{-b}). \qquad (10.7)$$

Its experimental validation supports the idea that all earthquakes or seismic activity, whatever their magnitude and whether or not they are aftershocks of a bigger earthquake, result from the same mechanisms and are involved in the same complex multiscale dynamics.

10.2.4 Inflation of the Lungs

The first studies on lung inflation simply measured the duration T of inspiration (intake of breath). The resulting histogram $P(T)$ seems to follow a power law or, equivalently, the spectral density seems to obey the scaling law $S(f) \sim f^{-0.7}$. But the small range of values observed for T means we cannot see these results as more than an encouragement to conduct a more detailed experimental study on how air fills the lungs.

The end branches of the lungs close during expiration and gradually reopen during inspiration. The dynamics of their (aerodynamic) resistance whilst the lungs fill at constant flux has been precisely measured (locally at the scale of an alveolus). It is observed that discrete variations superimpose on the continuous decrease of this resistance. The probability distribution $P(\tau)$ of the time τ separating two jumps, as well as the distribution $P(A)$ of jumps show power law behaviour [29]:

$$P(\tau) \sim \tau^{-\alpha} \text{ where } \alpha = 2.5 \pm 0.2, \qquad P(A) \sim A^{-\mu} \text{ where } \mu = 1.8 \pm 0.2. \quad (10.8)$$

A theoretical model, based on the hierarchical structure of the pulmonary system and describing the cascade of opening of its branches (analogue of avalanches on a sand pile) explains these scaling laws [6]. Changes in their exponents is a signature of certain respiratory diseases.

10.2.5 Ecosystems and Evolution

Fossil studies have yielded some information about the extinction of ancestral species [24]. These have shown that extinctions do not occur continuously but in the form of events lasting very short times on the evolutionary timescale. In the context of ecosystems, criticality means that species are highly interdependent, for example via food chains or mutualism (interaction between organisms in which each gains a fitness benefit). From the fact that all coupled species disappear at the same time, we understand that major extinction events can occur [22].

A model, by Bak and Sneppen [5], was designed to try to capture the basic mechanisms behind this feature of evolution.[6] This model, which does not try to describe reality but only to suggest a plausible mechanism, is as follows. We place species on a virtual lattice (that is to say without an inherent physical space) representing the network of interactions. Neighbouring species on the lattice are those that are coupled, for example by a trophic relationship (pertaining to food). The value of the *fitness* $f(i)$ of a species i is a number between 0 and 1 such that the characteristic lifetime of this species, if it was isolated, would be:

$$\tau_i = e^{b(f_i - f_{\text{ref}})}. \tag{10.9}$$

A species whose fitness value becomes less than the fixed threshold f_{ref} dies out. The fact that species are interdependent causes changes in the fitness of one to have repercussions on the fitness of others and leads to a reorganisation of the ecosystem, first locally and then, little by little, globally. The evolutionary algorithm consists in taking the species i with the lowest fitness and replacing it with a random value r_i. The neighbours are also modified by replacing $f_{i\pm1}$ by $(f_{i\pm1} + r_{i\pm1})/2$. Extinction cascades and a spontaneous evolution of fitness values towards a value f_c are observed. One this global steady state has been reached, we observe an alternating series of extinctions and rest phases, in which the lifetime T follows the statistics $N(T) \sim 1/T$. We call this "punctuated equilibrium" to describe the intermittency of extinction observed in the histogram reconstructed from fossil records. More sophisticated models have been developed since taking the basic idea that we have just presented but taking into account better the evolution and adaptive dynamics of ecosystems [18, 23].

---

[6]We do not discuss here the reliability of observations or how real the phenomenon the model intends to reproduce and explain actually is.

10.2.6 Other Examples

In some ways fully developed turbulence (Sect. 9.5) is a self-organised critical phenomenon since at large Reynolds numbers $Re \gg 1$, there is a large separation between the (macroscopic) scale at which energy is injected and the (microscopic) scale at which viscous dissipation takes place. The connection between these two scales is spontaneously established by a spatiotemporal fractal organisation, ensuring optimal energy transfer from the injection scale to the dissipation scale.

The concept of self-organised criticality has also been invoked with more or less relevance in social sciences. For example, Richardson (who we have already come across in the context of his work on turbulence) showed that the frequency of wars follows a power law as a function of their intensity, measured by the number of people killed [26]. More recent studies, relating the number of deaths to the total population involved in the conflict, again gives a power law with exponent -1.39. A general reason for caution in this type of analysis is firstly that there is far less data compared to sample sizes available in physics and biology, and secondly that the experiments are not reproducible. From one observation to another the context has changed and this non-stationary can greatly affect the phenomenon being studied (in a way that is impossible to assess quantitatively), not to mention the affect of actual causes of historical events. Finally it is difficult to determine relevant quantitative and absolute indices. The cited example is significant in the sense that is it the "right" way to measure the intensity of a war? Taking into account the advancement of weaponry, population dynamics and even the change in "targets", the number of deaths is questionable, to say the least.

We can therefore question the relevance and potentially even danger of extrapolating a concept rooted in simple dynamical laws to domains in which the existence of universal and fixed organising principles is by no means established.

10.3 Conclusion

10.3.1 Towards an Explanatory Picture: Feedback Loops and Marginal Stability

The first studies and even the emergence of the concept of self-organised criticality were done from numerical simulations face to face with experimental observations. They were done with the focus on the relevant *observables* and the *hallmarks* of self-organised criticality detectable in the data, namely a power law form of the distribution $P(A)$ of amplitude A of events ("avalanches") as well as the distribution $P(T)$ of their characteristic times T. This scale invariance is also seen in a dependence $S(f) \sim f^{-\beta}$ of the spectral density, called "$1/f$ noise", although it should not be seen as noise but as the expression of long range temporal correlations (see also Sect. 11.3). In each context, numerical studies of minimal models have identified the essential *ingredients* of the phenomenon.

The question that remains to be addressed is that of the *mechanisms* at work in self-organised criticality. The example of forest fires is perhaps the most instructive. Let us put the conclusions reached in Sect. 10.1.2 in dynamic terms:

- The local response of the system shows a *threshold* (here the local density of trees). Below this threshold an external perturbation has no noticeable effect. A step function response is a necessary ingredient, in other words there can be no self-organised criticality if the local response is gradual.
- The timescale on which energy or material is injected is very slow compared to that of the local dynamics. Due to this, the evolution of the state of the system and that of the control parameter do not take place over the same timescale. At each moment in time the state of the system rapidly adapts to the value $p(t)$ as if it was a fixed value of the parameter.
- The global state, quantified by the order parameter $P(p)$, controls the response of the system to spontaneous ignition. The key mechanism is then a *feedback of the order parameter on the control parameter*, which controls the local state and consequently the local response properties of the system.
- The value $p = p_c$ is stable for the global dynamics.

The take home message of all this is that the collective dynamics feeds back on the local dynamics in such a way that the critical state is found to be the attractor of the system [12]. This idea can be checked experimentally by artificially introducing a coupling between the order parameter and the control parameter in various phase transitions, for example liquid–vapour transitions, superfluidic transition of helium 4 or superconductivity. The idea is to design an apparatus sensitive to the correlation length ξ and be able to tune the temperature to increase this length ξ. In the case of the liquid–vapour transition, this objective is realised using light diffraction, sensitive to the isothermal compressibility, itself related to ξ; the result is then coupled to the thermostat regulating the temperature of the medium [11, 27].

10.3.2   Success and Reservations

Very schematically, the term self-organised criticality applies to dissipative systems spontaneously exhibiting a spatiotemporal scale invariance, in particular a dissipation of energy on all scales. This phenomenon is "complex" in the sense that it involves a large number of elements whose collective global behaviour cannot be simply deduced from individual components. We find here concepts described throughout this book, which could be summarised under the name of "physics of exponents". Justifying "self-organised criticality" in a given concrete situation would require having a definition that has a consensus and above all is usable. The different aspects highlighted in the examples in previous sections show that such a definition does not yet exist (and might never exist). There are precedents, for example the concept of fractal is not strictly defined either, except via the scaling law characterising the fractal dimension. Here, the scaling law relating the size of observed events to their probability could play the same role.

The reservation that we can express is that, at the moment, this concept is not accompanied by a working methodological framework. Self-organised criticality as it is observed in numerical models (cellular automata) is therefore not very fertile, at best purely descriptive, since it is not associated with scientific tools that would help to explain the underlying mechanisms of the phenomenon. So identifying self-organised criticality in a phenomenon could be seen as simply a summary, admittedly a concise and elegant one, of a set of striking scaling properties.

More positively, self-organised criticality is an interesting phenomenon to detect in so far as it reveals a universality in the mechanisms at work in the system and shows that it is useless to look for one explanation for "normal" events and another one specific to "catastrophes". It puts the emphasis on the hierarchical organisation of correlations and stresses. The interest of the concept is to unify a category of critical phenomena and to build an intuitive picture by offering some models which are both striking and familiar. As for deterministic chaos, it is a new paradigm but has not yet reached the same conceptual or functional maturity. Quantitative analysis of the models presented in this chapter is an initial step and *many* others remain to be gone through in this still little explored field of complex systems. We emphasize the importance of the ideas of *marginal stability*, which explains the random nature of response to a perturbation, and of *feedback loops* (or "control circuits"), explaining the possibility of cascade amplification and response on all scales. It seems clear that only a global *multiscale* view could further the study of complex systems. Therefore we should seek to understand multiscale organisation, global control of flows, frustrations and competition present in the system in order to determine the different trade-offs each giving a possible version of the spatiotemporal behaviour of the system.

For more examples and more extensive discussions we refer to the many existing articles and works on the subject, for example among the seminal ones [1, 17] and [28].

References

1. P. Bak, *How Nature Work: The Science of Self-Organized Criticality* (Springer, Berlin, 1996)
2. P. Bak, C. Tang, K. Wiesenfeld, Self-organized criticality: an explanation of $1/f$ noise. Phys. Rev. Lett. **59**, 381 (1987)
3. P. Bak, C. Tang, K. Wiesenfeld, Self-organized criticality. Phys. Rev. A **38**, 364 (1988)
4. P. Bak, K. Chen, C. Tang, A forest-fire model and some thoughts on turbulence. **147**, 297 (1990)
5. P. Bak, K. Sneppen, Punctuated equilibrium and criticality in a simple model of evolution. Phys. Rev. Lett. **71**, 4083 (1993)
6. A.L. Barabasi, S.V. Buldyrev, H.E. Stanley, B. Suki, Avalanches in the lung: a statistical mechanical model. Phys. Rev. Lett. **76**, 2192 (1996)

360   10   Self-Organised Criticality

7. K. Chen, P. Bak, Self-organized criticality. Sci. Am. 46 (1991)
8. B. Chopard, M. Droz, *Cellular Automata Modeling of Physical Systems*. Collection "Aléa
   Saclay" (Cambridge University Press, Cambridge, 1998)
9. K. Christensen, L. Danon, T. Scanlon, P. Bak, Unified scaling law for earthquakes. *Proc. Natl.
   Acad. Sci USA* **99**, 2509 (2002)
10. B. Drossel, F. Schwabl, Self-organized critical forest-fire model. Phys. Rev. Lett. **69**, 1629
    (1992)
11. N. Fraysse, A. Sornette, D. Sornette, Critical phase transitions made self-organized: proposed
    experiments. J. Phys. I France **3**, 1377 (1993)
12. L. Gil, D. Sornette, Landau–Ginzburg theory of self-organized criticality. Phys. Rev. Lett. **76**,
    3991 (1996)
13. J.F. Gouyet, *Physics and Fractal Structures* (Springer, New York, 1996)
14. B. Gutenberg, C.F. Richter, Bull. Seismol. Soc. Am. **64**, 185 (1944)
15. B. Gutenberg, C.F. Richter, *Seismicity of the Earth and Associated Phenomenon* (Princeton
    University Press, Princeton, 1954)
16. D. Helbing, M. Treiber, Jams, waves and clusters. Science, **282**, 2001 (1998)
17. H.J. Jensen, *Self-Organized Criticality* (Cambridge University Press, Cambridge, 1998)
18. S.A. Kauffman, *The Origins of Order: Self-Organization and Selection in Evolution* (Oxford
    University Press, 1993)
19. H.Y. Lee, H.W. Lee, D. Kim, Origin of synchronized traffic flow on highways and its dynamic
    phase transitions. Phys. Rev. Lett. **81**, 1130 (1998)
20. K. Nagel, M. Schreckenberg, A cellular automaton model for freeway traffic. J. Phys. I France
    2, 2221 (1992)
21. S.R. Nagel, Instabilities in a sandpile. Rev. Mod. Phys. **64**, 321 (1992)
22. M.E.J. Newman, Evidence for self-organized criticality in evolution. Physica D **107**, 293
    (1997)
23. M.E.J. Newman, Simple models of evolution and extinction. Comput. Sci. Eng. **2**, 80 (2000)
24. M.E.J. Newman, G.J. Eble, Power spectra of extinction in the fossil record. Proc. R. Soc.
    London B **266**, 1267 (1999)
25. M. Paczuski, K. Nagel, in *Traffic and Granular Flow*, ed. by D.E. Wolf, M. Schreckenberg,
    A. Bachem. Self-organized criticality and 1/f noise in traffic (World Scientific, Singapore,
    1996) pp. 73–85
26. L.F. Richardson, Frequency of occurrence of wars and other fatal quarrels. Nature **148**, 598
    (1941)
27. D. Sornette, Critical phase transitions made self-organized: a dynamical system feedback
    mechanism for self-organized criticality. J. Phys. I France **2**, 2065 (1992)
28. D. Sornette, *Critical Phenomena in Natural Sciences. Chaos, Fractals, Self-Organization and
    Disorder: Concepts and Tools* (Springer, Berlin, 2000)
29. B. Suki, A.L. Barabasi, Z. Hantos, F. Peták, H.E. Stanley, Avalanches and power-law behaviour
    in lung inflation. Nature **368**, 615 (1994)
30. D.L. Turcotte, Self-organized criticality. Rep. Prog. Phys. **62**, 1377 (1999)
31. D.E. Wolf, M. Schrenckenberg, A. Bachem (eds.), *Traffic and Granular Flow* (World Scien-
    tific, Singapore, 1996)

Chapter 11
Scale Invariance in Biology

11.1 Introduction

In this chapter we will show that biology cannot escape the omnipresence of scaling laws and self-similar structures underlying them. Some authors go as far as to suggest that there are scaling laws that could be specific to living systems and their particular multiscale organisation due to the result of evolution.

The most obvious scaling properties are those associated with the numerous fractal structures seen in living organisms, such as the branched structure of vascular systems and neurones, alveolar structure of lungs, porous structure of bones etc. We should not forget the plant kingdom, with plant root networks and branches of leaf veins, flowering of a cauliflower or leaves of a fern. The purpose of these fractal structures, established by successive trials over the course of evolution, is that they optimise exchange surfaces (either internal or with the environment) for a given volume (of material that has to be constructed and maintained so is costly) and maximise currents and metabolic flows.

Other, less obvious scale invariances exist. We will mainly present *allometric* scaling laws describing the observed similarity, after changes in scale, between shapes but also metabolisms of different organisms belonging to sometimes very different species. Although their origin, or even their exponents, remain controversial, if correct they would reflect a definite structural as well as functional universality in the organisation of living creatures [12, 51].

Other scaling laws are observed inside living organisms themselves, appearing in the form of *long range correlations* and reflecting the existence of collective phenomena. We will present a few typical examples known by biologists, namely DNA sequences, heat beat and brain activity as observed by electroencephalography (EEG). We will describe in detail in these particular cases a few general methods, namely spectral analysis, detrended fluctuation analysis (fluctuation analysis after correction for non-stationarity) and wavelet transforms. If positive, long range correlations reflect the emergence of collective behaviour, leading to radically new properties on larger scales. As for negative long range correlations

A. Lesne and M. Laguës, *Scale Invariance*, DOI 10.1007/978-3-642-15123-1_11,
© Springer-Verlag Berlin Heidelberg 2012

(anti-correlations), they are crucial for homeostasis, i.e. the stability and robustness of physiological processes. It is therefore highly likely that long range correlations will be found in many biological functions. We suggest that they result from a delicate balance between on the one hand local dynamics and on the other hand constraints on resources or regulation mechanisms taking place at a much larger scale.

Turning finally from the level of an organism to that of an ecosystem, we again find scaling properties reflecting quantitatively the complexity of their spatiotemporal dynamics. We saw an example in Sect. 9.4.2 with intermittent rarity of some species.

Themes in this chapter connect with those covered in Chap. 10. The concept of self-organised criticality and life share major ingredients of self-organisation and the open dissipative and multiscale nature of systems under consideration. Relying on scaling properties of living systems and using tools developed in physics in this context, may be relevant in describing, measuring and understanding their complexity.

11.2   Universality in the Metabolism of Living Systems

11.2.1   Observation of Allometric Scaling Laws

The first family of scaling laws we will consider expresses a scale invariance called *allometric*, which is observed when animals (different species or individuals within a species) of different masses M are compared. The scaling laws are of the form $Y = Y_0 M^b$, for various observables Y. Such laws are brought to light by experimental data (M, Y) represented on a log-log plot. Since Y_0 is not universal or robust we will rather write $Y \sim M^b$.

The most studied quantity is the metabolic rate B at rest, that is the quantity of energy the organisms needs each day simply to remain alive. As this is to do with the metabolism at rest, we could think that this metabolic rate is proportional to thermal dissipation. We then intuitively expect $B \sim M^{2/3}$, due to the argument that heat dissipates through the body's surface, which goes as $V^{2/3}$, taking the density as constant in a first approximation. Since the founding work of Kleiber, this idea has been superseded by that of an "anomalous" scaling law [34, 35]:

$$ B \sim M^{3/4} \qquad \text{(Kleiber's law)} \qquad (11.1) $$

which now carries his name. The energy required per day, per unit mass (specific metabolic rate) therefore behaves as $M^{-1/4}$, meaning that the larger the organism the less it costs to keep a kilogram of itself alive [36]. This idea had been pointed out earlier by d'Arcy Thompson [15] who said that no mammal smaller than a shrew exists. This limitation is stronger the more dissipative the external environment, for example, there are no small marine mammals. A large organism still consumes more in absolute terms, requiring abundant resources, as well as other constraints enforcing an upper limit on the size, for example skeletal strength or blood pressure. An unconventional example is that given by Rothen [47] of the Lilliputians described in *Gulliver's Travels*. Using Kleiber's law, the amount of food needed for a Lilliputian can be calculated to be more than that given by Jonathan Swift, who did not know this $M^{3/4}$ law.

Allometric scaling laws extend to biological clocks and characteristic times such as lifetime ($\sim M^{-1/4}$), time for an embryo to develop (gestation time in mammals) and heart rate ($\sim M^{1/4}$). Large animals live longer and at a slower pace. Observation of these other power laws, whose exponents are also simple multiples of $1/4$ (integers or half integers), reinforces the plausibility of this relation to metabolism.

The rise of the concept of fractals and scaling theories has resurfaced the question of the origin of the universality of Kleiber's law, initially observed in mammals. Scientists claim (maybe exaggerating) that this law for metabolism can be extended to not only plants but also single celled organisms and even organelles involved in the control of energy flux, for example mitochondria (site of cellular respiration in animals and plants) and chloroplasts (site of photosynthesis in plants), which are sort of basic units of the "metabolic factory" (see Fig. 11.1) [54]. The idea is that the same principles of transport and energy transformation are at work at the level of organs (to feed the cells) and inside cells (to feed mitochondria or chloroplasts).

These scaling laws do not include the prefactors and therefore cannot in any case be used to determine the value of the quantities Y considered above knowing the typical mass M of the organisms, but only to describe or predict how these quantities *vary* when M changes. These laws are also interesting not so much for their, as it happens limited, predictive power, but since they indicate universal

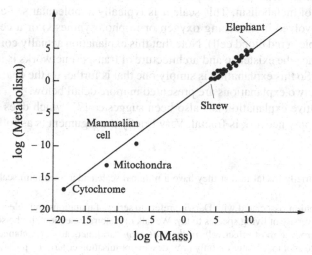

Fig. 11.1 Kleiber's generalised law expressing a $M^{3/4}$ dependence of resting metabolism, observed at different scales and in different species (after [54])

principles in the structure and function of living systems. Via the value of their exponent they are a quantitative intermediary between experimental measurements and more theoretical schemes of the organisation of living beings. Let us emphasise now, before looking at explanations, that these laws must be prudently considered and maybe their experimental validity discussed or at least limits to their validity carefully defined (Sect. 11.2.3).

11.2.2 Proposed Explanations

The first explanations of Kleiber's law simply rest on dimensional analysis of materials, physiological times and diffusion in the system. They involve many specific hypotheses and therefore cannot make up a valid explanation if the law proves to be as universal as some results let us think [8].

- Among the most recent advances, one explanation rests on the fact that the metabolism of living beings depends on the transport of substances (nutrients, oxygen etc.) through branched (fractal[1]) networks, filling all space and therefore the hierarchical structure ends with tiny branches (capillaries) that have the same size in all species. It then rests on a hydrodynamic (quite rudimentary, or even unrealistic) description of transport and dissipation in these networks [55].
- A second explanation was then proposed by the same authors drawing on their first arguments in a simpler and more general way [56]. The major difference is that it no longer explicitly makes reference to nutrient fluid transport (blood, sap, etc.) through a network, making it applicable to single celled organisms. However, it remains based on the existence of a universal (identical in all species) minimal scale a and a hierarchical organisation for metabolism, starting from this elementary scale a and shaped over the course of evolution[2] to optimise the efficiency of metabolism. This scale a is typically a molecular scale (that of an enzyme involved in absorbing oxygen or in photosynthesis) or a cellular scale (for example, a red blood cell). Note that this explanation actually coincides with the first since the existence and architecture of transport networks is a product of evolution. So this explanation is simply one that is further up the chain of causes. These first two explanations are presented in more detail below.
- An alternative explanation has also been suggested [7], which does not assume that the supply network is fractal. Very briefly the argument is as follows. Let L

---

[1] They are not strictly fractal in that they have a minimal scale and a maximum scale (the size of the organism).

[2] The term evolution, associated with Darwin, refers to series of mutations and selections that have produced the diversity of living species today. We talk of "natural selection" in the sense in which species that reproduce most effectively *in the long term* "arithmetically" spontaneously emerge. It is important to spot that evolution only provides an optimisation criterion a posteriori and there is no anticipation nor finality in evolution. Nevertheless, looking back retrospectively, *everything happens as if living systems have an additional organising principle* that inert systems do not have.

be the average distance travelled by a nutrient or gas molecule before reaching the site where it will be "consumed". As the network serves the whole volume of the organism, the number of sites goes as L^3. The total amount of nutrients then behaves as L^4, since at a given moment in time we should count the fraction being transported to active sites. This total amount, directly related to the volume of transporter fluid (blood or sap), is proportional to the mass M of the organism, hence $L \sim M^{1/4}$ and $B \sim M^{3/4}$ since the metabolism is controlled by the number of consummation sites. This argument, generalised to river systems,[3] shows that these non Euclidean anomalous power laws are a general feature of optimal network architecture, optimal in that they ensure the most efficient transport possible.

Opinion remains very divided on the validity of one or another of these explanations. The question is far from being resolved! In addition, an argument to reject approaches of this type[4] is that, in these models, the quantity of blood contained in large vessels (and therefore not useful for exchange of oxygen or metabolites) is much larger than that contained in capillaries.

Hydrodynamics in a network or metabolism optimisation
Here we enter into details of the quantitative arguments put forward by West, Brown and Enquist. Their first explanation is based on the following assumptions:

1. Energy dissipation controlling B occurs during transport through nutrient supply networks.
2. These networks fill all space such that all cells are supplied.
3. Terminal branches (capillaries) of these networks have the same geometric and hydrodynamic features in all species.
4. They are self-similar in a bounded range of scales, with the upper bound being the size of the organism and lower bound the universal size of terminal branches.
5. Organisms have evolved to minimise the energy necessary for their survival, i.e. energy dissipated during fluid (blood or sap) transport through the network.

We will see that the combination of these geometric, dynamic, and energetic constraints is sufficient to explain empirical scaling laws, in particular Kleiber's law. We will use the vocabulary of the cardiovascular system, but the reasoning stays valid for the respiratory system (where the fluid is gas) or the plant vascular system (where the fluid is sap).

---

[3] These networks are written in a plane ($d = 2$) so it explains the existence of scaling laws with exponents that are multiple of $1/(d + 1) = 1/3$.

[4] We thank Pierre–Gilles de Gennes for suggesting this argument to us.

We denote network "levels" by indices $k = 0, \ldots, K$, k being also the number of branch points encountered since the aorta ($k = 0$). A branch of level $k-1$ splits into n_k branches.[5]So we have $N_k = n_0 n_1 \ldots n_k$ (with $n_0 = 1$ so $N_0 = 1$) and consequently $n_k = N_{k+1}/N_k$. The self-similarity of the network (assumption 4) ensures that $n_k = n$ is independent of k, so that $N_k \sim n^k$. A level k blood vessel has a radius r_k and length l_k. The blood in it flows with velocity u_k (averaged over the tube cross section) and the pressure drop between its ends is Δp_k. We define $\gamma_k = l_{k+1}/l_k$ and $\beta_k = r_{k+1}/r_k$. The flow rate in the branch is $\phi_k = \pi r_k^2 u_k$. The total flow rate in the system being constant (no accumulation of fluid, steady state regime), we have:

$$N_k \phi_k \equiv \pi N_k r_k^2 u_k = \text{constant} = \phi_0 \sim B \sim M^a \quad \text{so} \quad \frac{u_{k+1}}{u_k} = n\beta_k^2.$$

(11.2)

Here the scaling form $B \sim M^a$ is assumed from experimental results, the aim being to determine the value of its exponent a. According to assumption 3 the final level ($k = K$) is universal, which means that l_K, r_K and u_K are independent of M and therefore also of ϕ_K as well as $B \sim N_K$. As this total number N_K of branches behaves as n^K, we deduce that the number K of levels varies with the mass M of the organism as $K \sim a \log M / \log n$. Assuming the network is space filling (assumption 2), its fractal dimension is equal to 3 so $N_k \sim l_k^3$. Also, since $N_k \sim n^k$, we have $\gamma_k = \gamma \sim n^{-1/3}$.

One way to continue is to rely on the self-similarity (assumption 4) of the structure to assume that the total area at a given level k is independent of k. It then follows that $\beta_k = \beta \sim n^{-1/2}$ is independent of k and therefore the velocity $u_k = u$ is independent of k. As $n\gamma\beta^2 \sim n^{-1/3} < 1$ and $K \gg 1$, the total volume of fluid, proportional to the mass, is written:

$$V_K \sim \frac{(\gamma\beta^2)^{-K}}{1 - n\gamma\beta^2} \sim M,$$

(11.3)

where $K \sim -\log M / \log(\gamma\beta^2)$. Comparing this with the previous expression of K, we see[6]$a = -\log n / \log(\gamma\beta^2)$. Substituting in the values $\gamma \sim n^{-1/3}$ and $\beta \sim n^{-1/2}$, we finally find $a = 3/4$. This reasoning, relying on the *assumption* of conservation of total network cross section when we change

---

[5]To implement the fundamental hypotheses of the model simply, we add an assumption of network uniformity: branches starting at a given level $k - 1$ are all identical and characterised by a single branching number n_k. In the same way, branches of the same level k are described by the same parameters r_k, l_k, u_k. This assumption can be relaxed by introducing a statistical dispersion, without fundamentally changing the result. In this case the network is fractal in a statistical sense.

[6]This formula is true when n_k, γ_k and β_k do not depend on k.

levels, is correct for plants, where the result $u_k = u = \text{constant}$ is actually observed. However it is incorrect for cardio-vascular systems in mammals as it is in disagreement with experimental observations showing the slowing down of blood at the level of capillaries, thereby allowing the absorption of nutrients and gases.

Therefore we should abandon the hypothesis of conservation of total area and rely on the constraint of minimisation of dissipated energy (assumption 5). This boils down to minimising the total hydrodynamic resistance of the system.[7] In this way we obtain $\beta_k \sim n^{-1/3}$, which gives an incorrect exponent $a = 1$.

The answer to this problem adopted by West, Brown and Enquist is to resort to a hybrid model, in which the dependence of β_k on k takes the simple form of a crossover. We still have $\beta_k \sim n^{-1/2}$ in the first levels (conservation of total area for small values of k). However the expression for β_k in the later levels is determined by minimising the hydrodynamic resistance (which is actually essentially due to the capillaries). This then gives $\beta_k \sim n^{-1/3}$ at large values of k. In this way a ratio $u_0/u_K \approx 250$ agreeing with reality can be reproduced and a value of $a = 3/4$ obtained for the exponent. It can similarly be shown that the diameter $2r_0$ of the aorta grows with mass as $M^{3/8}$.

Criticisms of this explanation concern on one hand the *ad hoc* nature of this hybrid model, increasing the number of arguments and hypotheses to match experimental reality, and on the other hand the assumption (4) of self-similarity, from which it was deduced that $n_k = n$ did not depend on k. This is completely unnecessary and optimised networks can actually be constructed with n_k dependent on k [7, 16].

The second explanation, introduced to avoid appealing to (contested) hydrodynamic arguments, is more abstract. The reasoning is as follows. The metabolic system has scales $l_1 \ldots l_n$, varying with the size of the organism and a universal minimal scale l_0 (e.g. cross section of capillaries). The area of gas or nutrient exchange is written by simple dimensional analysis:

$$A(l_0, l_1, \ldots, l_n) = l_1^2 \, \widetilde{A}\left(\frac{l_0}{l_1}, \ldots, \frac{l_n}{l_1}\right). \tag{11.4}$$

---

[7] The hydrodynamic resistance of a branch of level k is given by Poiseuille's formula: $R_k = 8\mu l_k/\pi r_k^4$ where μ is the viscosity of the fluid. The resistances of each level add in series, whereas within a level branches are parallel and it is the inverse of their resistances which add, giving $R_{tot} = \sum_{k=0}^{K} R_k/N_k \approx R_K/[N_K(1 - n\beta^4)]$. However we note that this reasoning is questionable in the case of blood, which does not behave at all like a simple liquid (its circulation in capillaries that are adapted to the size of red blood cells is much faster).

Imagine a scaling transformation, which transforms l_i to λl_i, unless $i = 0$ as l_0 does not change. Consequently,

$$A(\lambda) \equiv A(l_0, \lambda l_1, \ldots, \lambda l_n) = \lambda^2 l_1^2 \, \widetilde{A}\left(\frac{l_0}{\lambda l_1}, \ldots, \frac{l_n}{l_1}\right). \tag{11.5}$$

In contrast to what we would have in a scale invariant network, $A(\lambda) \neq \lambda^2 A(1)$, i.e. there remains an explicit dependence on λ in $\widetilde{A}$. The hierarchical nature of metabolic systems (in the broad sense in that there may not necessarily be an actual distribution network) justifies a power law form for $\widetilde{A}$:

$$\widetilde{A}(x_0, x_1, \ldots, x_n) \sim x_0^{-\epsilon_A} \text{ with } 0 \leq \epsilon_A \leq 1. \tag{11.6}$$

Therefore,

$$A(\lambda) \sim \lambda^{2+\epsilon_A} A(1). \tag{11.7}$$

The biological volume V involved in metabolism is written $V \equiv AL \sim M$. For the characteristic length L we follow the same reasoning as for the exchange area A, leading to:

$$L(\lambda) \sim \lambda^{1+\epsilon_L} L(1) \tag{11.8}$$

where:

$$M(\lambda) \sim \lambda^{3+\epsilon_L+\epsilon_A} M(1) \tag{11.9}$$

and therefore:

$$A \sim M^{(2+\epsilon_A)/(3+\epsilon_L+\epsilon_A)}. \tag{11.10}$$

The exponents ϵ_A and ϵ_L are determined by writing that the metabolism is optimal (as a result of evolution). At fixed M, maximising A with respect to ϵ_A and ϵ_L leads to $\epsilon_A = 1$ and $\epsilon_L = 0$ from which we deduce that:

$$B \sim A \sim M^{3/4}. \tag{11.11}$$

The exponent of the exchange area is $2 + \epsilon_A = 3$, showing that the network is space filling and does involve all the cells of the organism.

11.2.3 Objections to the Reliability of the Scaling Law

However, the origin of allometric scaling laws, not to mention their validity, remains controversial. A statistical re-analysis of experimental data from Kleiber and his contemporaries seems to show that we cannot reject the intuitive value $\alpha = 2/3$

in favor of the more surprising value $\alpha = 3/4$ [16]. We suggest readers look at the original articles [11, 34] to form their own opinion. We draw out a few points that we consider to be indicative of the difficulty in experimentally establishing reliable scaling laws and hence of the need to at the same time rely on theoretical arguments to construct a coherent, solid and productive set of laws. First of all, the exponent obtained varies with how the data is separated a priori into groups (e.g. mammals/birds, large mammals/small mammals). It also varies if we exclude a priori data assumed to show a deviation from the law due to a specific anatomical reason or the nature of a particular environment. Finally, data points are few and far between and are obtained in different, sometimes indirect, ways, introducing a systematic bias with respect to the metabolic rate B found theoretically.

There are too many and contradictory models and theoretical justifications too settle the debate. Each rests on a set of more or less restrictive assumptions, which are more or less unfounded and can often be questioned [16]. Nevertheless, even if this idea of anomalous scaling laws arising from the hierarchical structure of metabolic networks is not as simple as the arguments detailed above claim, it remains an interesting paradigm to analyse living systems and even more to identify organising principles of their own. This, now famous, example shows how rigorous we must be in the experimental demonstration of a scaling law. We will return to this point in Chap. 12. For comparison, let us just recall the time and work spent by physicists to be convinced that the exponent β of the liquid–vapour transition was not $1/3$ but 0.322.

11.2.4   Other Examples

In the same vein, a simple mechanical model of the skeleton, taking into account the weight and elastic forces acting on bones subjected to constraints, predicts that the length L and diameter D of bones should vary as $L \sim M^{1/3}$ and $D \sim M^{3/8}$ [10].

Another example is observed in the cortex of mammals. Anatomy studies show that the proportion of grey matter G (the highly convoluted surface layer where local processes take place) and white matter (the innermost part, in particular containing fibres providing long distance connections through the cortex) follows a scaling law:

$$B \sim G^{1.23}. \tag{11.12}$$

One explanation proposed is the constraint of minimisation of average length of axons crossing the white matter. It agrees with the similarity in anatomy of mammals brains in terms of the organisation of white and grey matter [57].

A final example, in this case a counter example, is encephalisation (the amount of brain mass exceeding that related to the organism's total body mass). A statistical model determines (in vertebrates) the scaling relation between the mass m_{th} of the brain and the total mass M of the animal:

$$m_{th} \sim M^{0.76}. \tag{11.13}$$

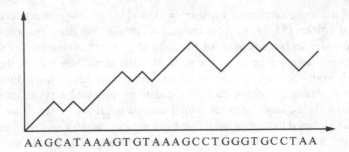

AAGCATAAAGTGTAAAGCCTGGGTGCCTAA

Fig. 11.2 Random walk representing a DNA sequence. In this example (of a real sequence) we see the necessity to correct the walk by locally eliminating the trend (drift) for sequences that have strong inhomogeneities in the proportion of base types (purines A and G or pyrimidines T and C)

It describes the way in which the size of the brain increases simply because the animal is larger [20]. The deviation from this law, more precisely the ratio m/m_{th} between the mass m_{th} estimated in this way and the true mass is the measure of the encephalisation, that is to say the *relative* development of the brain that can be interpreted in terms of increased cognitive abilities.

11.3 Long Range Correlations

Complex and hierarchical structures result in anomalies in correlation functions in general. The divergence of the range ξ of correlations (e.g. spatial) is reflected in the transition from an exponential decay $C(r) \sim e^{-r/\xi}$ to a power law decrease $C(r) \sim r^{-\alpha}$. So the presence of such long range correlations in observations related to the functioning of a living creature is nothing surprising [49, 51].

11.3.1 Coding and Non Coding DNA Sequences

As we have already mentioned in Sect. 8.1, a technique to read the four letter (A, C, G, T) "text" a DNA molecule is made up of makes it possible to not only detect long range correlations but also to associate them with non coding regions, that is regions whose sequence is never translated into proteins [27].

   The idea is to interpret the genetic sequence as a random walk in one dimension. The purines A and G correspond to a step up whereas their complementary bases (pyrimidines) T and C correspond to a step down. Starting from 0, the position of this fictitious walker is $y(N)$ after having read the Nth letter[8] (see Fig. 11.2). So we

---

[8]A being paired with T and G with C in the DNA double helix, the two complementary strands making up the molecule will give walks with $y \leftrightarrow -y$ symmetry, which is satisfying (and justifies the choice of the displacement rule!).

Table 11.1 Analysis of correlations in a signal u. Remember that the three exponents are related by $\beta = 1 - \alpha = \gamma - 1$. The exponent γ of increments w of u is $\gamma_w = \gamma - 2$ and that for the integrated system y is $\gamma_y = \gamma + 2$

Long range anti-correlations	$0 < \gamma < 1$	$1 < \alpha < 2$	*
White noise	$\gamma = 1$	$\alpha = 1$	$\beta = 0$
Power law correlations	$1 < \gamma < 2$	$0 < \alpha < 1$	$0 < \beta < 1$
Long range correlations	$2 \le \gamma < 3$	*	$1 \le \beta < 2$
Wiener process	$\gamma = 3$	*	$\beta = 2$

determine the diffusion law $\langle y^2(N) \rangle \sim N^\gamma$ for the sequence under consideration. In the absence of correlations between successive steps, or if correlations are short ranged, we have $\gamma = 1$ (Sect. 4.3.1). A value of $\gamma \ne 1$ with $\gamma < 2$ reveals the presence of persistent long range correlations[9] if $\gamma > 1$. So correlations between steps, i.e. between bases, decay with a power law $C(t) \sim t^{-\alpha}$ with $\alpha = 2 - \gamma$ (see Table 11.1). The remarkable point, whose biological meaning is still not well understood, is that the exponent γ observed here depends on whether the sequence considered is coding (when it is $\gamma \approx 1$) or non coding (when it is clearly $\gamma > 1$). Coding sequences only have very short ranged correlations (<10 base pairs).

A preliminary explanation is that selection pressure (natural selection over the course of evolution) is much weaker in non coding regions, allowing repeated and abnormally correlated sequences to be present. In support of this, it was shown that γ grew over evolution by adding non coding sequences and several models of sequence evolution have been proposed along these lines. One direction, which is very interesting but still very much a debated hypothesis, is that non coding sequences could contribute to controlling the tertiary structure of DNA in the cell nucleus. The correlations would then reflect their role at this higher level of organisation [4, 6].

Many studies are currently in progress to clarify this property, its origins and possible biological interpretations. For example, statistical methods are used to eliminate bias and non-stationarity (drift) in the walk (see below). Other procedures, coming directly from physics, are also proving fruitful, namely multifractal analysis (Sect. 9.5.6), wavelet transform and local spectral decomposition carried out at an adjustable scale (see Sect. 11.3.2) [3, 4].

Detrended fluctuation analysis
Usual statistical analysis of fluctuations of a temporal signal $u(t)$ assumes that this signal is stationary, that is to say statistically invariant to transformations in time. The autocorrelation function is estimated according to the formula:

---

[9]Other potential causes of anomalous diffusion (broad step distributions, trapping etc.) are absent here.

$$C(t) = \frac{1}{N-1} \sum_{s=0}^{N-1} \left[u(t+s)u(s) - \left(\frac{1}{N} \sum_{s=0}^{N-1} u(s) \right)^2 \right]. \tag{11.14}$$

Equivalently, the spectrum $S(f)$ of fluctuations (Fourier transform of the autocorrelation function $C(t)$), known by the name of *power spectrum*, can be studied. But the assumption of stationarity sometimes turns out to be very wrong. To overcome this difficulty, a correction method has been developed consisting of removing the drift or trend (that is to say the movement of the mean). It is known by the name *detrended fluctuation analysis* (DFA). The novelty and power of this method is to correct the signal by subtracting its *instantaneous* mean (average signal) calculated *at an adjustable scale* [28,44].

The elementary time step is fixed once for all, usually prescribed by the experimental device (or even by the system itself in the case of DNA sequences). Let u be the signal, recorded over a time N. First of all its integral $y(k) = \sum_{i=1}^{k} u_i$ is calculated. Then the observation time interval N is subdivided into time windows of duration n where n is adjustable. In each time window, the local trend of the time series $y(k)$ is calculated, i.e. the straight line that best fits the series $y(k)$ (found by a method of least squares, Fig. 11.3). In this way a series of segments representing the local "deterministic" (linear with respect to time) trend of the signal is constructed. At each moment in time k, the point on the corresponding segment is $y_n(k)$. This *local* trend will be used to correct the non-stationarity of the integrated signal (note that if the signal u is actually stationary, with average $\langle u \rangle$, the series of segments reduces to a single line $y_n(k) = k\langle u \rangle$ for any n large enough that fluctuations from the local mean are negligible). We will then analyse the standard deviation of this integrated and "detrended" signal:

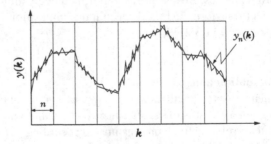

Fig. 11.3 Analysis of non-stationary signal fluctuations. *Thin lines* represent the local linear trend $y_n(k)$, obtained for each window of the integrated signal $y(k)$ by a method of least squares fit. The length of the windows n is an adjustable variable which will affect the standard deviation $F(n)$ of fluctuations $y(k) - y_n(k)$ [44]

$$F(n) = \sqrt{\frac{1}{N} \sum_{k=1}^{N} [y(k) - y_n(k)]^2} \sim n^{\gamma/2}. \qquad (11.15)$$

This standard deviation depends on the scale n at which the average is calculated (it grows with n). Fractal fluctuations lead to a power law dependence of $F(n)$, with exponent $\gamma/2$. To better understand this quantity $F(n)$, let us imagine a stationary signal x for which it makes sense to calculate the temporal autocorrelation function $C(t)$ and power spectrum $S(f)$. In the case of a signal with long range correlations, we would have:

$$\begin{cases} C(t) \sim t^{-\alpha} \\ S(f) \sim 1/f^{\beta} \qquad \text{with} \quad \beta = 1 - \alpha = \gamma - 1. \\ F(n) \sim n^{\gamma/2} \end{cases} \qquad (11.16)$$

We have long range correlations following a power law only if $\alpha > 0$, or equivalently if $\gamma < 2$. This criterion $\gamma < 2$ is what will be used in the more general case of a non-stationary signal, where only the function $F(n)$ calculated from the "corrected" signal has any meaning. We have seen in Sect. 4.5.3 that typical self-similar processes [10]producing long range correlations are fractal Brownian motions. Their characteristic exponent H (Hurst exponent) is identified with the exponent $\gamma/2$ of the "diffusion law" $F(n) \sim n^{\gamma/2}$. The value of the exponents can be understood with reference to the case where y is such a process, in which case $\gamma > 1$ corresponds to a persistent motion whereas $\gamma < 1$ corresponds to anticorrelations. The borderline case $\gamma = 1$ corresponds to the case where y is a Wiener process (simple Brownian motion). The signal x is then a white noise and its correlation function is identically zero. The limiting case $\alpha = 0$ corresponds to a $1/f$ noise and therefore $\beta = 1$ and $\gamma = 2$. If $2 < \gamma \leq 3$, we still have long range correlations but they no longer follow a power law. In particular, if the signal x is a Wiener process (also called "Brownian noise") we have $\gamma = 3$ and $\beta = 2$ and it is correlations of the increments that asymptotically follow a power law. These different cases are summarised in Table 11.1. The exponent γ can also be seen as a roughness exponent of the line making up the integrated signal y (see Sect. 8.1.2), i.e. the larger γ, the "smoother" this curve.

---

[10]More specifically, these are the exactly self-similar representatives (fixed points of renormalisation transformations) of the universality classes associated with different anomalous diffusion laws.

11.3.2   Heart Rate

A question with obvious medical applications is to extract the maximum amount of information about the functioning of the heart just from a recording of the heart rate (electrocardiogram). One method to analyse this signal is to consider the time interval $u(j)$ between the jth and the $(j+1)$th beat. Note that here the discretisation is intrinsic to the system and so takes the basal heart rate as the (discrete) reference time. The series of intervals is extremely irregular and looks like a very noisy signal. To go beyond this observation quantitatively, we analyse the fluctuations of u or the fluctuations of its increments $w(n) = u(n+1) - u(n)$ [5]. The DFA method presented in the previous section is used to correct potential non-stationarity in the recorded signal [14]. It shows long range correlations in u as well as in w. Their correlation functions are scale invariant, following a power law: $C(t) \sim t^{-\alpha}$. The exponents corresponding to w are related to those characterising u according to $\gamma_w = \gamma - 2$, $\alpha_w = \alpha + 2$ and $\beta_w = \beta - 2$ (with $\beta = 1 - \alpha = \gamma - 1$ for both sets of exponents). As in the context of DNA sequences, better quality results are obtained by multifractal analysis (Sect. 9.5.6) or by wavelet transform (see below).

The remarkable point for the biologist is that these quantitative characteristics show significant differences with age, activity level (awake or asleep, e.g. [33]) and various pathologies. The disappearance of anti-correlations observed in w is associated with a pathological functioning (although it is still not possible to establish a causal link). For example, the exponent γ is 2.10 in a waking state, 1.7 during sleep and 2.40 in a heart attack, whereas we would find $\gamma = 3$ if u was a Wiener process (w being then white noise). From a physiological point of view, u and w give information about heart rate regulation. Identification of these long range correlations in a healthy individual indicates that this rhythm is regulated over a large range of temporal scales.

It was recently shown that heart rate was multifractal [30, 31]. It seems then that the collective dynamics of the set of cardiac cells and nerve cells controlling their activity is very complex and shows, as well as their (vital!) property of global synchronisation, a hierarchical spatiotemporal structure involving many temporal scales and responsible for the scale invariance observed in the fluctuations. The loss of this complexity is associated with heart diseases.

Wavelet transform

The basic tool to carry out a spectral analysis is the Fourier transform. It decomposes the signal $u(t)$, for example a sound signal, into purely sinusoidal components $\hat{u}(\omega)e^{i\omega t}$. The Fourier transform $\hat{u}$ of u therefore indicates which frequencies ω are present in the signal and with what weight $\hat{u}(\omega)$ they contribute. It therefore proposes a completely transverse view of the signal and we also talk about "conjugate" space to designate the frequency space (or wavevector space in the spatial case).

Fig. 11.4 An example of a wavelet g. It is even and integrates to zero

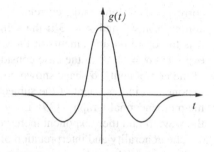

The aim of wavelet transform is to carry out a *local* spectral decomposition, keeping track of the temporal sequence. For a sound signal it consists in reconstructing the musical score. We want to determine not only the notes (frequencies) involved but also the moment when they are played (their position in the music) and their duration (minims, crotchets, quavers etc.). This objective is achieved by carefully choosing the kernel of the transformation. The trigonometric function ($\sin \omega t$ and $\cos \omega t$, or $e^{i\omega t}$) is replaced by a function g that is *localised* (zero outside an interval) and integrates to zero (for example a function in the form of a "Mexican hat", see Fig. 11.4). The result will depend on this "wavelet" g, the time t_0 around which we are placed and an adjustable scale factor λ, enabling signal analysis at an appropriate scale, as you would do with a microscope of adjustable magnification:

$$U(g, t_0, b) \equiv \frac{1}{\lambda} \int u(t) \, g\left(\frac{t - t_0}{\lambda}\right) dt. \tag{11.17}$$

This transformation is commonly used in signal processing in very diverse contexts when the phenomenon observed has a multiscale structure, for example turbulence, fractal structure and signals from complex dynamics.

11.3.3 Electroencephalography (EEG)

Analysis of electroencephalograms (EEG) can be done following statistical methods analogous to those presented in the context of DNA sequences and heart rate [37]. Here also, long range temporal correlations have been shown. For example, in spontaneous activity of the brain (when the eyes are closed), certain recordings show a power law spectrum $S(f) \sim 1/f^\beta$ with $\beta = 1.52$, corresponding to an exponent $\gamma = 2.52$ (using the same notation as the previous section) [52,53]. This corresponds to long range correlations but decreasing faster than a power law (for comparison we had $\gamma = 3$ if the signal is Brownian noise i.e. a Wiener process). It is therefore the increments of the signal that show power law correlations (with exponent $\alpha = 1.48$ in the case considered). Other recordings, focused on analysis of the component[11] α, have shown power law correlations $C(t) \sim t^{-0.6}$, where the exponent is independent of the subject and the recording method (EEG or MEG – magnetoencephalography) [39]. Long range correlations have also been detected in the wave β and their exponent increases with the level of vigilance [46].

The generality and interpretation of long range correlations observed in EEG are still debated. Their observation suggests that the brain functions around a critical state [25]. A more certain conclusion is that the existence of such correlations strongly calls into questions many statistical methods of EEG analysis founded on the assumption that no such correlations exist. For example, representing EEG as a linear superposition of white or coloured[12] ($\gamma = 1$) noise is invalid. Either it involves fractal Brownian motions, or we must abandon the linear approach. From a dynamic perspective, the interpretation of long range correlations is that the observed signal reflects collective dynamics of the underlying network of neurons, in which emergent properties have a large range of time and most probably spatial scales. A spatiotemporal exploration is therefore necessary to support this view.

11.4 Biological Networks: Complex Networks

11.4.1 A New Statistical Mechanics

Determining the "macroscopic" behaviours arising from the assembly of a large number of known and well described "microscopic" elements is the very heart of statistical physics. However, in the systems we have seen treated successfully by statistical physics methods, interactions between elements remained relatively simple and in any case fully determined, including at critical points. These systems include ferromagnetic interactions (Chap. 1) and excluded volume interactions between monomers in a polymer (Chap. 6). In such systems, interaction networks are simple and regular, being square or cubic networks (or triangular or hexagonal variants) where each element is connected to its nearest neighbours, or "infinite dimensional" networks where each element is connected to all the others. Often, they are simply the result of a discretisation of the physical space in which the phenomenon takes place.

---

[11]The EEG signal is traditionally subdivided into 7 components whose frequencies lie in separate well determined bands, which we call the waves δ, θ, α_1, α_2, β_1, β_2 and γ (in order of increasing frequency). The component α lies between 8 and 13 Hz and β between 13 and 20 Hz.

[12]Noise whose amplitude depends on time but always without temporal correlations.

However, interactions are not always as basic and can form networks of irregular architecture. This is often the case outside of the realm of physics, for example gene networks, metabolic networks inside a cell [32], neural networks [2], ecological networks [45], communication networks and social networks [50].

Statistical theory of networks is a way of approaching the study of these systems. It extends percolation models to much more complex topologies and in which physical space and the associated concept of proximity may be absent. The main aim is to formulate quantities to quantitatively describe the topological properties of a (model or real) network and to define statistical ensembles of networks sharing the same features, thus providing simple and robust models to represent natural networks. We suggest [43] for a complete review. As in the case of percolation, knowing and modelling the network structure is crucial to predict and quantify transport properties (e.g. propagation of information and viruses by the Internet, infectious diseases in a population structured by social connections, drugs in a metabolic system, etc.) and understand all phenomena in which local dynamics of a node is coupled to that of neighbouring nodes in the network.

11.4.2 Real Networks: Random Graphs?

The simplest model, developed by Erdös and Rényi [19], consists of randomly establishing, with a probability p, a bond between *any* two sites. It is therefore an extension of the bond percolation model (Sect. 5.1) where bonds only connect nearest neighbour sites, with probability p. Here, *there is no longer an underlying real space*. The topology of the network is entirely determined by the set of connections, i.e. the neighbourhood of site i is defined by the sites directly connected to it. The distance between two sites is given by the number of bonds in the shortest path connecting them. The diameter of the network will be the average distance between two sites.

A real network of N nodes (sites) and K bonds may be compared to a random graph having on average the same number of bonds, that is to say the graph defined by the probability $p = 2K/N(N-1)$. Unlike regular grids, the Erdös–Rényi model reproduces well a feature common to many real networks: their small diameter. We talk about "small world" networks in reference to the experiment conducted by the sociologist Milgram in 1967, showing that, on average, only 6 links of acquaintance connected any two people living in the United States of America (sometimes referred to the "six degrees of separation"). More specifically, it can be shown that the diameter of a random graph of N nodes grows only as $\log N$, whereas it grows as $N^{1/d}$ for a regular grid of dimension d. Due to this, the Erdös–Renyi model has long been used to model real networks. However, over the past decade or so the arrival of experimental data on large networks (communication networks, biological networks, etc.) has shown that it did not at all reproduce several other basic properties of real networks, primarily their degree distribution $P(k)$ (probability that a site is connected to k other sites). It can be shown that the

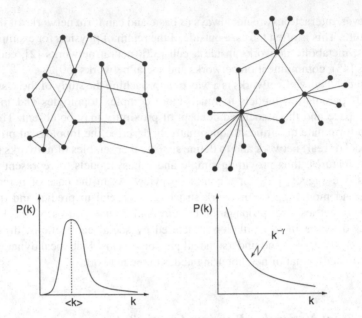

Fig. 11.5 Comparison between random graphs (*left*) and scale-free networks (*right*). Random graphs are relatively homogeneous, whereas in scale-free networks there exist "hub" sites with a large number of connections. Random graphs have a degree distribution $P(k)$ peaked at $\langle k \rangle$ and decaying exponentially, whereas scale-free networks have a power law distribution $P(k) \sim k^{-\gamma}$

Erdös–Renyi model leads to a Poisson distribution peaked at $k = \langle k \rangle$ and decaying *faster than exponential* at large values of k:

$$P(k) = \frac{\lambda^k e^{-\lambda}}{k!} \qquad \text{with} \quad \lambda = \langle k \rangle. \qquad (11.18)$$

In contrast, observed networks have a very broad distribution $P(k)$ decaying much more slowly, typically *as a power law* (Fig. 11.5):

$$P(k) \sim k^{-\gamma} \qquad (k \text{ large}). \qquad (11.19)$$

In such networks, there is no typical degree.[13] We therefore call them *scale-free networks*. The power law decay of $P(k)$ reflects the large heterogeneity of degrees (their variance diverges if $\gamma < 3$) and consequently the non negligible probability that highly connected nodes exist. Another feature of real networks that is not shared with random graphs is the strong clustering coefficient, that is to say the fraction of

---

[13]The average number of bonds per node can be infinite (if $\gamma < 2$) but even if it is finite, it does not correspond to the typical degree.

links present between the neighbouring sites of a given site (it measures to what extent "my friends are friends with each other").

11.4.3 Scale-Free Networks

The correct paradigm to describe real networks then seems to be that of scale-free networks with a very broad degree distribution, or even the narrower class of networks in which the degree distribution follows a power law $P(k) \sim k^{-\gamma}$ at large k. When the connections are directed we must distinguish between $P_{in}(k)$ (probability that k connections enter a given site) and $P_{out}(k)$ (probability that k connections leave a given site), defining two exponents γ_{in} and γ_{out}. For example, metabolic networks (observed on 43 species belonging to three kingdoms [32]) show this power law, with $\gamma_{in} = 2.4$ and $\gamma_{out} = 2.0$. This is also the case for the Internet and the World Wide Web (with different exponents) [43].

Differences observed between properties of these networks and those of random graphs suggest that real complex networks are not constructed at random but by following organisational principles and that the topological differences reflect this different creation. What realistic mechanism of self-organisation can account for this absence of characteristic scale and the form observed for $P(k)$? The simplest model, known by the name of "preferential attachment model" supposes that connections are created preferentially towards sites that are already very connected: "the rich get richer" [1]. This process of growth actually leads to a power law degree distribution with $\gamma = 3$. The number of publications on the subject has grown and is still growing exponentially and many variations and extensions have been proposed, which can be found in [18] and [43].

Another property of scale-free networks is the existence of key nodes that are extremely well connected (called *hubs*). Their presence significantly increases the efficiency of communication in the network and contributes to decreasing the diameter. Network properties are greatly affected by the distribution of such hubs. On the other hand, these networks are very robust to random loss of nodes. Ninety percent of nodes chosen at random can be dammaged without significantly affecting communication in the network [43]. In this model we reproduce the ambivalent behaviour of real networks, which are robust to random failures but fragile with respect to targeted attacks on hubs.

However in many real networks (food chain, metabolic networks, etc.), due to the limited number of nodes and the relatively small maximum number of connections they can establish, their degrees are intrinsically and so irredeemably bounded and their distribution only extends over a limited span, often less than two decades [26]. It is therefore impossible to reliably show, certainly not definitively, a power law behaviour of the degree distribution (a power law over two decades can easily be reproduced by a superposition of two or three exponentials). We should therefore only keep the features of the ideal model that remain significant, namely the large heterogeneity of degrees, reflected in a very broad distribution, as opposed to the very peaked law of regular networks and the exponential decay of random graphs. A critical and historical analysis of this concept of scale-free network is presented in [21].

11.4.4 In Practice: Reconstruction of a Complex Network

A major problem is that of the reconstruction of interaction networks from experimental data, for example gene networks (a connection $i \rightarrow j$ meaning that the protein coded for by gene i is involved in the regulation of the expression of gene j) or protein networks (a link $i \rightarrow j$ meaning that the proteins i and j bind to each other). This issue is central in biology today and it motivates two types of difficult and costly experiments. The first is to record in parallel a large number of variables intended to describe the state of all the nodes (for example the transcriptome, which is the expression level of all genes in a genome, or the proteome, which is the set of all proteins expressed at a given instant). The second is a series of separate experiments determining for each pair of nodes whether or not there is an interaction.

The first type of approach has the advantage of giving a global picture of the system, but it can only access correlations between nodes (by comparing data obtained at different moments in time or under different conditions) resulting in a correlation matrix C_{ij}. It could be *represented* by a network by placing with weight C_{ij} a connection between nodes i and j, but this must not be confused with the network of direct interactions between nodes i and j. In particular, a path has no meaning in the network representing the correlation matrix.

Pairwise reconstruction methods will give the set of possible connections, however the network actually active at a given instant in time will only be a subset of this. It will then be necessary to determining the "true" configuration of the network and how it varies during functional activity. Many experimental and theoretical efforts are underway today to develop reconstruction and analysis methods for biological networks. The difficulty is even greater in that it is not only to obtain the statistical properties of the network but also to describe and understand its functional dynamics. The impact of network topology on spatiotemporal phenomena taking place in it is then a crucial but still wide open question [9, 13, 29, 40].

11.4.5 Avenues to Explore the Multiscale Structure of Networks

One way to describe and exploit the possible multiscale structure of a network is to identify *modules*, defined as sub-networks in which the nodes are more densely interconnected than they are with the rest of the network. They are also referred to as *communities* in the context of social networks. Many algorithms have been developed to identify these communities, based for example on methods of hierarchical clustering [41] or using an auxiliary random walk remaining trapped for long durations in different communities [24,38]. In biological contexts, modules are often interpreted as functional units, but this identification is a bit hasty. Since biological processes are essentially out of equilibrium and active, a definition of function based on flow through the network seems more relevant [42].

A concept of self-similarity was defined for networks by Gallos et al. [23] according to a simple extension of the concept of self-similarity introduced in fractal geometry (Chap. 2). However, its scope is limited to networks written naturally in a metric space (that is to say provided with a distance), typically a plane or three dimensional physical space. But we have already pointed out that for many networks such as the food chain, metabolic networks or biochemical regulation networks, the only intrinsic distance is that associated with the connectivity [48].

The generalisation to complex networks of renormalisation methods developed for percolation networks will hit the major difficulty of defining super nodes and super links.[14] By adapting a coarse-graining procedure identifying the super nodes with modules, we expect three typical behaviours: (i) convergence under the action of renormalisation towards a complete network in which each node is connected to all the others; (ii) convergence towards a regular fractal network for example a tree or linear network (i.e. $P(k) = \delta(k - k_0)$ with k_0 equal to a few units); (iii) convergence towards a network with a broad degree distribution, the typical (but ideal) case being a power law distribution $P(k) \sim k^{-\gamma}$.

As in the case of percolation, the aim of renormalisation will be to accentuate and purify the large scale features of the network to better discriminate different categories of behaviour. The objective is to justify the use of very simple hierarchical and exactly self-similar reference models and to adjust their parameters by applying the renormalisation operator to networks reconstructed from experimental data (if there are enough nodes). Then we will be able to more fully exploit renormalisation transformation in the study of dynamic properties, for example to determine the statistical and topological properties of a network that will influence the scaling laws describing transport phenomena and change the value of their exponents compared to those observed in a regular or random network [22].

11.5   Conclusion: A Multiscale Approach for Living Systems

Living organisms, even the simplest, are complex systems in the sense that new global behaviour emerges from the assembly of simple elements and feeds back on the properties of these elements giving the system the capacity for adaptation. Scaling laws that can be satisfied by this global behaviour provide access to

---

[14]We must be careful not to limit the study to network models intrinsically adapted to a renormalisation process [17]. The action of the renormalisation operator would then be clearly defined only on this particular network and be therefore difficult to exploit.

valuable quantitative information on the organisation of the system and on the way in which the whole and the parts are connected structurally and functionally. A good indication of the possible existence of such scaling laws is the presence of underlying fractal structures, themselves resulting from an optimisation process, typically optimisation of exchange surface areas or fixed volume interactions (e.g. lungs, vascular system or nervous system). Structural and dynamical aspects can rarely be separated and scaling laws can also be detected in the analysis of temporal data that can be obtained on various aspects of function of a living being (as we have seen in the case of electrocardiograms and EEG) or even at the scale of populations. Multiscale approaches are essential to understand the functional connections between different levels of organisation of a living creature, from the molecular scale to that of the organism. Such approaches are particularly needed to explain the emergence and persistence of the creatures over the course of evolution. The scale invariances we have just met provide an initial pointer to set out on this path crossing different levels of observation and ultimately leading to a global description of living systems.

References

1. R. Albert, A.L. Barabasi, Statistical mechanics of complex networks. Rev. Mod. Phys. **74**, 47 (2001)
2. D. Amit, *Modelling Brain Functions, the World of Attractor Neural Networks* (Cambridge University Press, 1989)
3. A. Arneodo, E. Bacry, P.V. Graves, J.F. Muzy, Characterizing long-range correlations in DNA sequences from wavelet analysis. Phys. Rev. Lett. **74**, 3293 (1995)
4. A. Arneodo, C. Vaillant, B. Audit, F. Argoul, Y. d'Aubenton-Carafa, C. Thermes, Multi-scale coding of genomic information: From DNA sequence to genome structure and function. Physics Reports **498**, 45 (2011)
5. Y. Ashkenazy, P.Ch. Ivanov, S. Havlin, C.-K. Peng, A.L. Goldberger, H.E. Stanley, Magnitude and sign correlations in heartbeat fluctuations. Phys. Rev. Lett. **86**, 1900 (2001)
6. B. Audit, C. Thermes, C. Vaillant, Y. d'Aubenton-Carafa, J.F. Muzy, A. Arneodo, Long-range correlations in genomic DNA: a signature of the nucleosomal structure. Phys. Rev. Lett. **86**, 2471 (2001)
7. J.R. Banavar, A. Maritan, A. Rinaldo, Size and form in efficient transportation networks. Nature **399**, 130 (1999)
8. J.J. Blum, On the geometry of four-dimensions and the relationship between metabolism and body mass. J. Theor. Biol. **64**, 599 (1977)
9. S. Boccaletti, V. Latora, Y. Moreno, M. Chavez, D.U. Hwang, Complex networks: Structure and dynamics. Physics Reports **424**, 175 (2006)
10. K. Bogdanov, *Biology in Physics* (Academic Press, 2000)
11. S. Brody, *Bioenergetics and Growth* (Reinhold, New York, 1945)
12. J.H. Brown, G.B. West (eds), *Scaling in Biology* (Oxford University Press, 2000)
13. A.R. Carvunis, M. Latapy, A. Lesne, C. Magnien, L. Pezard, Dynamics of three-state excitable units on random *vs* power-law networks: simulations results. *Physica A* **367**, 595 (2006)
14. Z. Chen, P.Ch. Ivanov, K. Hu, H.E. Stanley, Effect of nonstationarities on detrended fluctuation analysis. Phys. Rev. E **65**, 041107 (2002)
15. W.T. D'Arcy, *On Growth and Form* (Cambridge University Press, 1917)

16. P.S. Dodds, D.H. Rothman, J.S. Weitz, Re-examination of the "3/4-law" of metabolism. J. Theor. Biol. **209**, 9 (2001)
17. S.N. Dorogovtsev, Renormalization group for evolving networks. Phys. Rev. E **67**, 045102 (2003)
18. S.N. Dorogovtsev, J.F.F. Mendès, Evolution of networks. Adv. Phys. **51**, 1079 (2002)
19. P. Erdös, A. Rényi, On the evolution of random graphs. Publ. Math. Inst. Hung. Acad. Sci. **5**, 17 (1960)
20. R.A. Foley, P.C. Lee, Ecology and energetics of encephalization in hominid evolution. Phil. Trans. Roy Soc. London B **334** 223 (1991)
21. E. Fox Keller, Revisiting "scale-free" networks. BioEssays **27**, 1060 (2005)
22. L.K. Gallos, C. Song, S. Havlin, H.A. Makse, Scaling theory of transport in complex biological networks. Proc. Natl. Acad. Sci. USA **104**, 7746 (2007)
23. L.K. Gallos, C. Song, H.A. Makse, A review of fractality and self-similarity in complex networks. Physica A **386**, 686 (2007)
24. B. Gaveau, L.S. Schulman, Dynamical distance: coarse grains, pattern recognition and network analysis. Bull. Sci. Math. **129**, 631 (2005)
25. D.L. Gilden, T. Thornton, M.W. Mallon, $1/f$ noise in human cognition. Science **267** 1837 (1995)
26. N. Guelzim, S. Bottani, P. Bourgine, F. Képès, Topological and causal structure of the yeast transcriptional regulatory network. Nature Genetics **31**, 60 (2002)
27. S. Havlin, S.V. Buldyrev, A. Bunde, A.L. Goldberger, P.Ch. Ivanov, C.K. Peng, H.E. Stanley, Scaling in nature: from DNA through heartbeats to weather. Physica A **273**, 46 (1999)
28. K. Hu, P.Ch. Ivanov, Z. Chen, P. Carpena, H.E. Stanley, Effect of trends on detrended fluctuation analysis. Phys. Rev. E **64**, 011114 (2001)
29. M. Hütt, A. Lesne, Interplay between topology and dynamics in excitation patterns on hierarchical graphs. Front. Neuroinform. **3**, 28 (2009)
30. P.Ch. Ivanov, L.A.N. Amaral, A.L. Goldberger, S. Havlin, M.G. Rosenblum , Z. Struzik, H.E. Stanley, Multifractality in human heartbeat dynamics. Nature **399**, 461 (1999)
31. P.Ch. Ivanov, L.A.N. Amaral, A.L. Goldberger, S. Havlin, M.G. Rosenblum, H.E. Stanley, Z. Struzik, From $1/f$ Noise to Multifractal Cascades in Heartbeat Dynamics. Chaos **11**, 641 (2001)
32. H. Jeong, B. Tombor, R. Albert, Z.N. Oltvai, A.L. Barabasi, The large-scale organization of metabolic networks. Nature **407**, 651 (2002)
33. J.W. Kantelhardt, Y. Ashkenazy, P.Ch. Ivanov, A. Bunde, S. Havlin, T. Penzel, J.-H. Peter, H.E. Stanley, Characterization of sleep stages by correlations of heartbeat increments. Phys. Rev. E **65**, 051908 (2002)
34. M. Kleiber, Body size and metabolism. Hilgardia **6**, 315 (1932)
35. M. Kleiber, Body size and metabolic rate. Physiol. Rev., **27**, 511 (1947)
36. M. Kleiber, *The Fire of Life: An Introduction to Animal Energetics* (Wiley, New York, 1961)
37. J.M. Lee, D.J. Kim, I.Y. Kim, K.S. Park, S.I. Kim, Detrended fluctuation analysis of EEG in sleep apnea using MIT/BIH polysomnography data. Comput. Biol. Med. **32**, 37 (2002)
38. A. Lesne, Complex networks: from graph theory to biology. Lett. Math. Phys. **78**, 235 (2007)
39. K. Linkenkaer-Hansen, V.V. Nikouline, J.M. Palva, R.J. Ilmoniemi, Long-range temporal correlations and scaling behavior in human brain oscillations. J. Neurosci. **21**, 1370 (2001)
40. C. Marr, M.T. Hütt, Similar impact of topological and dynamic noise on complex patterns. Phys. Lett. A **349**, 302 (2006)
41. M.E.J. Newman, J. Girvan, Finding and evaluating community structure in networks. Phys. Rev. E **69**, 026113 (2004)
42. J. Papin, J.L. Reed, B.O. Palsson, Hierarchical thinking in network biology: the unbiased modularization of biochemical network. Trends Biochem. Sci. **29**, 641 (2004)
43. R. Pastor-Satorras, A. Vespignani *Evolution and Structure of the Internet* (Cambridge University Press, Cambridge, 2004)
44. C.K. Peng, S. Havlin, H.E. Stanley, A.L. Goldberger, Quantification of scaling exponents and crossover phenomena in nonstationary heartbeat time series. Chaos **5**, 82 (1995)

45. S.L. Pimm, *Food Webs* (Chapman & Hall, London, 1982)
46. L. Poupard, R. Sartene, J.C. Wallet, Scaling behavior in beta-wave amplitude modulation and its relationship to alertness. Biol. Cybern. **85**, 19 (2001)
47. F. Rothen, *Physique générale. La physique des sciences de la nature et de la vie*, Chap. 17, (Presses polytechniques et universitaires romandes, Lausanne, 1999)
48. C. Söti, P. Csermely, Aging cellular networks: chaperones as major participants. Exp. Gerontol. **42**, 113 (2007)
49. H.E. Stanley, S.V. Buldyrev, A.L. Goldberger, Z.D. Goldberger, S. Havlin, R.N. Mantegna, S.M. Ossadnik, C.K. Peng, M. Simons, Statistical mechanics in biology: how ubiquitous are long-range correlations? Physica A **205**, 214 (1994)
50. S.H. Strogatz, Exploring complex networks. Nature **410**, 268 (2001)
51. T. Vicsek (Editor), *Fluctuations and Scaling in Biology* (Oxford University Press, Oxford, 2001)
52. P.A. Watters, Fractal structure in the electroencephalogram. Complex. Int. **5** (1998)
53. P.A. Watters, Time-invariant EEG power laws. Int. J. Syst. Sci. (Special Issue on Emergent Properties of Complex Systems) **31**, 819 (2000)
54. G.B. West, The origin of universal scaling laws in biology, Physica A **263**, 104 (1999)
55. G.B. West, J.H. Brown, B.J. Enquist, A general model for the origins of allometric scaling laws in biology. Science **276**, 122 (1997)
56. G.B. West, J.H. Brown, B.J. Enquist, The fourth dimension of life: fractal geometry and allometric scaling of organisms. Science **284**, 1677 (1999)
57. K. Zhang, T.J. Sejnowski, A universal scaling law between gray matter and white matter of cerebral cortex. Proc. Natl. Acad. Sci. USA **97**, 5621 (2000)

Chapter 12
Power and Limits of Scaling Approaches

12.1 Criticality and Scaling Laws

After the wide panorama presented in this book, we can draw a few general conclusions on scaling approaches. We have seen that they apply in situations when the system no longer has a characteristic scale, as for example observed for a temperature $T \to T_c$ in the case of phase transitions, for a number of monomers $N \to \infty$ in polymers, for a filling density $p \to p_c$ in percolation, for a time span $t \to \infty$ in growth or diffusion, at the onset of chaos in the transition to chaos by period doubling or for a Reynolds number $Re \to \infty$ in turbulence. This absence of characteristic scale is quantitatively reflected in the *divergence of the range ξ of correlations*. So we have scale invariance generically. A signature, which is both fundamental theoretically and exploitable experimentally (or numerically), is the behaviour of correlation functions, i.e. the exponential decay $(C(r) \sim e^{-r/\xi}$ in the spatial case) observed away from critical points becomes a slow *power law* decay $(C(r) \sim r^{-\alpha}$ with $\alpha > 0)$ at points where $\xi = \infty$, which are called *critical* points. We should however point out that there are at least three types of criticality:

- *Criticality crossing a threshold,* related to a marginal stability that allows fluctuations and responses at all scales. This is the case of percolation, bifurcations and critical phase transitions. The system is only exactly critical at infinite size (or infinite duration if it is a dynamical system) and when the control parameter takes its threshold value.
- *"Constitutive" criticality* met in polymers (self-avoiding random walks), anomalous diffusion taking place on a fractal structure, or more generally in systems with highly connected elements. The system is then exactly critical when its size (duration for dynamical phenomena) is infinite, in a wide range of values of control parameters.
- *Self-organised criticality* where feedback mechanisms from the global state on the local dynamics spontaneously bring the system to a marginally stable state that is critical in the first sense stated above.

A. Lesne and M. Laguës, *Scale Invariance*, DOI 10.1007/978-3-642-15123-1_12,

Beyond this distinction, signatures of criticality are always the same, namely that phenomena have a large number of characteristic scales *which do not decouple*. One observable consequence of this is the presence of fluctuations on all scales such that the distribution of fluctuations follows a power law and not an exponential law. For the same reason, criticality accompanies anomalous response properties since the system is extremely sensitive and an infinitesimal perturbation can be followed by effects on all scales, bringing about a complete reorganisation of the state of the system. Therefore, only a global approach can help to understand such systems. More than that, the crux of the phenomenon is contained in the *connection between scales*, typically in the way in which the flow of material, energy and information is organised hierarchically between different levels and not in the details at any particular scale. It is this that explains the existence of scaling laws, self-similar structures and universal properties. Anomalous exponents are the signature of *emergent* phenomena, in which properties are not directly visible in the constituents nor in the basic laws of construction or evolution. These properties emerge, at a larger scale, from the particular organisation of a large number of elements.[1]

12.2 Experimental Determination of Scaling Laws

Scaling approaches most often begin by experimental demonstration of scaling laws. This does not seem to pose any methodological difficulty a priori. Two observables X and Y, suspected of scaling, are plotted on a log-log plot, which gives a straight line in the case of a scaling law $Y \sim X^a$. This procedure both shows the existence of a scaling law and determines its exponent. But, in practice difficulties arise, which are not apparent in this ideal description and these should be kept in mind to avoid artifacts and misinterpretations.

The first caveat is that it is quite easy to produce a reasonably linear section of a curve formed by a few experimental points, or from a scattered cluster of points, even more so on a log-log plot which "squashes" deviations. It is therefore not enough to show an exponent, we should also carefully *estimate its accuracy and reliability*. A necessary condition is that the data (X_i, Y_i) actually cover different scales, in other words that $X_{\max}/X_{\min} \gg 1$, in practice several orders of magnitude.

An important problem arises: should the data be represented by *one* scaling law $Y \sim X^a$, or by *two* different scaling laws depending on the region of the scale of X, typically $Y \sim X^{a_1}$ for small values of X connected by a *crossover* to another scaling law $Y \sim X^{a_2}$ for large values of X (something we have already encountered in the experimental determination of fractal dimensions)? This is particularly dangerous if the second regime is not developed or sampled enough to be detected (it only gives a few "aberrant" points with respect to the single scaling law $Y \sim X^{a_1}$). This leads

---

[1]P.W. Anderson, More is different. *Science*, **177**, 393–396 (1973).

Fig. 12.1 Wrong estimation of the exponent α of a scaling law $Y \sim X^{\alpha}$ (*thick black line*) if a *crossover* in the scaling law is not detected

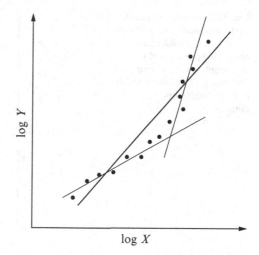

to a wrong estimation of the exponent and the law $Y \sim X^{a}$, where a is actually an effective exponent resulting from the two "real" scaling laws and is between a_1 and a_2 (Fig. 12.1). The risk of making this mistake can be increased if the accuracy and reliability of measurements varies with the scale. This effective exponent a has no real meaning in terms of the hierarchical organisation of the system observed and does not even have phenomenological interest for example in extrapolating the data or comparing several systems. On the other hand, it is even more impossible to reliably extract *two* exponents of this type and thereby identify a crossover.

In the same vein, scaling laws can only exist in a certain range of values of X and we must choose a priori how to truncate the experimental data. Also, data may belong to different independent families and the determination of scaling laws needs to be done separately for each (see Fig. 12.2). The prior classification, whether it has been done by statistical methods or according to arguments from an additional qualitative understanding we have of the system under consideration, can greatly affect the exponents obtained in the scaling analysis (see for example Sect. 11.2.3). There is then clearly a risk of using an *ad hoc* classification to obtain a desired result or show an ideal scale invariance.

Other difficulties can crop up such as the system has finite size effects or more generally we may have to consider more complex scaling laws with more than one variable.

Finally, there are situations in which the scaling behaviour has fluctuations around the deterministic scaling law $Y \sim X^{a}$. To correctly describe such behaviour, the fact that the observable Y is random must be taken into account and it is now the probability distribution of Y which will show a scale invariance:

$$P(Y|X) \sim X^{-a} \; f(YX^{-a}). \tag{12.1}$$

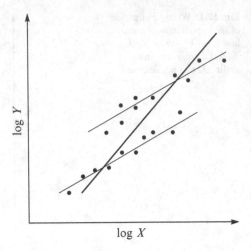

Fig. 12.2 Wrong estimation of the exponent α of a scaling law $Y \sim X^{\alpha}$ (*thick black line*) if the presence of two sub-populations, satisfying the same scaling law but with different prefactors, is not detected

In practice, such scale invariance is confirmed by showing that the histograms obtained for various values of X superimpose when they are multiplied by X^{a} (after normalisation) and plotted as a function of YX^{-a}. The universal curve obtained in this way gives the experimental scaling function f. It is clear that this kind of approach requires a large amount of data and that its statistical validity must be carefully established.

With these precautions, experimentally determined scaling laws provide significant information about the organisation of the system. They are guides towards a global approach to multiscale systems. They open up powerful theoretical approaches, namely those of scaling theories and renormalisation methods. Theoretical developments in return provide support to analyse experimental data and test the assumptions of scale invariance made in the first place.

12.3   Renormalisation and Status of Models

We have seen throughout this book that critical phenomena can be grouped into universality classes, within which critical exponents take well determined values which are identical for each member of the class. The very existence of these universality classes reflects the fact that the critical exponents do not depend on the microscopic details of the system but only on the geometric properties of the space and the type of order. For example we have seen that universality classes of critical phase transitions are categorised according to the spatial dimension d and the number of components of the order parameter. Microscopic physical properties are only involved insofar as they control these geometric properties. In this sense, *a scale invariant phenomenon is robust* in that the scaling laws describing it are not affected by changes in the microscopic details, provided that this change does not push it into another universality class (in other words provided that it is not an essential perturbation inducing a crossover). For the same reason, *its model will be equally robust* (if of course we are only interested in its scaling properties). All we need to do it associate the phenomenon with any model belonging to its universality class. In particular, we can just analyse the simplest representative model of the universality class as this minimal model will give access to the scaling properties as well as, if not better than, a model incorporating all the microscopic details.

The renormalisation approach rigorously establishes the features of this minimal model, by classifying the components as a function of their "relevance" or "irrelevance". It leads to a classification of models, dividing the space of models into universality classes and thereby demonstrating the very unspecific nature of asymptotic scaling properties vis-à-vis the small scale constitutive properties. Finally, by focusing on the hierarchical organisation of phenomena, renormalisation quantitatively reveals the scale invariance and the value of the associated exponents. Renormalisation therefore has a *demonstrative* and *predictive* power far beyond those of phenomenological scaling theories. However it is not always possible to implement technically and construction of the renormalisation transformation itself comes with some reservations, as we have discussed in Chap. 3.

We hold onto the fact that scaling approaches encapsulate and exploit the idea that in the case of a scale invariant phenomenon, describing the *connection* between observations at different scales is enough to explain and predict statistical laws that describe their global behaviour (collective or asymptotic). To do so, models that are simplified to the extreme, even abstract in the case of percolation and self-avoiding chains, are sufficient. It is even precisely because it is legitimate to use such, we could say "skeletal", models, that it is possible to access, at least numerically, global properties, which are inaccessible to an exhaustive model due to the number of degrees of freedom.

In conclusion, the universality of critical phenomena has changed the way we think of, construct and use a model, in particular giving an unexpected interest to models as rudimentary as the Ising model, percolation or logistic maps.

12.4 Open Perspectives

The panorama presented in this book is far from being exhaustive and we finish by mentioning a few perspectives emerging from scaling approaches, criticality and more generally, identification of systems whose hierarchical multiscale organisation is enough to determine the macroscopic behaviour without having to consider the details of the constituents and their interactions. Generalisation of scaling approaches developed for critical phenomena to these more general systems called *complex systems* faces several challenges:

- The elementary constituents are not necessarily identical, which is the case in *disordered systems* and even more so in *living systems*.

- The systems are usually maintained *far from equilibrium* by their interaction with their environment (e.g. flow of material and energy) and tools from statistical mechanics no longer apply. This is the field of *dissipative systems*, addressed in Sect. 9.5 with the example of turbulence. The transfer of energy between the scale at which it is injected and that where dissipation mechanisms become significant generates a complex spatiotemporal organisation, which in this case is hierarchical and even self-similar. Other examples are found in certain chemical reactions taking place in vessels continually supplied with reactants (e.g. Belousov–Zhabotinski, Sect. 9.2.1) and in living systems. We also mention one dimensional models, studied under the name of traffic theories and used to model real situations as well as to develop in model system the new concepts and tools demanded by these systems far from equilibrium (Sect. 4.6).
- If the systems are isolated, their complexity can give rise to *metastable* and *aging* phenomena. They are out of equilibrium because their dynamics is too slow for a steady state to be reached over the observation timescale. These phenomena are observed in glasses, spin glasses and more generally in all systems in which a combination of conflicting influences (we talk of *frustration*) generates energy landscapes with very rich topology and many local minima.
- The systems are generally extended in space and dynamical systems tools must be generalised and adapted to *spatiotemporal dynamics*. A preferred model, coming from a cross between chaotic dynamical systems and the Ising model, is that of networks of coupled oscillators. Here a rich variety of behaviours are observed including local and regional synchronisation, global synchronisation, spatiotemporal chaos and turbulence. They are used to address all sorts of spatiotemporal pattern formation problems encountered in physics, chemistry or biology, for example to model neural networks.
- The presence of feedback loops between different levels brings an element of *circular causality* and leads to self-organised phenomena, or even *self-organised critical phenomena*.

Throughout this book we have emphasized the qualitative role of correlations and in particular their range, compared on different scales of space and time in the system. We have seen that this opened up questions of universality, importance (or not) of specific microscopic details and the role and status of models. The focus was then put on the connection between different scales of observation and description and, more generally, on the hierarchical and dynamical organisation of the system. In all the fields we have referred to, these same, still very open, questions arise.

Index

A

Aging 154
Allometry 362
Alloys 11, 29
Angle doubling 316
Asymptotic (regime) 295, 302
Atmosphere 308
Avogadro (number) 112, 140

B

Backausen 25
Baker's (transformation) 312, 327
Bardeen 234
Barenblatt (equation) 127
BCS theory 234, 235
Belousov–Zhabotinski (reaction) 309
Bernoulli 22
Bethe 179
Biased (diffusion) 129
Bifrucation
    saddle-node 300
Bifurcation 6, 11, 298, 385
    diagram 298, 322
Binary liquids 11
Birkhoff (ergodic theorem) 315
Bjerrum (length) 209
Blob 228
Bloch Felix 25, 232
    functions 235
Boiling 19
Boltzmann
    approximation 137
    equation 137

Bosons 234
Box counting 46
Boyle 22
Brownian motion 110, 132
    fractal 133, 152, 373

C

Cantor (set) 44
Catastrophe
    IR 78
    UV 77, 78
Cauchy (distribution) 150
Cayley-Bethe 169
Cellular automaton 353
Central limit theorem 130, 148
Chain
    Gaussian 203
    ideal 227
    self-avoiding 203, 207
Chaos 77, 135, 146, 293
    Hamiltonian 319
    molecular 137, 143, 146, 324, 325
Chaotic transport 151
Characteristic time 303
Chemical potential 280
Cluster
    dominant 176
    infinite 169, 185, 189
    percolation 189
Coarse-graining 79
Coast of Britain 45
Colloidal limit 119
Community 380
Complex (system) 381